LASER ABLATION - FROM FUNDAMENTALS TO APPLICATIONS

Edited by **Tatiana E. Itina**

Laser Ablation - From Fundamentals to Applications
http://dx.doi.org/10.5772/67961
Edited by Tatiana E. Itina

Contributors

Cristian Focsa, Stefan Andrei Irimiciuc, Silviu Gurlui, Michael Ziskind, Petru-Edward Nica, Maricel Agop, Nisar Ali, Umm-I- Kalsoom, Shazia Bashir, Narjis Begum, Ion N. Mihailescu, Gianina-Florentina Popescu-Pelin, Carmen Ristoscu, Maria Badiceanu, Keiichi Nakagawa, Takakazu Suzuki, Fumihiko Kannari, Valentin Ion, Nicu D. Scarisoreanu, Andreea Andrei, Maria Dinescu, Daniel Nieto, Alberto Jorge-Mora, Naroa Imaz, Nekane Frutos, Rodolfo Gomez, Jose Bartolomé, Jesus Pino, Gerard O'Connor, Chung-Wei Cheng, Jinn-Kuen Chen, Bülend Ortaç, Elif Uzcengiz Şimşek, Canan Kurşungöz, Angel Aragón, Carmen Bao-Varela, Alejandro Fernazdez, Gerard Oconnor, Vladimir Mazhukin, Alexandra Palla Papavlu, Valentina Dinca, Mihaela Filipescu, Abdalla Darwish, Sergey Sarkisov, Paolo Mele, Shrikant Saini, Ruiqi Shen

Published by InTech
Janeza Trdine 9, 51000 Rijeka, Croatia

Publishing Process Manager Romina Rovan
Technical Editor SPi Global
Cover InTech Design team

First published December, 2017
Printed in Croatia
Legal deposit, Croatia: National and University Library in Zagreb

Additional hard copies can be obtained from orders@intechopen.com

Laser Ablation - From Fundamentals to Applications, Edited by Tatiana E. Itina
p. cm.
Print ISBN 978-953-51-3699-6
Online ISBN 978-953-51-3700-9

For EU product safety concerns:
IN TECH d.o.o., Prolaz Marije Krucifikse Kozulić 3, 51000 Rijeka, Croatia,
info@intechopen.com or visit our website at intechopen.com.

Meet the editor

Prof. Tatiana E. Itina is one of the world's leading experts in the field of laser-matter interactions and, particularly, in numerical modeling of laser ablation for numerous applications ranging from surface micro- and nanostructuring to nanoparticle synthesis, thin-film deposition, material analysis, laser-assisted plasmonic and magneto-plasmonic nanostructure formation, etc. She currently holds a CNRS research director position at the Hubert Curien Laboratory, CNRS/University Jean Monnet/University of Lyon (Saint Etienne, France), where she is also an experienced team leader. Before this, she worked at the Pennsylvania State University (USA) and at LP3 Laboratory in Marseille (France). Also, she was invited as a researcher or a research professor to the Virginia State University (USA), Australian National University (Australia), ITMO University (Saint Petersburg, Russia), and other institutions. Prof. Itina has authored more than 100 scientific publications, took part in many international conferences, initiated numerous international collaborations, and received several awards.

Contents

Preface

In 1964, Charles Hard Townes, Nikolai G. Basov, and Alexander M. Prokhorov were awarded the Nobel Prize in Physics for their "fundamental work in the field of quantum electronics, which has led to the construction of oscillators and amplifiers based on the maser-laser principle." These very important studies had been accomplished around 10 years before resulting in the appearance of what is called *laser* today. Shortly after the demonstration of the first laser, the most intensely studied theoretical topics dealt with laser-matter interactions. Many experiments were undertaken to clarify the major ablation mechanisms. At the same time, numerous theoretical studies, both analytical and numerical, were proposed to describe these interactions. It was demonstrated that interaction mechanisms depend both on the main parameters of laser systems, such as pulse wavelength, temporal and spatial shape, duration, repetition frequency, and fluence, and on a wide range of material parameters determining their optical, thermal, and mechanical properties. Many experiments were undertaken to clarify the mechanisms of the extremely complex process involved in laser ablation. At the same time, numerous models, both analytical and numerical, were proposed to describe laser interactions. In these models, different experimental conditions were considered, and several terms were proposed to denote the processes occurring during laser action on different materials. Thus, "laser ablation," "evaporation," "desorption," or "sputtering"—all these terms are relevant to the interaction of a laser beam with a solid (or a liquid) surface that results into transition of the surface particles into a gas phase. For simplicity, here we will use mostly the term "laser ablation." One of the main advantages of this technique is the simplicity of the experimental setup. The other advantage is the possibility of adjusting the experimental conditions in order to obtain the desirable treatment quality.

Both experimental and theoretical studies paved the ways toward the development of numerous laser applications, ranging from laser micro- and nanomachining to material analysis, nanoparticle and nanostructure formation, thin-film deposition, laser propulsion, etc. Some of them are based on unique properties provided by laser ablation, such as bare and chemically clean nanoparticles produced in liquids, clean and proper craters produced due to reduced thermal effects, extremely small and well-controlled 3D objects formed by two-photon polymerization, very precise material analysis including bacteria and explosive materials, etc. Among promising industrial applications, one can also mention laser cleaning, laser-induced forward transfer, surface nanostructuring, molecular mass spectrometry, laser surgery, etc. In addition, it was shown that lasers could be used not only to remove the material but also to modify material properties or even to create novel materials. Recently, even more promising laser applications have been proposed, including biomedicine, catalysis, photovoltaic, and optical memories. This field, thus, is under constant development, and even more impressive progress can be expected in the nearest future.

This book intends to provide the reader with a comprehensive overview of the current state of the art in laser ablation, from its fundamental mechanisms to several recent applications. The book contains 3 main parts and 13 chapters, as follows:

The first part deals with the fundamentals of laser ablation and contains the chapters focused both on the experimental studies and on the numerical modeling.

The second part is focused on several ablation analysis methods. Importantly, measurements of laser plasma properties are essential for better understanding of laser ablation experiments. Typically, plasma spectroscopy was used for this. However, new additional techniques have been recently developed and shown to be promising. The chapters presented in this part describe some of these methods.

The third part provides an overview of several applications. This part is the longest of the book. As it was noted above, laser applications can be very different, ranging from novel and developing medical applications to much more established thin-film deposition or the detection of explosives and other harmful substances.

It is worth underlining, finally, that this book can be helpful not only for the experienced researchers but also for students and engineers that have just started to work in the field of laser ablation. In fact, even if it is not aimed at providing a complete picture, it gives an overview of the main mechanisms and applicative issues of laser-matter interactions.

Tatiana E. Itina
CNRS Research Director/Research Professor,
Hubert Curien Lab., University of Lyon,
Saint Etienne, France

Section 1

Fundementals of Laser Ablation

Chapter 1

Experimental and Theoretical Studies on the Dynamics of Transient Plasmas Generated by Laser Ablation in Various Temporal Regimes

Petru-Edward Nica, Stefan Andrei Irimiciuc, Maricel Agop, Silviu Gurlui, Michael Ziskind and Cristian Focsa

Additional information is available at the end of the chapter

http://dx.doi.org/10.5772/intechopen.70759

Abstract

During the last decade, our groups have performed systematic experimental studies on the characterization of plasma plumes generated by laser ablation in various temporal regimes (ns, ps, fs) on materials ranging from simple metals (Al, Cu, Mn, Ni, In, W, …) to more complex compounds (ceramics, chalcogenide glasses, ferrites). Optical (fast imaging and space- and time-resolved emission spectroscopy) and electrical (mainly Langmuir probe) methods have been applied to experimentally investigate the dynamics of the plasma plume and its constituents. Influence of the target physical (thermodynamic and electrical) parameters on the plasma dynamics has been studied. A mathematical correlation between the local and global plasma parameters and the physical properties of the target was proposed for the first time. Peculiar behaviors like plume splitting or plasma oscillations have been evidenced for high laser fluence ablation in vacuum. Along with results from the literature, our findings provide convincing arguments for the existence of multiple double-layers in the laser ablation plasma plume, in a scenario including two-temperature electrons. New fractal-based theoretical approaches have been developed to qualitatively and quantitatively account for the observed phenomena. The space and time evolution of expansion velocity, particle number, current density and plasma temperature were theoretically investigated.

Keywords: laser ablation, transient plasma dynamics, plasma oscillations, Langmuir probe, optical emission spectroscopy, plasma simulation, fractal, hydrodynamics

1. Introduction

Despite its widespread use in a large number of applications (e.g. pulsed laser deposition [1], generation of nanoparticles [2, 3], chemical analysis [4, 5] or cleaning of delicate artwork [6]), a comprehensive picture of laser ablation remains a challenge for both experimentalists and theoreticians. The difficulty arises from the multi-physics nature of the ablation process, coupling optics and electrodynamics (absorption of light by the target material), thermodynamics (heating, phase transitions, cooling), gas dynamics (expansion of ablation plume into vacuum or background gas), plasma physics (collisions, electric interactions) and laser-plume interaction (plasma heating by absorption of laser photons, inverse Bremsstrahlung, multi-photon ionization), some of these evolving on very short time scales, which can make it challenging to find the adequate resolving probe. Moreover, the fundamental mechanisms involved in the ablation process and the properties and dynamics of the subsequent laser-produced plasmas depend strongly on the laser beam parameters (pulse duration, fluence, wavelength or beam profile) and also on the properties of the irradiated material (thermal/electrical conductivity, reflectivity, heat of vaporization, binding energy, etc.).

Fundamental differences can be revealed when investigating the role of the laser pulse duration with respect to the specific timescales of the irradiated material response [7]. For instance, in nanosecond ablation regime, the laser pulse is significantly longer than the usual electron cooling time ($\sim$10 fs) and the lattice heating time ($\sim$ps). In this case, the energy absorbed by the electrons has enough time to be transferred to the lattice. The electrons and the lattice can further reach thermal equilibrium, and the main energy loss is the heat conduction into the solid target. Consequently, the mechanisms involved in this ablation regime are mainly thermal (e.g. phase explosion, normal vaporization, etc.) [8, 9]. Most notably, the subsequent plasma generated in this ablation regime absorbs a significant percentage of the beam energy (pulse "tail", during several nanoseconds), leading to an important heating of the plume. In the case of ultra-fast laser ablation ($\sim$fs), when the laser pulse duration is shorter than (or on the same scale as) the electron cooling time, the electrons in the surface layer suffer cooling by heat diffusion and by heat transfer to the lattice ions. This stage continues for several picoseconds. The picture changes in the case of a semiconductor target which is heated by an ultra-short pulse. In the latter case, laser energy is transferred into the solid by creating a "bath" of hot electrons and holes [10]. Hot carriers subsequently transfer energy to the lattice by creating optical and acoustic phonons. In the case of both metals and semiconductors, the thermalization of laser energy in the hot carrier bath takes place within a few femtoseconds, while the typical time-scale for lattice heating falls within the 1–10 ps range, where thermal conduction is negligible [11]. For low fluence fs-laser irradiation, the Coulomb explosion [12] is the dominant ejection mechanism, while at sufficiently high laser intensities the phase explosion is followed by non-thermal vaporization of the bulk material and becomes the main mechanism for material removal [13]. A particular case of ultra-fast laser ablation is represented by ps-laser ablation. This temporal regime acts like a bridge between the previous cases manifesting characteristics from both regimes. The pulse duration is long enough so that some thermal damage occurs due to the heating of the lattice. If the laser pulse width is in the 1–10 ps range, the particle ejection is still dominated by the Coulomb explosion with minimal contribution from the thermal mechanism and no interaction between the plasma plume and the incoming laser beam. For longer pulse durations (∼100 ps) the balance is "tipped" in favor of thermal mechanisms coupled with a brief absorption of the laser beam by the ejected cloud [7].

Extensive efforts have been fostered by many research groups in the last decades for unveiling this complexity. Most of them were dedicated to characterizing the laser-produced plasma dynamics and to establishing a link between the local/global plasma parameters and the fundamental ablation mechanisms, some also addressed the link with the physical properties of the irradiated material (see, e.g., [14–48]). In order to draw a comprehensive picture of the laser-produced transient plasma evolution, one needs detailed space- and time-resolved information on the chemical and electrical composition of the expanding plume (ions, electrons, neutral atoms, molecules, clusters, with associated number densities, ionization stages, etc.), the plasma dynamics—expansion regime (plasma as a whole), kinetic energies (of every individual species), etc. Optical (fast photography, shadowgraphy, interferometry, optical emission and absorption spectroscopies, laser-induced fluorescence, resonant ionization spectroscopy, Thomson scattering) and electrical (Langmuir probes (LP), Faraday cups, electrostatic analyzers, mass spectrometry) methods [49, 50] are available for undertaking this complex investigation. The difficulty in getting a complete description of the laser ablation plasma plume arises mostly from its transient character, with a lifetime typically on the 10 μs scale, but with inner rapid phenomena (e.g. oscillations) which can exhibit sub-nanosecond timescales. Consequently, extreme care must be taken for the specific application of well-known steady state plasma characterization tools (e.g. Langmuir probes) to this transient case. Moreover, not all of the above-mentioned techniques can be applied for probing the great variety of laser ablation plasmas (nor even one given plasma plume along its spatial and temporal evolution). For instance, higher electron number densities are needed to get an effective response when using interferometry or Thomson scattering than when Langmuir probes are used. Typical irradiation conditions (in the range of GW/cm^2) will result in laser ablation plasmas which can be considered as cold (electron temperature in the eV range, number densities roughly in the 10^{13}–10^{18} cm^{-3} range). Higher irradiation values (exceeding 1 PW/cm^2) can lead to hot plasmas with higher electron temperature and number densities, usually studied in a laser inertial confinement nuclear fusion context [51].

During the last decade, our groups have performed systematic experimental and theoretical studies on the fundamental characterization and applications (mainly pulsed laser deposition of thin films) of plasma plumes generated by laser ablation in various temporal regimes (ns, ps, fs) on materials ranging from simple metals (Al, Cu, Mn, Ni, In, W, etc.) to more complex compounds (ceramics, chalcogenide glasses, ferrites) [52–79]. Optical (fast gate Intensified Charge Coupled Device (ICCD) camera imaging and space- and time-resolved emission spectroscopy) and electrical (mainly Langmuir probe) methods have been applied to experimentally investigate the dynamics of the plasma plume and its constituents. The analysis of probe current-voltage characteristics at various delays after the laser pulse gave access to the temporal evolution of ion density, electron temperature and plasma potential. The recording timescales are analyzed by coupling the distribution functions of electrons and ions through an effective mass. The space and time evolution of expansion velocity, particle current density and plasma temperature were theoretically investigated by a fractal hydrodynamic model. We present here a short overview of these experimental and theoretical studies, with a special focus on the characterization of transient laser-produced plasmas by electrical methods and on recent developments of the fractal hydrodynamic model.

2. Experimental details

A schematic view of the experimental set-up often used in our studies is given in **Figure 1**. The solid targets (usually 20 mm diameter, 1 mm thick disks) of various chemical composition were placed on a translation-rotation stage in vacuum or controlled atmosphere and irradiated by ns, ps or fs laser pulses at various wavelengths (usually 532 nm Nd:YAG and 800 nm Ti:Sa, Quantel, Continuum, Spectra Physics). We used laser fluences spanning the 10^{-1}–10^{3} J/cm^2 range, corresponding to irradiances in the 10^{6}–10^{14} W/cm^2 limits for laser spot dimensions on the target in the range 0.1–1 mm and laser pulse durations in the range 40 fs–10 ns.

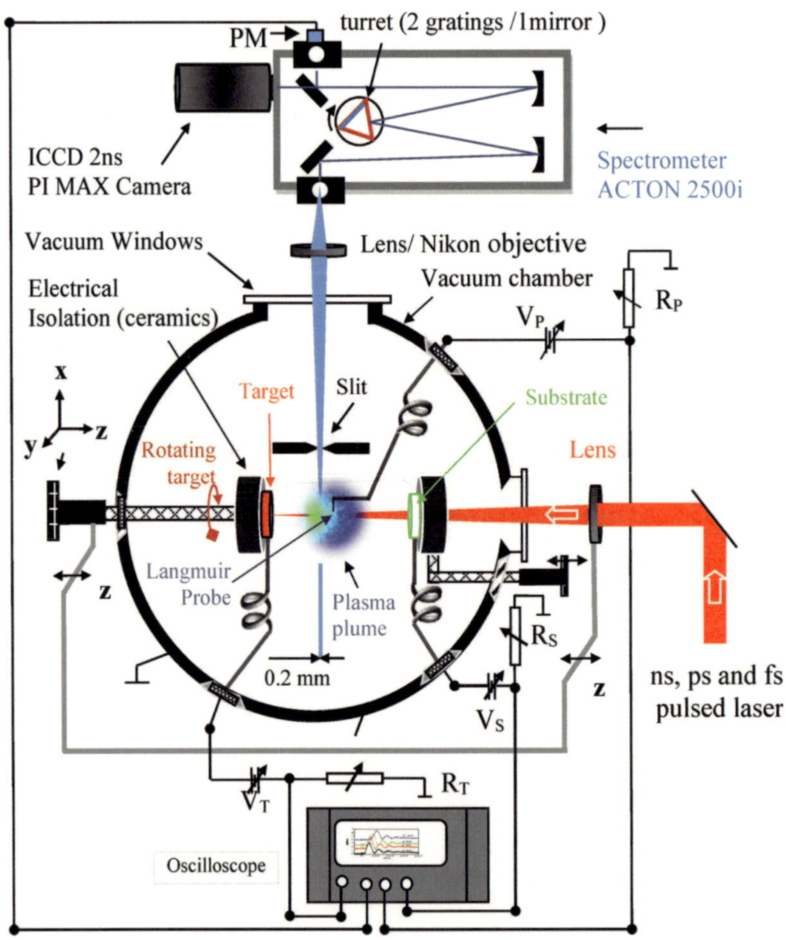

Figure 1. Schematic view of the experimental set-up at the University of Lille.

The electric diagnostics mainly used were Langmuir probes immersed at various positions in the plasma plume to record the ionic or electronic currents, depending on the probe biasing voltage (V_P) [52, 53, 55, 58, 60, 77, 79]. Using bunches of Langmuir probes we obtained the angular distribution of charge carriers at various distances from the target [52]. The Langmuir probes were 0.8 mm diameter, 5 mm length stainless steel wires. When using metallic targets, they were also biased to observe the influence on the probe current, and to correlate it with the transitory electrical signal recorded from the target [65]. The transitory signals were registered by digital oscilloscopes with 0.5–2 GHz bandwidth (LeCroy) using 50 Ω or 1 MΩ impedances.

To record the plasma optical emission, three configurations have been often used: plasma plume imaging (ICCD fast photography), space- and time-resolved optical emission spectroscopy (OES) and temporal evolution of a given spectral line intensity [54–57, 62, 64, 66–68, 73, 75]. For the imaging experiments, ICCD gate widths of 5 ns were usually employed in order to catch as much as possible sharp temporal snapshots in the space-time evolution of the plume. For space-resolved OES, a 1 mm × 5 mm translating slit was placed in the vacuum chamber, at 40 mm from the normal to the target, to observe plasma plume "slices" of 1 mm width [57]. Finally, the temporal evolution of a given spectral line intensity was recorded with a fast (sub-ns rise-time) photomultiplier tube (PMT, Hamamatsu) placed on the second output port of the monochromator (Acton Princeton Instruments), once the spectral line has been selected and isolated with appropriate entrance and exit slit widths and diffraction grating positions [54, 57].

3. Experimental results

Both our optical and electrical investigations on the dynamics of the transient plasmas produced by laser ablation in vacuum revealed some peculiar phenomena as the plume splitting in (at least) two components or the occurrence of oscillations in the currents recorded by the Langmuir probe or on the irradiated target. We have recently presented an overview of these peculiar findings, along with similar results from the literature, in a review paper [78]. We will therefore orient the presentation below more on the extraction of significant plasma parameters (to be compared with the theoretical model predictions) from the time-of-flight profiles of the currents recorded by the LP.

The typical time-evolution of the ion current recorded by the probe placed at various distances (axial, radial) with respect to the center of the laser irradiation spot shows that it extends in the μs range and it generally consists in a fast part having an oscillatory behavior, and a slower tail which arrives at longer times (see some examples in **Figure 2**, for ns-ablation and $V_P = -30$ V probe biasing). As it was expected, the time at which the first current maximum is recorded (t_{max}) significantly increases with the distance from the target (z), while it is only slightly increasing with the radial distance (r). Moreover, for $r > 1$ cm and $z > 3$ cm the oscillations disappear. In a simple estimation, the z-t_{max} dependence, displaying a linear evolution, gives an axial expansion velocity $v_z = 1.25 \times 10^5$ m/s. Also, a decrease of t_{max} when increasing the laser pulse energy was evidenced in our previous work [55], as a consequence of the higher expansion velocity. For comparison, from the time-evolution of the optical maximum signals recorded by ICCD imaging, one can derive the center-of-mass velocities of the plasma structures [55,56,62]. In all our studies, we found values in the range of 10^4 m/s for the first (fast) structure and of 10^3 m/s for the second (slow) one, in agreement with experimental results given in the literature [80–82] and rough calculations performed in simple thermodynamic framework [58, 59]. As an example, for the experimental conditions in which the data from **Figure 2** were registered [55], we have obtained $v_{fast} = 4.66 \times 10^4$ m/s which is half of the previous value, as consequence of ion acceleration/oscillation in the probe electric field. Moreover, by smoothing the probe signal to remove the oscillations, the temporal trace is well described by a shifted Maxwellian velocity distribution function [83, 84], with center-of-mass (or "drift") velocity in the range of 10^4 m/s [77].

The currents induced in the target by the ablation process (displayed in **Figure 3a** for various laser energies/pulse) can be correlated with the probe signal, as the positive charging arises

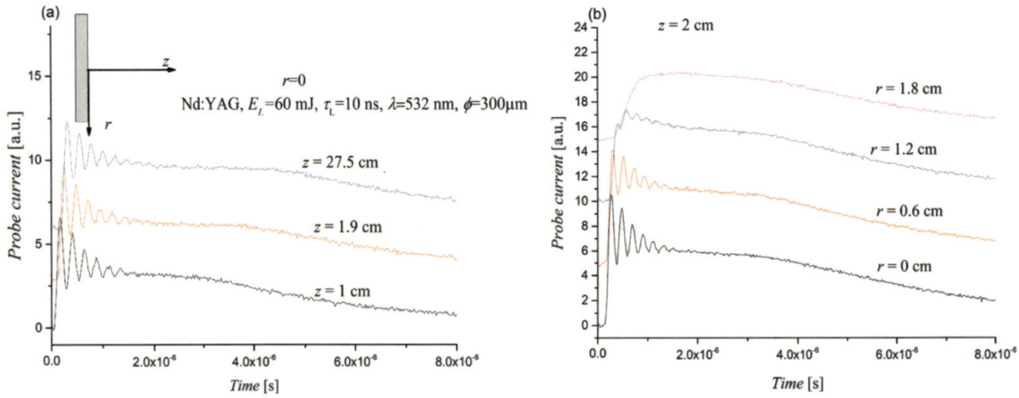

Figure 2. Typical time-dependence of the ionic current recorded by the Langmuir probe (biased at $V_P = -30$ V) placed at various axial (a) and radial (b) distances from the center of the laser irradiation spot (Al target, ns ablation [55]).

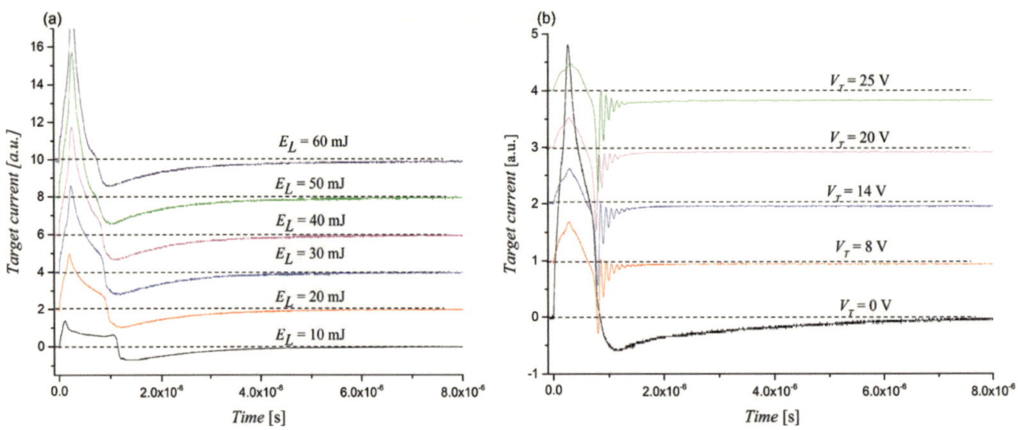

Figure 3. Currents induced in the ns-irradiated target (Al) for (a) various laser beam energies ($V_T = 0$ V) and (b) target biases ($E_L = 60$ mJ/pulse).

through the electrons escaping from the expanding plasma to the grounded chamber, while the negative charging is given by the ions escaping from the target. The asymmetry of the laser intensity distribution and non-uniform target absorptivity can also lead to the generation of considerable currents along a conductive target [85]. Experiments revealed two possible mechanisms: induction due to the magnetic dipole moment of the plasma and a second mechanism resulting from the phenomena at the plasma-target interface [85–87]. From these experimental results, we observed that the amplitudes of the fast peak increase rapidly with the laser energy and the fast electron contribution becomes dominant above E_L = 16 mJ/pulse (~20 J/cm^2 fluence) [65]. Biasing the target (V_T) by an external voltage stabilized source, the negative part of the target current, which is given by the ion contribution, acquires an oscillatory behavior of the same frequency as previously recorded for the Langmuir probe current (**Figure 3b**). We therefore deduced that such periodic fluctuations were induced by the probe/target electric field, with a target bias threshold for their occurrence.

When extracting the oscillatory part from the original signal (**Figure 2**), that is, subtracting the smoothed temporal trace, the resulted time-dependence revealed a good fitting with a usual damp oscillator [48, 60]. Thus, one can assume that in the electric field near the probe the ion equation of motion is $\ddot{x} + \nu_{ei}\dot{x} + \omega_{pi}^2 x = 0$, where the dissipative term is given by the electron-ion collision frequency, $\langle \nu_{ei} \rangle \sim 1.5 \times 10^{-6}\ z^2 n_e \ln \Lambda / T_e^{3/2}$ [Hz], and $f_{pi} = \omega_{pi}/2\pi = 210z\sqrt{n_i/A}$ [Hz] is the plasma ion frequency [49, 60]. We applied this procedure also for studying the LP current oscillations in the case of fs-laser ablation of Al. In a more detailed analysis, we observed that biasing the target can influence the oscillation frequency, two regimes being observed, corresponding to the fast and slow components, respectively. For example, in **Figure 4a** and **b** we observed good fitting for V_T = +1.6 V and V_T = +35 V, resulting in oscillation frequencies υ = 8.37 MHz and υ' = 6.46 MHz. For an intermediate value, V_T = +19.5 V, we distinguished two parts: (i) at short recording times (<400 ns), a fast plasma structure oscillating with a frequency of υ = 8.93 MHz, which is similar with the previous value for V_T = +1.6 V; (ii) a slow structure for longer times (>400 ns), oscillating with a frequency of υ' = 6.74 MHz, which is similar with the previous value for V_T = +35 V. This is in good agreement with the existence of two types of particles, that is, plasma structures, which are formed during the initial expansion process: a fast (hot) one consisting in highly charged particles promptly ejected in an expansion process characterized by the acceleration through the electric field given by initial charge separation (ambipolar diffusion); and a slow (cold) tail consisting in thermalized particles with low average charge state. Such double-structures are often reported in the literature for both fs- and ns-laser ablation, being observed for plume propagating in an ambient gas [14, 15, 20, 88] and in vacuum for high-fluence irradiation [56, 57, 67, 75]. Reports from literature usually present a difference of one order of magnitude between the expansion velocities of the two plasma components. The Fourier transform of the probe current (**Figure 4c**) confirms the existence of two oscillatory regimes, through the existence of two peaks which are evidenced differently at various target voltages. Their frequencies are plotted in **Figure 4d** vs. V_T as obtained from the above-mentioned sine damp fitting and from the Fourier transform, respectively. Let us observe that both methods gave similar results, two groups being computed with average values υ = 8.5 MHz and υ' = 6.5 MHz.

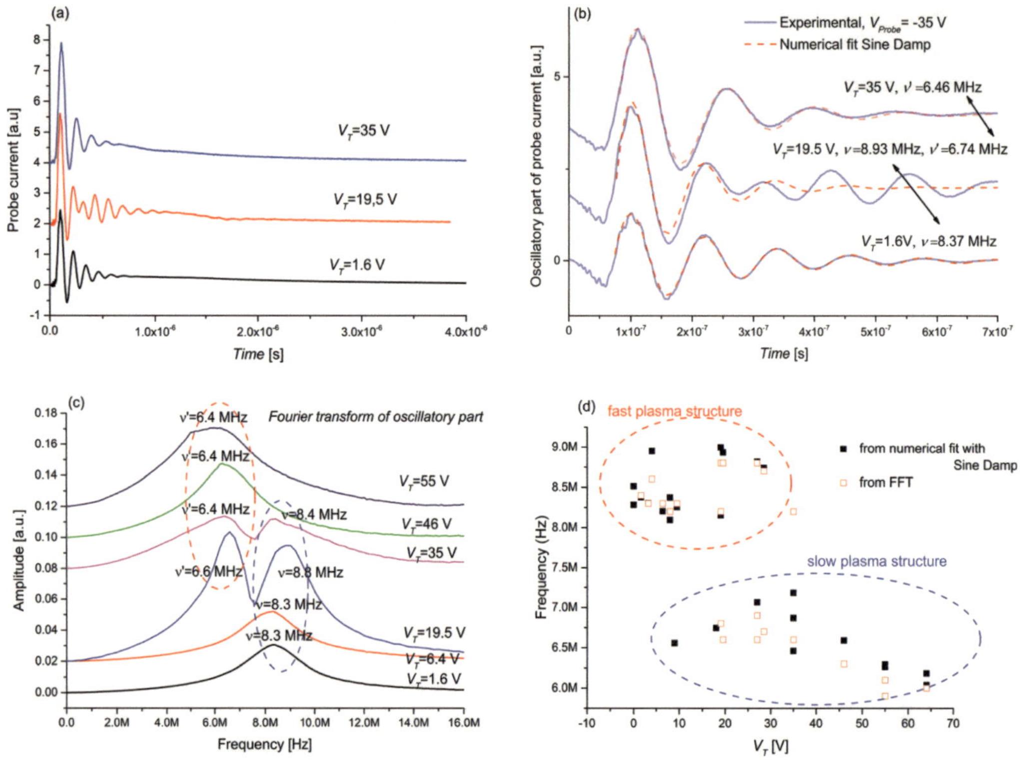

Figure 4. Typical temporal evolution of the current recorded by the negatively biased probe, $V_P = -35$ V (a), oscillatory part obtained by smoothing the temporal trace and subtracting from the original one (b), its Fourier transform (c) and frequencies obtained for various target biasing V_T (d), for Al fs-laser-produced plasma.

There is no clear dependence of the oscillation frequencies on the probe biasing, their variations being in the limits of the determination error bars. A similar behavior was observed when using a Cu target: at short recording times, the probe current oscillates with a frequency of $\upsilon = 10.5$ MHz, while it shifts later to $\upsilon' = 6.4$ MHz, the values being averaged over the various target biases. Moreover, both oscillating regimes are observed in the probe current regardless of probe biasing, the behavior being slightly different when compared with the case of Al target, where they occurred only for intermediate values $V_T = +(19\text{–}35)$ V.

During these measurements, a discussion has arisen on the use of low or high oscilloscope impedance for recording the transient signals. Both configurations have advantages and disadvantages. Using low impedance ensures good temporal resolution, while using a high input impedance has the advantage of improving the signal amplitude, although the temporal trace $i(t)$ is artificially extended through its convolution with the apparatus function,

$$i_C(t) = \int_{-\infty}^{\infty} i(\tau)h(t-\tau)d\tau \tag{1}$$

where t is the time and $h(t)$ the convolution function, which for our experiments was of exponential decay type, $h(t) = \exp(-t/\tau_0)$, with $\tau_0 = 16 \times 10^{-6}$ s, the time constant deduced from

http://dx.doi.org/10.5772/intechopen.70759

the electrical circuit parameters. In such experimental conditions, the probe current is given in **Figure 5a**, where we can observe that the oscillations are preserved at short recording times, since the convolution of a periodic function is also periodic. However, through the inverse mathematical procedure, the numerical deconvoluted signals fit well with the probe current recorded using a low oscilloscope input impedance. For example, in **Figure 5b** such comparison is made for Al targets and $V_T=0$, $V_P=-30$ V. We therefore used one or the other configuration, with easy conversion between them through convolution/deconvolution procedures.

The consequence of applying a positive voltage on the metallic target, $V_T=+35$ V, is the appearance of an initial hump at short recording times (see **Figure 6a**). Such biasing is influencing the plasma expansion at the initial stages by accelerating the ions and retarding the electrons. Therefore, an excess of fast positive electric charges results, in the form of a residual ionic current. The fast plasma structure acceleration is given by the ambipolar electric field within a thin layer in the plume periphery, where quasi-neutrality of the plasma is broken [34], and the resulting double layer is influenced by the target biasing, at the early stages of

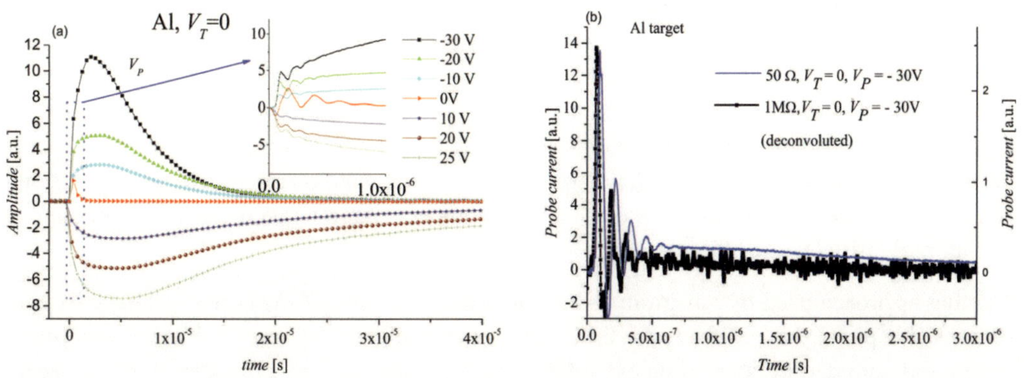

Figure 5. Typical temporal profiles of the current recorded by the LP probe at 3.5 mm from the target and various values of the probe biasing voltages: Al target, fs-laser ablation, high input impedance (a) and numerically deconvoluted signal compared to the current recorded using a low oscilloscope input impedance (b).

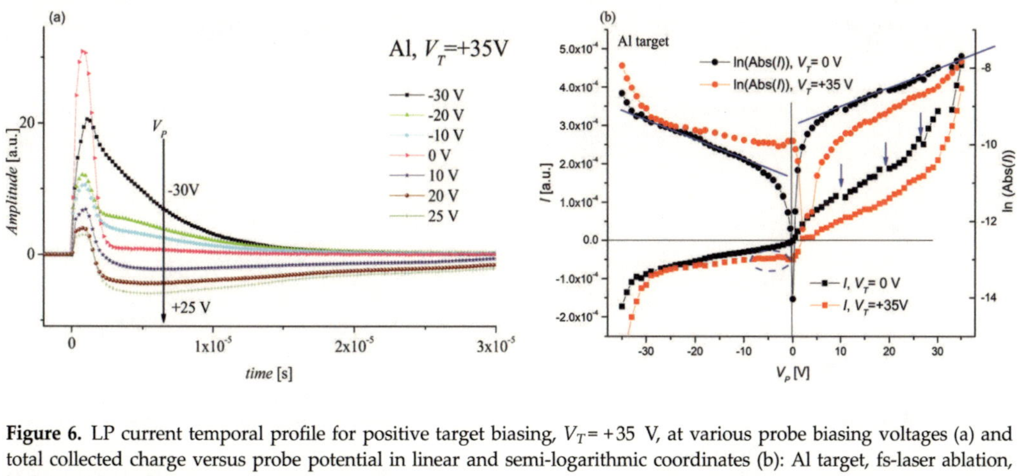

Figure 6. LP current temporal profile for positive target biasing, $V_T=+35$ V, at various probe biasing voltages (a) and total collected charge versus probe potential in linear and semi-logarithmic coordinates (b): Al target, fs-laser ablation, high input impedance.

expansion. The time-extension of the initial hump is longer in the case of higher atomic mass targets and it indicates a dependence on the inertial properties, the light ions being faster repelled by the positively biased target. These measurements can be successfully used to calculate the global temperature and average charge state for the fast plasma structure, by time-integration of probe current intensity to obtain the total collected charge dependence versus probe potential (**Figure 6b**). It was observed that for both electronic and ionic branches, corresponding to positive or negative LP biasing, for small values of the probe potential ($|V_P| < 10$ V), the dependence is of $\sqrt{|V_P|}$ type. This is given by the repelling/attracting of charges from the cold (slow) structure, and it shows a saturation trend as the potential increases because the slow structure center-of-mass velocity is easily influenced by the probe field. Between ±10 and ±35 V, an exponential dependence results, the collected charge is mainly influenced by the behavior of the hot (fast) plasma structure and a semi-logarithmic plot (**Figure 6b**) shows a linear dependence. The positive exponential branch corresponding to the fast electrons allows the computing of electron temperatures from the slope of the semi-logarithmic representation, the resulting values being time-averaged. Moreover, in the assumption of an isothermal plasma, taking also into account the ionic branch, the average ion charge state can be derived as the ratio of the two regions slopes. The results are given in **Table 1** for various metallic targets. We observed that a high positive correlation exists between the deduced average charge states and the plasma temperatures. This result is in agreement with plasma standard theoretical models, for example, local thermodynamic equilibrium or collisional radiative [89], where average charge state or ion fractional populations can be easily computed in the case of a homogenous plasma, for given electron temperature and atom density (e.g. see **Figure 5** of [90]).

Another approach used by our group is based on the treatment of the current-voltage characteristics (*I-V* plots) derived from the time-of-flight current profiles recorded at various probe biases and sampled at specific delays after the laser pulse [77]. In principle, this "classical" method [40, 91–93] considers only the thermal movement of the particles (without streaming) and a Maxwellian distribution function. Space-time evolution of electronic temperature, thermal velocity, plasma potential and particle density can be derived from the *I-V* characteristic based on the results of the "classical" LP theory. In the framework of the LP method, all main plasma parameters are derived from the electronic part of the *I-V* characteristic:

Target	Atomic weight	T_e (eV)	Average charge state (z)
Al	27	14.21 ± 2.15	2.63 ± 0.74
Mn	55	8.15 ± 0.79	1.08 ± 0.42
Ni	59	6.72 ± 1.03	1.29 ± 0.37
Cu	64	6.54 ± 0.45	1.59 ± 0.23
In	115	8.66 ± 1.29	2.23 ± 0.94
Te	128	8.04 ± 0.84	1.44 ± 0.24
W	184	5.29 ± 0.53	0.92 ± 0.17

Table 1. Plasma temperature for various metallic targets.

$$I_{Probe} = I_{e0}\exp\left[-e(V_{Plasma} - V_{Probe})/k_B T_e\right] \; - \; I_{i0}, \;\; V_{Probe} < V_{Plasma} \tag{2}$$

where I_{Probe} is the probe current, I_{e0} and I_{i0} are the electron and ion saturation currents, V_{probe} is the bias applied on the probe, V_{Plasma} is the plasma potential, e is the electron charge, k_B is the Boltzmann constant, T_e is the electron temperature.

The electron temperature (T_e) can be further determined from the slope of the semi-logarithmic plot of the probe current from the saturation region (I_e) versus probe bias (V_p), the plasma potential (V_{Plasma}) representing the inflection point separating the linear increase from the saturation region, while the ion density and thermal velocity are given by the following equations:

$$I_{i0} = \frac{1}{4} e A n_i \sqrt{\frac{8 k_B T_i}{\pi m_i}} \tag{3}$$

$$v_{thermal} = \frac{1}{4}\sqrt{\frac{8 k_B T_i}{\pi m_i}} \tag{4}$$

where A is the probe collecting area, n_i is the ionic density, m_i is the mass of the particle and T_i the ion temperature which is considered to be equal to the electron temperature, assuming the local thermodynamic equilibrium (LTE) hypothesis.

We applied this particular method to study the dynamics of the plume at relatively long delays (>1 μs) after the laser pulse. Based on our previous ICCD fast camera imaging and space- and time-resolved optical emission spectroscopy measurements, we know that these times are characteristic for the observation of the slow plasma component [55, 56, 67]. As expected, it resulted that all studied parameters have a significant space-time decrease, due to the cooling process and rarefaction during expansion (**Figure 7**). Comparing the values for the electron temperatures determined using time-sampling methods with the ones given by the previous method, some notable differences occur. The time-resolved method only captures the cold tail arriving at the probe surface as the sampling is done after 1 μs, and thus the values of the

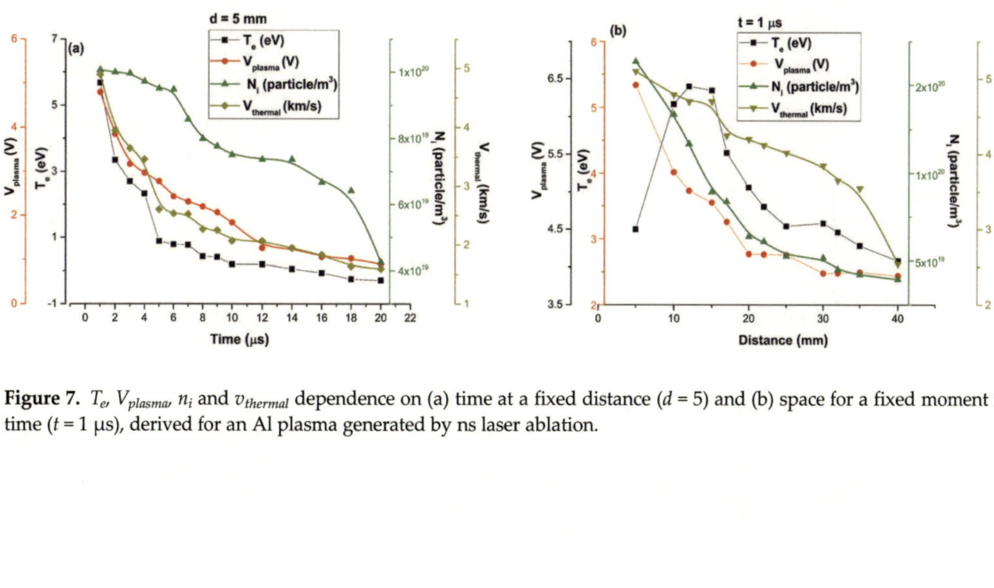

Figure 7. T_e, V_{plasma}, n_i and $v_{thermal}$ dependence on (a) time at a fixed distance (d = 5) and (b) space for a fixed moment in time (t = 1 μs), derived for an Al plasma generated by ns laser ablation.

electron temperatures are considerably lower than the ones derived from the total collected charge versus probe potential representation.

The same technique has been implemented for the study of transient plasmas generated by laser ablation in three different temporal regimes (ns, ps, fs) on a series of six metallic targets (Al, Ti, Mn, Ni, Cu, Zn). **Figure 8a** shows the decrease of the saturation ion density (recorded at 5 mm from the target and after 1 μs with respect to the laser pulse) with the melting point of the target material. Other groups have reported similar evolutions for ns-laser-produced plasmas. A significant decrease of the ablation efficiency (estimated as a function of the ablated crater depth) with the increase of the melting point was reported in [94]. A similar influence of the melting point on the ablation yield was found by Schou et al. [29, 95], who discussed the decrease in the ablation yield as a consequence of the target cohesive energy increase. Both melting point and cohesive energy are considered as a measure of the degree of volatility. The same dependence proposed by previously mentioned authors was also found by our group for fs-laser ablation of various metallic targets (W, Te, In, Cu, Ni, Mn, Al) [77]. The decrease of the ionic density with the increase of the melting point (or cohesive energy) appears to be a general characteristic of the laser-produced plasmas as it is confirmed by our current systematic study on the ns, fs and ps ablation for a wide range of metallic targets.

Figure 8b displays the evolution of the electronic temperature with the target electrical conductivity for the three ablation regimes. The values derived at $t = 1$ μs and $d = 5$ mm present a steep increase for low conductivity elements reaching a quasi-saturation regime after 10 MS/m. The experimental data were fitted with the same function [77] for the three ablation regimes: $T(\sigma) = c_T - \frac{b_T}{\sigma + a_T}$ where a_T, b_T, c_T are constants, characteristic for each ablation regime. Considering the wide range of target conductivities, spanning approximately two orders of magnitude from 0.62 MS/m for Mn to 59 MS/m for Cu, and the values of the electronic temperature (1–9 eV, commonly encountered for plasmas generated by laser ablation at low-moderate fluence), a generalization is somehow appealing. However, we note that these particular values characterize only a particular plasma volume investigated at a specific moment in time

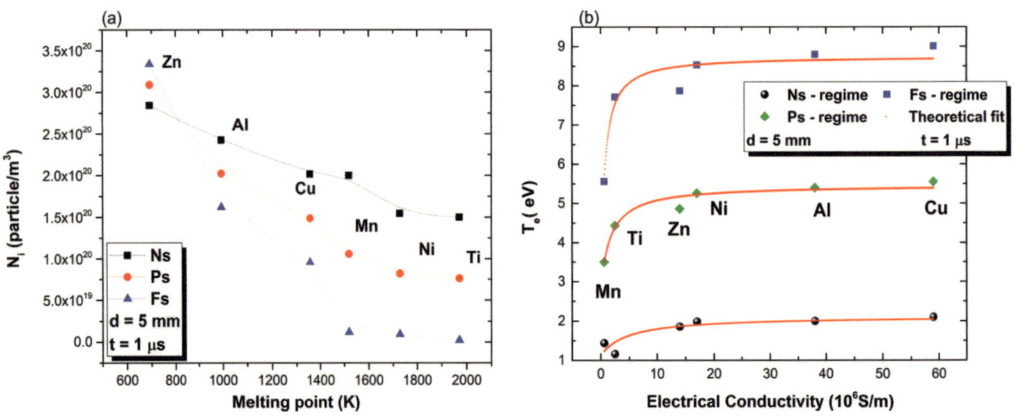

Figure 8. (a) Ion density dependence on the target melting point for three ablation regimes (ns, ps and fs) and (b) the electronic temperature dependence on the electrical conductivity of the metallic targets.

(1 μs delay and a distance of 5 mm from the target) and further investigations are thus required to establish the universality of the proposed dependence.

Let us note that in the results presented above only single-element targets were used to better understand the fundamental processes involved in laser-target interaction and subsequent plasma evolution. When using complex (multi-component) targets, the interpretation of the LP temporal profiles is more difficult. For example, we present in **Figure 9** the results for laser ablation of a chalcogenide glass ($Ge_{9.5}Sb_{28.6}Se_{61.9}$) [75], where three plasma structures are recorded, each of them oscillating with a specific frequency. These structures are also present in ICCD images (see the inset of **Figure 9**), displaying various expansion velocities.

An extensive investigation of plasmas generated by ns laser ablation of chalcogenide targets was presented in [75]. We reported there a strong evolution of the global expansion velocity of all three plasma components (derived from ICCD fast camera imaging) and of the excitation temperatures (determined through optical emission spectroscopy) with the thermal and electrical properties of the complex chalcogenide target. More precisely, the increase of the Sb_2Se_3 content led to a quasilinear increase of the expansion velocities and average plasma temperature. The results were interpreted in the frame of the target structural changes. Previous reports on the properties of similar systems revealed that the addition of Sb leads to the decrease of the bandgap energy [96], the increase of the glass transition temperature and

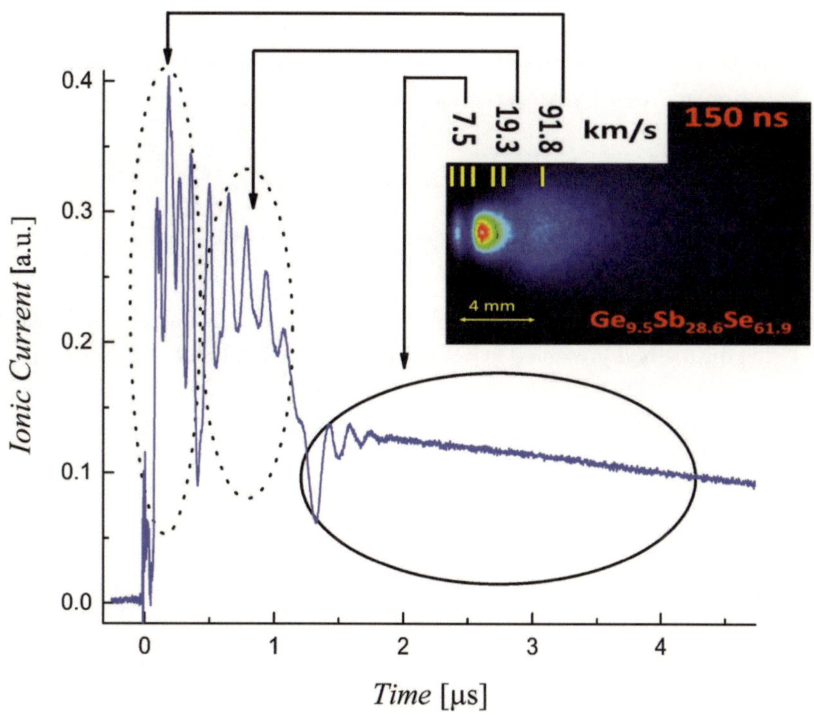

Figure 9. LP current for fs-laser-produced plasma using a chalcogenide glass target and ICCD image (inset) showing the occurrence of three plasma structures.

consequently the increase of the weak bonds concentration [97]. The increase in the weak bonds concentration leads to increase in the thermal and electrical conductivity of the glass. This is in good agreement with our reported results on pure metallic targets [77], where the increase in the electrical/thermal conductivity of the materials led to the increase of electron temperatures and ion drift velocities.

4. Theoretical investigations

Continuing our previous work [55, 58, 59, 63] on the fractal hydrodynamic model for laser ablation plasma dynamics, we recently proposed a compact version for the analysis of the spatial and temporal evolution of some plasma dynamic variables [76]. This version of our model was obtained by using normalized variables of the particle density, velocities, current density, etc., and by choosing adequate scale resolutions. In our initial model describing the evolution of the fractal fluid [59, 63, 98], we took into account a high number of factors (experimental ones by means of the width of the laser pulse Gaussian distribution, probe or target bias, etc., and theoretical ones by means of the fractal-non-fractal transition coefficient, resolution scale, fractal dimension of the movement curves, etc.), which increased the difficulty in performing a complete analysis of the plume dynamics. Through a viable choice of normalized dynamic variables with respect to the previous factors, we simplified the interpretation of the plume dynamics.

In the frame of fractal hydrodynamics with an arbitrary fractal dimension of the motion curves, D_F, we obtained for the one-dimensional case in absence of an external potential the specific momentum and density conservation laws [59, 63, 98]:

$$\partial_t v + v\partial_x v = -2\lambda^2 (dt)^{\left(\frac{4}{D_F}\right)-2} \partial_x \left(\rho^{\frac{-1}{2}} \partial_{xx} \rho^{\frac{-1}{2}}\right) \tag{4a}$$

$$\partial_t v + \partial_x (\rho v) = 0 \tag{4b}$$

with v the velocity, ρ the density, dt the scale resolution and λ a fractal-non-fractal transition coefficient. Using the method presented in [59, 63], the analytical solutions for Eqs. (4a) and (4b) are

$$v(x,t) = \frac{c\alpha^2 + \left[\frac{2\lambda (dt)^{(2/D_F)-1}}{\alpha}\right]^2 xt}{\alpha^2 + \left[\frac{2\lambda (dt)^{(2/D_F)-1}}{\alpha}\right]^2 t^2} \tag{5}$$

for the velocity field and

$$\rho(x,t) = \frac{\pi^{-1/2}}{\left\{\alpha^2 + \left[\frac{2\lambda (dt)^{(2/D_F)-1}}{\alpha}\right]^2 t^2\right\}^{1/2}} \cdot \exp\left\{\frac{-(x-ct)^2}{\alpha^2 + \left[\frac{2\lambda (dt)^{(2/D_F)-1}}{\alpha}\right]^2 t^2}\right\} \tag{6}$$

for the particular initial and boundary conditions given by:

$$v(x, t = 0) = c, \quad \rho(x, t = 0) = \rho_0 e^{(-x/\alpha)^2} \tag{7}$$

$$v(x = ct, t) = c, \quad \rho(x = -\infty, t) = \rho(x = \infty, t) = 0 \tag{8}$$

Thus, we assumed that at $t = 0$ the center of the initial Gaussian distribution ρ is located at x $(t=0)>0$ and has the velocity $v(t=0)>c$. This is supported by the fact that the laser beam temporal profile is usually of Gaussian type, and consequently we can consider a similar distribution of the plasma plume along the expansion direction. Moreover, if the plasma parameters are investigated at time scales longer than the laser pulse width, one can also assume that the particles are ejected with a constant expansion velocity. In this approximation, a correlation could be established between the density distribution of the ejected particles and the laser pulse temporal distribution [55] (in the hypothesis of total laser energy absorption by the plasma plume). A similar outcome is obtained for <*x*> = *ct*. From such a perspective, with respect to the movement plane of any particle (<*x*> = *ct*) of the ablation plasma at any time $\neq 0$, the velocities of the ejected particles are constant during expansion (Eq. (8)) and the density becomes null at large distances from the target. By means of the conditions:

$$t \gg \frac{\alpha^2}{2\lambda(dt)^{\left(\frac{2}{D_F}\right)-1}} \tag{9a}$$

$$x \ll \frac{c\alpha^2}{2\lambda(dt)^{\left(\frac{2}{D_F}\right)-1}} \tag{9b}$$

the current density takes the approximate form:

$$j(x, t) = v(x, t)(x, t) = \pi^{-1}\frac{c}{\alpha}\left[\frac{\alpha^2}{2\lambda(dt)^{\left(\frac{2}{D_F}\right)-1}}\right]^2 \frac{1}{t^3} \times \exp\left\{-\left[\frac{\alpha^2}{2\lambda(dt)^{\left(\frac{2}{D_F}\right)-1}}\right]^2 \left(\frac{x}{t} - c\right)^2\right\} \tag{10}$$

which is similar with a shifted Maxwellian distribution [99, 100], often used for treating the temporal trace of the LP current:

$$j(t)\frac{1}{t^3}\exp\left[\frac{-m}{2k_B T_{ionic}}\left(\frac{d}{t} - v_{drift}\right)^2\right] \tag{11}$$

through the following identities:

$$x \equiv d, v_{drift} \equiv c, \frac{m}{2k_B T_{ion}} \equiv \left[\frac{\alpha^2}{2\lambda(dt)^{\left(\frac{2}{D_F}\right)-1}}\right]^2 \tag{12}$$

Given the dependences of the multiple dynamic variables (ρ, v, j, etc.) on the external factors ($ex : \alpha, \lambda, (dt)^{\left(\frac{2}{D_F}\right)-1}$, etc.), we choose an adequate normalization that will allow us to obtain more compact and simplified dependences. Basically, we want to reduce the explicit dependences on the external factors. Thus, by means of the condition $2\lambda = \alpha c$, we will choose the normalization:

$$\xi \rightarrow \frac{x}{\alpha}, \tag{13a}$$

$$\tau \rightarrow \frac{tc}{\alpha}, \tag{13b}$$

$$\mu \rightarrow (dt)^{(4/D_F)-2} \tag{13c}$$

This allows us to re-write the dependencies of the plasma dynamic variables on the external factors as follows [76]:

Normalized velocity:

$$V(\xi, \tau, \mu) = \frac{1 + \mu\xi\tau}{1 + \mu\tau^2} \tag{14}$$

Normalized particle density:

$$N(\xi, \tau, \mu) = \left[1 + \mu\tau^2\right]^{-1/2} \cdot \exp\left[\frac{-(\xi - \tau)^2}{1 + \mu\tau^2}\right] \tag{15}$$

Normalized current density:

$$J(\xi, \tau, \mu) = \frac{1 + \mu\xi\tau}{\left[1 + \mu\tau^2\right]^{3/2}} \cdot \exp\left[\frac{-(\xi - \tau)^2}{1 + \mu\tau^2}\right] \tag{16}$$

Let us analyze the influence of the μ parameter (named fractalization degree) which contains the contribution of all external factors, by means of the fractal-non-fractal transition coefficient, λ, and scale resolution, dt and fractal dimension, D_F, of the movement curves. From a physical perspective, fractalization implies different statistics (from Levy-type movements to Brownian movements, either by means of non-Markovian processes or Markovian ones [63, 98]) that are imposed by the fundamental processes involved in the plasma formation and expansion. In such context, we assumed that generally for laser ablation there are two main mechanisms responsible for the formation and expansion of two plasma structures [15, 16, 56, 67, 73, 80–82], which were previously experimentally evidenced. Thus, the first structure is a result of the electrostatic interactions, at very short time scale, the positive charge left on the target surface by electron laser excitation and detachment would accelerate the positive ions outwards the surface (Coulomb mechanism [13]). The second plasma structure that requires more time to form and expand is generated through thermal mechanisms (evaporation, phase explosion

[101]). Each mechanism is characterized in our model by a specific scale resolution, that is, an adequate choice of the fractalization degree, μ.

In this context, in **Figure 10a** and **b** the current densities given by Eq. (16) are plotted versus time (τ) for various normalized distances (ξ) and fractalization degrees: $\mu = 0.3$ (a) and $\mu = 7$ (b). Such distinct evolutions are associated with the two plasma structures, while their overlapping fits well the experimental LP temporal trace given in **Figure 2**, except the current oscillations which should be treated separately. The space-time evolution of the variables describing the plasma is in agreement with other theoretical models [102, 103]. Moreover, having in view the normalization used in our model, it is possible to control some dynamics of the plasma plume by means of the fractalization degree μ.

The validation of our model comes from the comparison with the time-dependence of plasma parameters obtained from LP measurements. In **Figure 11a** one observes that the time-decrease of the ion density is well fitted by the dependence (15), with the parameters

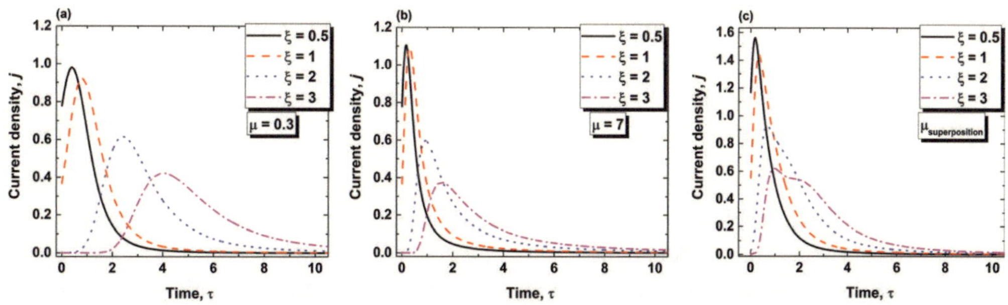

Figure 10. Current densities given by Eq. (16) dependence versus time (τ) for various normalized distances (ξ) and fractalization degrees: $\mu = 0.3$ (a) and $\mu = 7$ (b) and their overlapping (c).

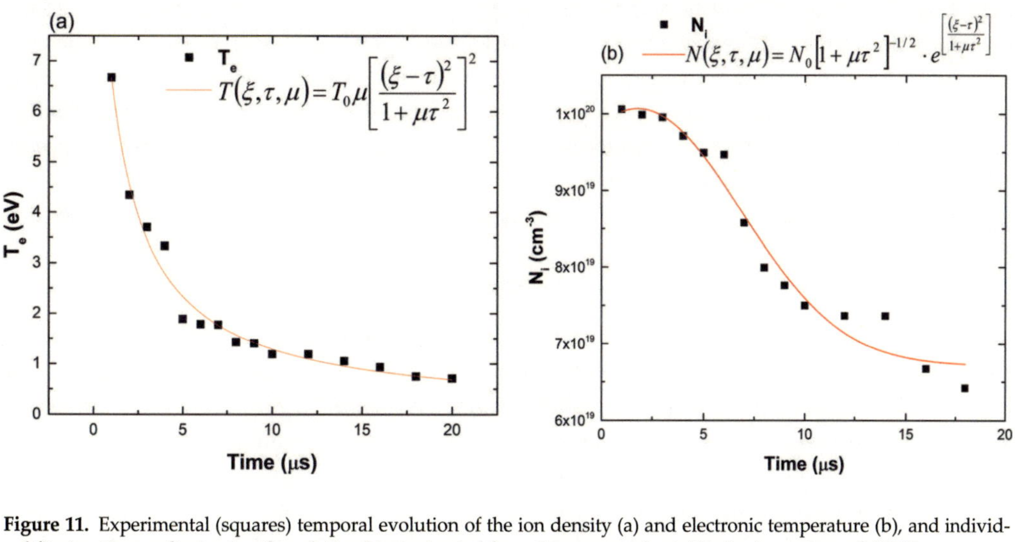

Figure 11. Experimental (squares) temporal evolution of the ion density (a) and electronic temperature (b), and individual fits (continuous line) using the relationships extracted from the compact fractal hydrodynamic model [76].

$N_0 = 4.25 \times 10^{19}\ \text{cm}^{-3}$, $\xi \sim 11.8$, $\mu \sim 2.95$, where N_0 is a normalization constant. Moreover, the right term of Eq. (4a) is the space derivative of a fractal potential,

$$Q = 2\lambda^2 (dt)^{\left(\frac{4}{D_F}\right)-2} \rho^{\frac{-1}{2}} \partial_{xx} \rho^{\frac{-1}{2}} \tag{17}$$

Then, using Eq. (17) and the normalizations (13a)–(13c), the fractal potential takes the normalized form:

$$q(\xi, \tau, \mu) = \mu \left[\frac{(\xi - \tau)^2}{1 + \mu\tau^2} \right]^2 \tag{18}$$

In the "classical" LTE model, the plasma temperature (T) is a measure of the thermal movement (described by the random movement of the plasma ions, electrons and neutrals), property which is reflected here by the plasma particles motion curves non-differentiability. In our model, since the fractal potential (q) is a measure of non-differentiability, this leads to a relation of proportionality between the two dynamic variables ($q \sim T$). In such context, we observed that the dependence (18) fits well the time evolution of the electronic temperature (**Figure 11b**), with the parameters $T_0 = 0.68\ \text{eV}$, $\xi \sim 11.5$, $\mu \sim 2.8$, where T_0 is a normalization constant.

The normalization of the plasma dynamic variables led us to a more compact and simple form of the fractal theoretical model, by allowing the use of a single control parameter that embodies the contributions of several external parameters on the dynamics of the ejected particles. In its compact form, we were able for the first time [76] to define some clear associations between fractal model variables and specific plasma parameters (electron temperature, thermal velocity, particle density). Moreover, when compared with experimental data depicting the temporal evolution of the plasma parameters determined though the Langmuir probe method, the model satisfactorily reproduces the experimental traces. From the theoretical fit, we determined a range of values for the fractalization degree which describes the laser-produced plasmas in the ns ablation regime. The success of this non-differential approach is also seen from the fact that the model is able to "recognize" the probe—target distance at which the experimental data was recorded. Further studies are required in order to test the generality of the model, by comparing it against other "classical" theoretical models and also against data extracted from ps and fs laser-produced plasmas in various experimental conditions (target-probe distance, laser fluence, background pressure).

In the classical concepts, the theoretical models (hydrodynamic, kinetic, etc.) are built assuming that the dynamics of individual elements are characterized by continuous and differentiable motion variables (energy, momentum, density, etc.). These variables are exclusively dependent on the spatial coordinates and time. In the real situation, the complex system dynamics is much more complicated and the classical theoretical models failed in the attempt to explain all the concerned aspects. These difficulties can be overcome in a complementary approach, using fractal concepts, describing "exotic" shapes that did not fit the patterns of

Euclidean geometry. Moreover, the depth analysis of different complex systems evolution showed that most of the phenomena are nonlinear and, therefore, new mathematical tools were required. These have been provided by the Scale Relativity Theory (SRT) and by Extended Scale Relativity Theory (ESRT) [98], that is, the SRT with an arbitrary constant fractal dimension. These theories consider that the motions of the complex systems structural units take place on continuous but non-differentiable curves (fractal curves). In this situation, Euclidean dynamics of a complex system subjected to external constraints is replaced by a fractal dynamics characterizing the same system free of any external constraints. More precisely, Euclidian constraints dependent motions, that is, on continuous but differentiable curves, are substituted by constraints independent motion in a fractal space, that is, on continuous but non-differentiable (fractal) curves (a free motion).

5. Conclusions

The dynamics of the transient laser-produced plasma in vacuum was investigated using electrical (Langmuir probe, target current) and optical (fast gate intensified CCD camera imaging) measurements for various ablation regimes and simple target materials. The typical time-evolution of the ion current recorded by the probe placed at various distances (axial, radial) with respect to the center of the laser irradiation spot shows that it extends in the μs range, and it generally consists in a fast part having an oscillatory behavior, and a slower tail which arrives at longer times. For the expansion velocities, values in the range of 10^4 m/s for the first (fast) structure and of 10^3 m/s for the second (slow) one were found, both from electrical and optical methods, which are in agreement with experimental results given in the literature and rough calculations performed in simple thermodynamic framework.

Measurements of the current induced in the target by the ablation process showed that it is correlated with the probe signal, as the positive charging arises through the electrons escaping from the expanding plasma to the grounded chamber, while the negative charging is given by the ions escaping from the target. Biasing the target by an external voltage source, the negative part of the target current, which is given by the ion contribution, acquires an oscillatory behavior of the same frequency as previously recorded for the Langmuir probe current. Consequently, such periodic fluctuations are assumed to be induced by the probe/target electric field, with a target bias threshold for their occurrence.

Extracting the oscillatory part from the original temporal trace of Langmuir probe, the resulted time-dependence revealed a good fitting with a usual damped oscillator, while its frequency is connected with the plasma ion frequency. Moreover, we observed that biasing the target can influence the oscillation frequency, two regimes being inferred, corresponding to the fast and slow components, respectively.

Electrical measurements can be successfully used to calculate the global temperature and average charge state for the fast plasma structure, through time-integration of probe current intensity to obtain the total collected charge dependence *vs.* probe potential. We observed a high positive correlation between these two parameters, the result being in agreement with plasma standard theoretical models (e.g. local thermodynamic equilibrium or collisional radiative).

Another approach used by us to study the dynamics of the plume at relatively long delays is based on the treatment of the current-voltage characteristics (*I-V* plots), that are derived from the time-of-flight current profiles recorded at various probe biases and sampled at specific delays after the laser pulse. Space-time evolution of electronic temperature, thermal velocity, plasma potential and particle density were computed, all of them having a significant decrease, due to the cooling process and rarefaction during expansion. The technique has been implemented for the study of transient plasmas generated by laser ablation in three different temporal regimes (ns, ps, fs) on a series of six metallic targets (Al, Ti, Mn, Ni, Cu, Zn). A decrease of the ionic density with the increase of the melting point (or cohesive energy) is observed, and it appears to be a general characteristic of the laser-produced plasmas, as it is confirmed by our current systematic study on the ns, fs and ps ablation for a wide range of metallic targets. The evolution of the electronic temperature with the target electrical conductivity for the three ablation regimes presents a steep increase for low conductivity elements, reaching a quasi-saturation regime after 10 MS/m. However, the obtained values characterize only a particular plasma volume investigated at a specific moment in time, and further investigations are required to establish the universality of the proposed dependence.

Theoretically, a compact version using normalized variables of our previous fractal hydrodynamic model was proposed for the analysis of the spatial and temporal evolution of some plasma dynamic variables. In this context, a new parameter named fractalization degree was introduced to account for the contribution of all external factors, that is, the fractal-non-fractal transition coefficient, the scale resolution, and the fractal dimension of the movement curves. When compared with experimental data depicting the temporal evolution of the plasma parameters determined through the Langmuir probe method, the model satisfactorily reproduces the experimental traces. In the compact form of the model, we were able to define some clear associations between fractal model variables and specific plasma parameters. From the theoretical fit, we determined a range of values for the fractalization degree which describes the laser-produced plasmas in the ns ablation regime, while more efforts are required to elucidate some features associated with femtosecond laser ablation (e.g. evolution of the oscillation period with the target atomic mass).

Acknowledgements

This work has been partially supported by the Agence Nationale de la Recherche through the LABEX CEMPI (ANR-11-LABX-0007), as well as by the Ministry of Higher Education and Research, Hauts de France Council and European Regional Development Fund (ERDF) through the Contrat de Projets Etat-Region (CPER Photonics4Society). S.A.I. thanks the Institut Français de Bucharest for a BGF cotutelle PhD grant.

http://dx.doi.org/10.5772/intechopen.70759

Author details

Petru-Edward Nica[1], Stefan Andrei Irimiciuc[2,3], Maricel Agop[1], Silviu Gurlui[2], Michael Ziskind[3] and Cristian Focsa[3]*

*Address all correspondence to: cristian.focsa@univ-lille1.fr

1 Department of Physics, Gheorghe Asachi Technical University, Iasi, Romania

2 Faculty of Physics, "Alexandru Ioan Cuza" University of Iasi, Iasi, Romania

3 Université de Lille, CNRS, UMR 8523, PhLAM—Physique des Lasers Atomes et Molécules, CERLA— Centre d'Etudes et de Recherches Lasers et Applications, Lille, France

References

[1] Eason R, editor. Pulsed Laser Deposition of Thin Films: Applications-Led Growth of Functional Materials. New Jersey: Wiley; 2011

[2] Miotello A, Ossi P, editors. Laser-Surface Interactions for New Materials Production: Tailoring Structure and Properties. New York: Springer; 2010

[3] Gerhard C, Wieneke S, Viöl W, editors. Laser Ablation: Fundamentals, Methods and Applications. New York: Nova Science Publishers; 2015

[4] Russo RE, Mao X, Gonzalez JJ, Zorba V, Yoo J. Laser ablation in analytical chemistry. Analytical Chemistry. 2013;**85**:6162-6177

[5] Cremers D, Radziemski L, editors. Handbook of Laser-Induced Breakdown Spectroscopy. Chichester: Wiley; 2006

[6] Phipps CR, editor. Laser Ablation and Its Applications. New York: Springer; 2007

[7] Rethfeld B, Sokolowski-Tinten K, von der Linde D, Anisimov SI. Timescales in the response of materials to femtosecond laser excitation. Applied Physics A: Materials Science & Processing. 2004;**79**:767-769

[8] Kelly R, Miotello A. On the role of thermal processes in sputtering and composition changes due to ions or laser pulses. Nuclear Instruments and Methods in Physics Research Section B. 1998;**141**:49-60

[9] Peterlongo A, Miotello A, Kelly R. Laser-pulse sputtering of aluminum: Vaporization, boiling, superheating, and gas-dynamic effects. Physical Review E. 1994;**50**: 4716-4727

[10] Shirk MD, Molian PA. A review of ultrashort pulsed laser ablation of materials. Journal of Laser Applications. 1998;**10**:18-28

[11] Leitz KH, Redlingshöfer B, Reg Y, Otto A, Schmidt M. Metal ablation with short and ultrashort laser pulses. Physics Procedia. 2011;**12**:230-238

[12] Bulgakova NM, Stoian R, Rosenfeld A, Hertel IV, Marine W, Campbell EEB. A general continuum approach to describe fast electronic transport in pulsed laser irradiated materials: The problem of Coulomb explosion. Applied Physics A: Materials Science & Processing. 2005;**81**:345-356

[13] Harilal SS, Freeman JR, Diwakar PK, Hassanein A. Femtosecond Laser Ablation: Fundamentals and Applications. In: Musazzi S, Perini U, editors. Laser-Induced Breakdown Spectroscopy Theory and Applications. New York: Springer; 2014. p. 143-166

[14] Geohegan DB, Puretzky AA. Dynamics of laser ablation plume penetration through low pressure background gases. Applied Physics Letters. 1995;**67**:197-199

[15] Geohegan DB, Puretzky AA. Laser ablation plume thermalization dynamics in background gases: Combined imaging, optical absorption and emission spectroscopy, and ion probe measurements. Applied Surface Science. 1996;**96–98**:131-138

[16] Wood RF, Chen KR, Leboeuf JN, Puretzky AA, Geohegan DB. Dynamics of plume propagation and splitting during pulsed-laser ablation. Physical Review Letters. 1997;**79**:1571-1574

[17] Wood RF, Leboeuf JN, Geohegan DB, Puretzky AA, Chen KR. Dynamics of plume propagation and splitting during pulsed-laser ablation of Si in He and Ar. Physical Review B. 1998;**58**:1533-1543

[18] Harilal SS, Issac RC, Bindhu CV, Nampoori VPN, Vallabhan CPG. Temporal and spatial evolution of C_2 in laser induced plasma from graphite target. Journal of Applied Physics. 1996;**80**:3561-3565

[19] Harilal SS, Issac RC, Bindhu CV, Nampoori VPN, Vallabhan CPG. Emission characteristics and dynamics of C_2 from laser produced graphite plasma. Journal of Applied Physics. 1997;**81**:3637-3643

[20] Harilal SS, Bindhu CV, Tillack MS, Najmabadi F, Gaeris AC. Plume splitting and sharpening in laser-produced aluminium plasma. Journal of Physics D: Applied Physics. 2002;**35**:2935-2938

[21] Harilal SS, Bindhu CV, Tillack MS, Najmabadi F, Gaeris AC. Internal structure and expansion dynamics of laser ablation plumes into ambient gases. Journal of Applied Physics. 2003;**93**:2380-2388

[22] Diwakar PK, Harilal SS, Hassanein A, Phillips MC. Expansion dynamics of ultrafast laser produced plasmas in the presence of ambient argon. Journal of Applied Physics. 2014;**116**:133301

[23] Bulgakov AV, Bulgakova NM. Gas-dynamic effects of the interaction between a pulsed laser ablation plume and the ambient gas: Analogy with an underexpanded jet. Journal of Physics D: Applied Physics. 1998;**31**:693-703

[24] Bulgakova NM, Panchenko AN, Zhukov VP, Kudryashov SI, Pereira A, Marine W, Mocek T, Bulgakov AV. Impacts of ambient and ablation plasmas on short- and ultrashort-pulse laser processing of surfaces. Micromachines. 2014;**5**:1344-1372

[25] Harilal SS, Tillack MS, O'Shay B, Bindhu CV, Najmabadi F. Confinement and dynamics of laser-produced plasma expanding across a transverse magnetic field. Physical Review E. 2004;**69**:026413

[26] Anoop KK, Harilal SS, Philip R, Bruzzese R, Amoruso S. Laser fluence dependence on emission dynamics of ultrafast laser induced copper plasma. Journal of Applied Physics. 2016;**120**:185901

[27] Anoop KK, Polek MP, Bruzzese R, Amoruso S, Harilal SS. Multidiagnostic analysis of ion dynamics in ultrafast laser ablation of metals over a large fluence range. Journal of Applied Physics. 2015;**117**:083108

[28] Anoop KK, Ni X, Wang X, Amoruso S, Bruzzese R. Fast ion generation in femtosecond laser ablation of a metallic target at moderate laser intensity. Laser Physics. 2014;**24**:105902

[29] Thestrup B, Toftmann B, Schou J, Doggett B, Lunney JG. Ion dynamics in laser ablation plumes from selected metals at 355 nm. Applied Surface Science. 2002;**197–198**:175-180

[30] Toftmann B, Schou J, Lunney JG. Dynamics of the plume produced by nanosecond ultraviolet laser ablation of metals. Physical Review B. 2003;**67**:104101

[31] Gonzalo J, Siegel J, Perea A, Puerto D, Resta V, Galvan-Sosa M, Afonso CN. Imaging self-sputtering and backscattering from the substrate during pulsed laser deposition of gold. Physical Review B. 2007;**76**:035435

[32] Schou J. Physical aspects of the pulsed laser deposition technique: The stoichiometric transfer of material from target to film. Applied Surface Science. 2009;**255**:5191-5198

[33] Canulescu S, Papadopoulou E, Anglos D, Lippert T, Montenegro MJ, Georgiou S, Döbeli M, Wokaun A. Nanosecond and femtosecond ablation of $La_{0.6}Ca_{0.4}CoO_3$: A comparison between plume dynamics and composition of the films. Applied Physics A. 2011;**105**: 167-176

[34] Bulgakova NM, Bulgakov AV, Bobrenok OF. Double layer effects in laser-ablation plasma plumes. Physical Review E. 2000;**62**:5624-5634

[35] Cleyssens F, Cheesman A, Henley SJ, Ashfold MNR. Studies of the plume accompanying pulsed ultraviolet laser ablation of zinc oxide. Journal of Applied Physics. 2002;**92**: 6886-6894

[36] Hansen TN, Schou J, Lunney JG. Angle-resolved energy distributions of laser ablated silver ions in vacuum. Applied Physics Letters. 1998;**72**:1829-1831

[37] Amoruso S, Armenante M, Berardi V, Bruzzese R, Velotta R, Wang X. High fluence visibleand ultraviolet laser ablation of metallic targets. Applied Surface Science. 1998; **127-129**:1017-1022

[38] Amoruso S, Bruzzese R, Spinelli N, Velotta R. Characterization of laser-ablation plasmas. Journal of Physics B: Atomic, Molecular and Optical Physics. 1999;**32**:R131-R172

[39] Schou J, Amoruso S, Lunney JG. Plume dynamics. In: Phipps C, editor. Laser Ablation and Its Applications. New York: Springer; 2007. p. 67-95

[40] Doggett B, Lunney JG. Langmuir probe characterization of laser ablation plasmas. Journal of Applied Physics. 2009;**105**:033306

[41] Esposito M, Lippert T, Schneider CW, Wokaun A, Donnelly T, Lunney JG, Tellez H, Vadillo JM, Laserna JJ. Pulsed laser ablation of silver: Ion dynamics in the plasma plume. Journal of Optoelectronics and Advanced Materials. 2010;**12**:677-680

[42] Toftmann B, Doggett B, Budtz-Jorgensen C, Schou J, Lunney JG. Femtosecond ultraviolet laser ablation of silver and comparison with nanosecond ablation. Journal of Applied Physics. 2013;**113**:083304

[43] Chen J, Lunney JG, Lippert T, Ojeda-G-P A, Stender D, Schneider CW, Wokaun A. Langmuir probe measurements and mass spectrometry of plasma plumes generated by laser ablation of $La_{0.4}Ca_{0.6}MnO_3$. Journal of Applied Physics. 2014;**116**:073303

[44] Amoruso S, Wang X, Altucci C, de Lisio C, Armenante M, Bruzzese R, Spinelli N, Velotta R. Double-peak distribution of electron and ion emission profile during femtosecond laser ablation of metals. Applied Surface Science. 2002;**186**:358-363

[45] Amoruso S, Armenante M, Bruzzese R, Spinelli N, Velotta R, Wang X. Emission of prompt electrons during excimer laser ablation of aluminum targets. Applied Physics Letters. 1999;**75**:7-9

[46] Sunil S, Kumar A, Singh RK, Subramanian KP. Measurements of electron temperature and density of multi-component plasma plume formed by laser-blow-off of LiF-C film. Journal of Physics D: Applied Physics. 2008;**41**:085211

[47] Kumar A, Singh RK, Thomas J, Sunil S. Parametric study of expanding plasma plume formed by laser-blow-off of thin film using triple Langmuir probe. Journal of Applied Physics. 2009;**106**:043306

[48] Singh SC, Fallon C, Hayden P, Mujawar M, Yeates P, Costello JT. Ion flux enhancements and oscillations in spatially confined laser produced aluminum plasmas. Physics of Plasmas. 2014;**21**:093113

[49] Hutchinson IH. Principles of Plasma Diagnostics. 2nd revised ed. Cambridge: Cambridge University Press; 2005

[50] Griem HR. Principles of Plasma Spectroscopy. Cambridge: Cambridge University Press; 1997

[51] Kirkwood RK, Moody JD, Kline J, Dewald E, Glenzer S, Divol L, Michel P, Hinkel D, Berger R, Williams E, Milovich J, Yin L, Rose H, MacGowan B, Landen O, Rosen M, Lindl J. A review of laser–plasma interaction physics of indirect-drive fusion. Plasma Physics and Controlled Fusion. 2013;**55**:103001

[52] Gurlui S, Sanduloviciu M, Strat M, Strat G, Mihesan C, Ziskind M, Focsa C. Dynamic space charge structures in high fluence laser ablation plumes. Journal of Optoelectronics and Advanced Materials. 2006;**8**:148-151

[53] Gurlui S, Sanduloviciu M, Mihesan C, Ziskind M, Focsa C. Periodic phenomena in laser-ablation plasma plumes: A self-organization scenario. AIP Conference Proceedings. 2006;**812**:279-282

[54] Focsa C, Ziskind M, Ursu C, Gurlui S, Pagnon D, Pellerin S, Pellerin N, Dudeck M. Laser-$BNSiO_2$ ceramics interaction: Simulation of the energy deposition on dielectric wall surfaces in Hall thrusters. Journal of Optoelectronics and Advanced Materials. 2008;**10**:2380-2385

[55] Gurlui S, Agop M, Nica P, Ziskind M, Focsa C. Experimental and theoretical investigations of a laser-produced aluminum plasma. Physical Review E. 2008;**78**:026405

[56] Ursu C, Gurlui S, Focsa C, Popa G. Space- and time-resolved optical diagnosis for the study of laser ablation plasma dynamics. Nuclear Instruments and Methods in Physics Research Section B. 2009;**267**:446-450

[57] Focsa C, Nemec P, Ziskind M, Ursu C, Gurlui S, Nazabal V. Laser ablation of As_xSe_{100-x} chalcogenide glasses: Plume investigations. Applied Surface Science. 2009;**255**:5307-5311

[58] Nica P, Vizureanu P, Agop M, Gurlui S, Focsa C, Forna N, Ioannou P, Borsos Z. Experimental and theoretical aspects of Aluminum expanding laser plasma. Japanese Journal of Applied Physics. 2009;**48**:066001

[59] Agop M, Nica P, Gurlui S, Focsa C. Fractal hydrodynamic model of high fluence laser ablation plasma expansion. AIP Conference Proceedings. 2010;**1278**:612-622

[60] Nica P, Agop M, Gurlui S, Focsa C. Oscillatory Langmuir probe ion current in laser produced plasma expansion. Europhysics Letters. 2010;**89**:65001

[61] Cimpoesu RH, Pompilian GO, Baciu C, Cimpoesu N, Nejneru C, Agop M, Gurlui S, Focsa C. Pulsed laser deposition of poly(l-Lactide) acid on nitinol substrate. Optoelectronics and Advanced Materials, Rapid Communications. 2010;**4**:2148-2153

[62] Ursu C, Pompilian OG, Gurlui S, Nica P, Agop M, Dudeck M, Focsa C. Al_2O_3 ceramics under high-fluence irradiation: Plasma plume dynamics through space-and time-resolved optical emission spectroscopy. Applied Physics A: Materials Science & Processing. 2010;**101**:153-159

[63] Agop M, Nica PE, Gurlui S, Focsa C, Paun VP, Colotin M. Implications of an extended fractal hydrodynamic model. European Physical Journal D: Atomic, Molecular, Optical and Plasma Physics. 2010;**56**:405-419

[64] Focsa C. Laser ablation transient plasma structures expansion in vacuum. IEEE Transactions on Plasma Sciences. 2011;**39**:Gurlui S, 2820-Gurl2821

[65] Nica P, Agop M, Gurlui S, Bejinariu C, Focsa C. Characterization of aluminum laser produced plasma by target current measurements. Japanese Journal of Applied Physics. 2012;**51**:106102

[66] Balika L, Focsa C, Gurlui S, Pellerin S, Pellerin N, Pagnon D, Dudeck M. Laser induced breakdown spectroscopy in a running Hall effect Thruster for space propulsion. Spectrochimica Acta Part B. 2012;**74–75**:184-189

[67] Pompilian OG, Gurlui S, Nemec P, Nazabal V, Ziskind M, Focsa C. Plasma diagnostics in pulsed laser deposition of GaLaS chalcogenides. Applied Surface Science. 2013;**278**: 352-356

[68] Balika L, Focsa C, Gurlui S, Pellerin S, Pellerin N, Pagnon D, Dudeck M. Laser ablation in a running Hall effect Thruster for space propulsion. Applied Physics A: Materials Science & Processing. 2013;**112**:123-127

[69] Dascalu G, Pompilian G, Chazallon B, Nica V, Caltun O, Gurlui S, Focsa C. Rare earth doped cobalt ferrite thin films deposited by PLD. Applied Physics A: Materials Science & Processing. 2013;**110**:915-922

[70] Dascalu G, Pompilian G, Chazallon B, Caltun O, Gurlui S, Focsa C. Femtosecond pulsed laser deposition of cobalt ferrite thin films. Applied Surface Science. 2013;**278**:38-42

[71] Irimiciuc SA, Mihaila I, Agop M. Experimental and theoretical aspects of a laser produced plasma. Physics of Plasmas. 2014;**21**:093509

[72] Irimiciuc S, Agop M, Nica P, Gurlui S, Mihaileanu D, Toma S, Focsa C. Dispersive effects in laser ablation plasmas. Japanese Journal of Applied Physics. 2014;**53**:116202

[73] Bulai G, Gurlui S, Caltun OF, Focsa C. Pure and rare earth doped cobalt ferrite laser ablation: Space and time resolved optical emission spectroscopy. Digest Journal of Nanomaterials and Biostructures. 2015;**10**:1043-1053

[74] Bulai G, Dumitru I, Pinteala M, Focsa C, Gurlui S. Magnetic nanoparticles generated by laser ablation in liquid. Digest Journal of Nanomaterials and Biostructures. 2016; **11**:283-293

[75] Irimiciuc S, Boidin R, Bulai G, Gurlui S, Nemec P, Nazabal V, Focsa C. Laser ablation of $(GeSe_2)_{100-x}(Sb_2Se_3)_x$ chalcogenide glasses: Influence of the target composition on the plasma plume dynamics. Applied Surface Science. 2017;**418B**:594-600

[76] Irimiciuc SA, Gurlui S, Nica P, Focsa C, Agop MA. Compact non-differential approach for modelling laser ablation plasma dynamics. Journal of Applied Physics. 2017;**121**: 083301

[77] Irimiciuc SA, Gurlui S, Bulai G, Nica P, Agop M, Focsa C. Langmuir probe investigation of transient plasmas generated by femtosecond laser ablation of several metals: Influence of the target physical properties on the plume dynamics. Applied Surface Science. 2017;**417**:108-118

[78] Focsa C, Gurlui S, Nica P, Agop M, Ziskind M. Plume splitting and oscillatory behavior in transient plasmas generated by high-fluence laser ablation in vacuum. Appl. Surf. Sci. 2017;**424P3**:299-309

[79] Nica P, Gurlui S, Osiac M, Agop M, Ziskind M, Focsa C. Investigation of Femtosecond Laser-Produced Plasma from Various Metallic Targets using Langmuir Probe Characteristic. Phys. Plasmas. 2017; in press

[80] Puretzky AA, Geohegan DB, Haufler RE, Hettich RL, Zheng XY, Compton RN. Laser ablation of graphite in different buffer gases. AIP Conference Proceedings. 1993;**288**: 365-374

[81] Harilal SS, Farid N, Freeman JR, Diwakar PK, LaHaye NL, Hassanein A. Background gas collisional effects on expanding fs and ns laser ablation plumes. Applied Physics A: Materials Science & Processing. 2014;**117**:319-326

[82] Ojeda-G-P A, Schneider CW, Döbeli M, Lippert T, Wokaun A. Plasma plume dynamics, rebound, and recoating of the ablation target in pulsed laser deposition. Journal of Applied Physics. 2017;**121**:135306

[83] Krása J, Jungwirth K, Gammino S, Krouský E, Láska L, Lorusso A, Nassisi V, Pfeifer M, Rohlena K, Torrisi L, Ullschmied J, Velyhan A. Partial currents of ion species in an expanding laser-created plasma. Vacuum. 2008;**83**:180-184

[84] Kelly R, Dreyfus RW. On the effect of Knudsen-layer formation on studies of vaporization, sputtering, and desorption. Surface Science. 1988;**198**:263-276

[85] Kabashin AV, Nikitin PI, Marine W, Sentis M. Experimental study of spontaneous electric field generated by a laser plasma. Applied Physics Letters. 1998;**73**:25-27

[86] Kabashin AV, Konov VI, Nikitin PI, Prokhorov AM, Konjević N, Vikor L. Laser plasma generation of currents along a conductive target. Journal of Applied Physics. 1990;**68**: 3140-3146

[87] Kabashin AV, Nikitin PI, Marine W, Sentis M. Electric fields of a laser plasma formed by optical breakdown of air near various targets. Quantum Electronics. 1998;**28**:24-28

[88] Wu J, Li X, Wei W, Jia S, Qiu A. Understanding plume splitting of laser ablated plasma: A view from ion distribution dynamics. Physics of Plasmas. 2013;**20**:113512

[89] Gornushkin IB, Panne U. Radiative models of laser-induced plasma and pump-probe diagnostics relevant to laser-induced breakdown spectroscopy. Spectrochimica Acta B. 2010;**65**:345-359

[90] Elsied AM, Termini NC, Diwakar PK, Hassanein A. Characteristics of ions emission from Ultrashort laser produced plasma. Scientific Reports. 2016;**6**:38256

[91] Weaver I, Martin GW, Graham WG, Morrow T, Lewis CLS. The Langmuir probe as a diagnostic of the electron component within low temperature laser ablated plasma plume. The Review of Scientific Instruments. 1999;**70**:1801-1805

[92] Hansen TN, Schou J, Lunney JG. Langmuir probe study of plasma expansion in pulsed laser ablation. Applied Physics A: Materials Science & Processing. 1999;**69**:S601-S604

[93] Dogar AH, Ilyas B, Ullah S, Nadeem A, Qayyum A. Langmuir probe measurements of Nd-YAG laser-produced copper plasmas. IEEE Transactions on Plasma Sciences. 2011;**39**:897-900

[94] Salle B, Chaleard C, Detalle V, Lacour JL, Mauchien P, Nouvellon C, Semerok A. Laser ablation efficiency of metal samples with UV laser nanosecond pulses. Applied Surface Science. 1999;**138-139**:302-305

[95] Schou J, Toftmann B, Amoruso S. Pulsed laser deposition: From basic processes to film deposition. Proceedings of SPIE. 2005;**5830**:1-10

[96] Olivier M, Němec P, Boudebs G, Boidin R, Focsa C, Nazabal V. Photosensitivity of pulsed laser deposited Ge-Sb-se thin films. Optical Materials Express. 2015;**5**:781-793

[97] Afifi MA, Labib HH, Fadel M. Electrical and thermal properties of chalcogenide glass system $Se_{75}Ge_{25-x}Sb_x$. Applied Physics A: Materials Science & Processing. 1992;**55**:167-169

[98] Merches I, Agop M. Differentiability and Fractality in Dynamics of Physical Systems. Singapore: World Scientific; 2015

[99] Krása J, Lorusso A, Doria D, Belloni F, Nassisi V, Rohlena K. Time-of-flight profile of multiply-charged ion currents produced by a pulse laser. Plasma Physics and Controlled Fusion. 2005;**47**:1339-1349

[100] Doria D, Lorusso A, Belloni F, Nassisi V. Characterization of a nonequilibrium XeCl laser-plasma by a movable Faraday cup. The Review of Scientific Instruments. 2004;**75**:387

[101] Kelly R, Miotello A. Comments on explosive mechanisms of laser sputtering. Applied Surface Science. 1996;**96–98**:205-215

[102] Anisimov SI, Luk'yanchuk BS. Selected problems of laser ablation theory. Physics-Uspekhi. 2002;**45**:293-324

[103] Murakami M, Kang YG, Nishihara K, Fujioka S, Nishimura H. Ion energy spectrum of expanding laser-plasma with limited mass. Physics of Plasmas. 2005;**12**:62706

Chapter 2

Nanosecond Laser Ablation: Mathematical Models, Computational Algorithms, Modeling

Vladimir I. Mazhukin

Additional information is available at the end of the chapter

http://dx.doi.org/10.5772/intechopen.70773

Abstract

The basic mathematical models, computational algorithms, and results of mathematical modeling of various modes of laser action on metals are considered. It is shown that for mathematical description and analysis of the processes of laser heating, melting, and evaporation of condensed media, various theoretical approaches are used: continuum, kinetic, atomistic, etc. Each of them has its own field of applicability, its advantages, and disadvantages. Mathematical description of ns-laser ablation is usually carried out within the framework of continuum approach in the form of hydrodynamic models that take into account reaction of irradiated material to varying density, pressure, and energy both in the target and in the vapor-gas medium. Within the framework of continuum approach, a multiphase, multifront hydrodynamic model and computational algorithm were constructed that were designed for modeling ns-PLA of metal targets embedded in gaseous media. It is shown that proposed model and computational algorithm allow to carry out the simulation of interrelated mechanisms of heterogeneous and homogeneous evaporation of metals manifested as a series of explosive boiling. Modeling has shown that explosive boiling in metals occurs due to the presence of a near-surface temperature maximum. It has been established that in ns-PLA, exposure regimes can be realized in which a phase explosion is the main mechanism of material removal. The verification of reliability of obtained results was carried out by comparing experimental data and calculations with atomistic models.

Keywords: nanosecond pulse, laser action, hydrodynamic model, mathematical modeling, explosive boiling, phase explosion, subsurface temperature maximum

1. Introduction

The pulsed laser ablation (PLA) of condensed media has been intensively studied over the past few decades [1, 2]. The increased interest in PLA is determined by the increasing possibilities of its use in a variety of applications, beginning with the already traditional ones: micro-processing [3, 4], pulsed laser deposition (PLD) [5, 6], laser-induced breakdown spectroscopy (LIBS) [7, 8], and new rapidly developing areas: production of nanomaterials [9, 10], surface nanostructuring [11], chemical, and physical synthesis [12]. The total effect of nanotechnology and photonics led to the emergence of a new direction—laser synthesis of colloids [13], which draws general attention to its extensive applicability, primarily in biomedicine [14, 15].

Numerous applications make PLA an attractive direction for fundamental investigations. Despite extensive studies of fundamental properties of laser ablation performed earlier, a number of important physical phenomena still remain insufficiently well studied and understood. Previous studies have established that the nature of interaction of laser radiation depends both on the modes of action: wavelength [16, 17], duration [18, 19], and laser pulse intensity [20, 21] and on thermophysical and optical properties of the target [22, 23], presence of the surrounding gas [24] and its pressure [25, 26]. The greatest differences in the physical mechanisms of laser ablation of metals are observed between short (ns) and ultrashort (ps, fs) pulsed modes [18, 19].

In ultrashort range (fs, pc) of influence, laser radiation freely reaches surface of the target. Absorption of laser radiation by a degenerate electron gas followed by a slowed-down energy exchange between electron and phonon components leads to strong deviation from locally thermodynamic equilibrium of the system as a whole. As a result, laser ablation and its accompanying processes develop after the end of the pulse.

Laser ablation in the nanosecond range is a more complex phenomenon involving many interrelated processes both during action and after the end of laser pulse. Such processes include laser target heating, heterogeneous, and homogeneous phase transformations, taking place in the evaporated matter, formation and expansion of plasma plume, heat transfer, laser and intrinsic plasma radiation transfer, generation and propagation of shock waves, and contact boundaries in the vaporized matter and the surrounding gas environment. In contrast to ultrashort regime, in nanosecond range, with a certain choice of parameters of the action, two experimentally observed and explored phenomena arise—volumetric boiling (phase explosion) of liquid phase of the target and formation of laser plasma in vaporized matter and surrounding gas.

Putting an irradiated target into an external gas environment, which is typical for most PLA applications, significantly complicates the situation. In these cases, the long-lived processes of laser-plasma plume and associated generation and propagation of interacting fronts of shock waves and contact boundaries in the vapor-gas medium play an important role in overall picture of laser ablation [27]. The presence of such a large number of interconnected physical processes creates additional difficulties in determination and investigation of the basic mechanisms of ablation. The information obtained by the methods of instrumental diagnostics is not sufficient due to the lack of data on fundamental phenomena associated with rapidly changing energy (thermal and laser radiation), hydrodynamic fields, and the kinetics of heterogeneous and homogeneous phase transformations in the solid and liquid phases, ionization of vaporized matter, and gas environment. At the same time, the understanding of fundamental physics of internal structure of plasma plumes and their spatiotemporal evolution in the process of plasma expansion in the background gas at atmospheric pressure is of decisive importance for many engineering applications. For this reason, nanosecond laser ablation continues to be an area of active research in which mathematical modeling plays an increasing role [28–31].

Any simulation begins with a choice of mathematical model, construction and development of which is given a paramount importance in the computing experiments. For theoretical description and analysis of PLA process of condensed media, various theoretical approaches are used: continuum, kinetic, atomistic (molecular dynamics, etc.). Each of them has its own field of applicability, its advantages, and disadvantages.

Atomistic models allow us to conduct research at the atomic level and obtain fundamental knowledge about structure, thermodynamic, and mechanical properties of crystalline materials [32, 33], about physical mechanisms of various processes [34, 35], including the kinetics of heterogeneous and homogeneous phase transitions [36, 37]. The basic methods of atomistic modeling—molecular dynamics (MD) and Monte Carlo (MC), which use as a rule, semiempirical interaction potentials, operate with tens and hundreds of millions of atoms, and allow calculations in the time range of nanosecond duration.

However, even with the use of parallel computer platforms, computational costs are enormous, and the space-time scales inherent for PLA processes are beyond the limits of accessibility for atomistic modeling methods. Therefore, in spite of constant progress in the field of designing interatomic potentials and increasing the power of computing systems, the final overcoming of computational constraints is hardly achievable, and continuum models will always remain relevant.

Continuous models based on the equations of continuous medium mechanics are realized, as a rule, in the form of hydrodynamic models [18, 24, 28, 30, 31, 38–42] and use the minimum of information and operate with average values of physical characteristics calculated on infinitesimal volume. The methods for solving them are more compact, they have higher accuracy and a relatively small amount of computation. The main shortcomings of continuum approach are manifested in the absence of the possibility of direct investigation of elementary processes in materials and limited possibilities of mathematical description of homogeneous mechanisms of phase transitions of the first kind and calculation of thermophysical, thermodynamic, optical, and other characteristics of matter in a wide range of parameters. These problems are much easier and more fully solved within the framework of atomistic modeling, the results of which can be used as input parameters in meso- and macrolevel models.

In this paper, the application of continuum approach to modeling in preplasma regime of processes dynamics and of main mechanisms of ns-PLA of metal target (Al) in air is considered. The mechanisms of heterogeneous and homogeneous phase transitions interacting with each other are analyzed in detail. Primary attention is paid to model the dynamics of phase explosion of liquid phase of aluminum and the expansion of its fragments in the air, since explosive boiling is considered to be one of the most effective thermal mechanisms of ns laser ablation of materials. Various aspects of this problem have been studied in a number of theoretical and experimental studies [43–56], but there is still no consensus on the mechanism of the phase explosion in metals. In order to obtain detailed information on interaction of heterogeneous and homogeneous mechanisms and data on laser plume morphology, simulation of laser heating, melting, surface evaporation, and evolution of plume in the vapor-gas medium is performed within the framework of new hydrodynamic model with temperature dependences of material properties of the target and explicit tracking of interphase boundary fronts, contact boundary, and shock wave. The release of liquid phase fragments into the atmosphere as a result of phase explosion is modeled by the procedure for introducing a quasinucleus of a new phase (vapor) in the region of near-surface maximum of temperature reaching the value $T_{max} \approx 0.9T_{cr}$ at time of reaching the maximum permissible overheating of metastable liquid phase.

2. Hydrodynamic model

Laser ablation of Al target placed in the air atmosphere is considered. In preplasma regimes of the action in ablation, the processes in three phase states (solid, liquid, and vapor) and two states (perturbed, unperturbed) of external gaseous medium are taken into account. Laser radiation propagates from right to left. The air for the selected action mode is completely transparent to the laser flow, which is partially reflected from the metal surface and partially absorbed in the near-surface layer of the target.

The mathematical description of ns laser ablation within the continuum approach is realized in the form of hydrodynamic models that allow one to take into account the reaction of condensed (target) and vapor-gas (vaporized matter, gas) media to varying density, pressure, and energy. The processes in each medium are described by a system of nonstationary equations of gas-hydrodynamics supplemented by equations of energy with thermal conductivity, equation of laser radiation transfer, and corresponding equations of state. The 1-D approximation is used for spatial variables. For laser action regimes under consideration, conditions of locally thermodynamic equilibrium (LTE) are assumed for all processes both in the irradiated target and in the vapor-gas medium. Accordingly, hydrodynamic model is formulated in one-temperature approximation.

Phase states and vapor-gas medium are separated among themselves by moving interphase boundaries solid-liquid $\Gamma_{s\ell}(t)$, liquid-vapor $\Gamma_{\ell v}(t)$, and by the fronts of contact boundary $\Gamma_{vg}(t)$ and shock-wave $\Gamma_{sh,g}(t)$. The right-hand boundary $\Gamma_g(t)$, running along unperturbed gas, is declared moving in order to improve economic efficiency of the computational algorithm. Schematically, their position and direction of motion are shown in **Figure 1**.

In order to obtain complete information on the kinetics of phase transformations, the morphology and dynamics of laser plume, all moving fronts and boundaries in the course of the solution are subjected to explicit tracking using appropriate relationships, which are simultaneously the boundary conditions for hydrodynamic and energy equations.

Complete system of equations is represented as:

http://dx.doi.org/10.5772/intechopen.70773

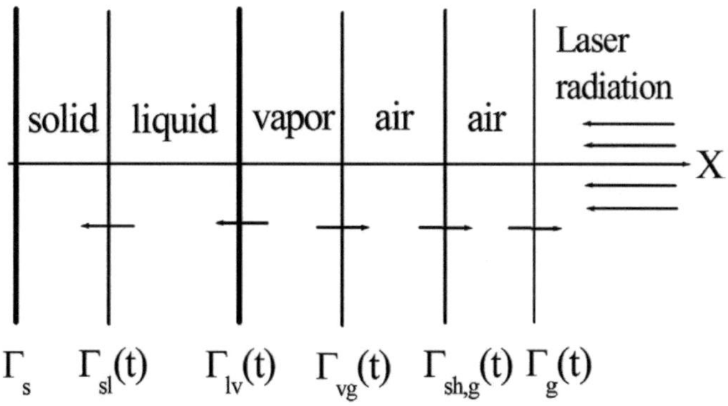

Figure 1. Scheme of spatial position of the phases and the direction of motion of interphase, contact boundaries, and the front of shock wave.

$$\left[\begin{array}{l} \frac{\partial \rho}{\partial t}+\frac{\partial(\rho u)}{\partial x}=0, \\ \frac{\partial(\rho u)}{\partial t}+\frac{\partial(\rho u^2)}{\partial x}+\frac{\partial P}{\partial x}=0, \\ \frac{\partial(\rho \varepsilon)}{\partial t}+\frac{\partial(\rho u \varepsilon)}{\partial x}=-\left(P\frac{\partial u}{\partial x}+\frac{\partial W_T}{\partial x}+\frac{\partial G}{\partial x}\right), \\ \frac{\partial G}{\partial x}+\kappa_L(\rho, T)G=0, \\ W_T=-\lambda(T)\frac{\partial T}{\partial x} \\ P=P(\rho, T), \quad \varepsilon=\varepsilon(\rho, T) \end{array}\right]_k \quad k=s, \ell, v, g \tag{1}$$

$$t>0, \quad \Gamma_s<x<\Gamma_{s\ell}(t)<x<\Gamma_{\ell v}(t)<\Gamma_{vg}(t)<\Gamma_g(t)$$

Here, ρ, u, ε, T, and P are the density, gas dynamic velocity, internal energy, temperature, and pressure of the substance, respectively, κ_L and G are the absorption coefficient and laser radiation fluence, W_T is heat flux density, and λ is coefficient of thermal conductivity. Indices s, ℓ, v, g denote the belonging of the quantities to solid, liquid, vapor, and air, respectively. In the condensed phase, the value ε_k has the meaning of enthalpy of liquid and solid phases H_k.

2.1. Boundary conditions

In the hydrodynamic models that easily combine with the kinetic ones, the heterogeneous mechanisms of phase transitions: melting/crystallization and evaporation/condensation are described naturally. A distinctive feature of heterogeneous phase transitions is the presence of sharp interfaces at which the main thermophysical and optical characteristics: enthalpy H, heat capacity C_p and thermal conductivity λ, density ρ, pressure p, and reflectivity R of the surface undergo a steplike change.

In the case of rapid phase transformations that typical for pulsed nanosecond action, a complete set of equations for mass, momentum, and energy flows is used to describe heterogeneous phase transitions. However, considering that the absence of phase equilibrium at interphase boundary is a necessary condition for phase transition to occur, the expressions for conservation laws must be supplemented by appropriate kinetic relations characterizing the degree of nonequilibrium of phase transition.

2.1.1. *Left fixed boundary,* x = Γ_s

The condition that flow of mass and heat be equal to zero is used as boundary conditions on the left fixed boundary:

$$x = \Gamma_s: \quad u = 0, \quad W_T = 0; \tag{2}$$

2.1.2. *Model of heterogeneous (surface) melting: crystallization,* õ = $\Gamma_{s\ell}(t)$

For fast phase transitions, heterogeneous melting model consists of a system of equations expressing three conservation laws: mass, momentum, and energy supplemented by kinetic condition for the velocity of melting front $v_{s\ell}$ [57] obtained from molecular-kinetic theory [58] and which is the main characteristic of the process melting-crystallization. In a stationary (laboratory) coordinate system, surface melting-crystallization model written on moving interface of melting $x = \Gamma_{s\ell}(t)$ can be represented as:

$$\begin{aligned} x = \Gamma_{s\ell}(t): \quad & \rho_s(u_s - v_{s\ell}) = \rho_\ell(u_\ell - v_{s\ell}) \\ & p_s + \rho_s(u_s - v_{s\ell})^2 = p_\ell + \rho_\ell(u_\ell - v_{s\ell})^2 \\ & \left(\lambda(T)\tfrac{\partial T}{\partial x}\right)_s - \left(\lambda(T)\tfrac{\partial T}{\partial x}\right)_\ell = \rho_s L_m^{ne} v_{s\ell} \end{aligned} \tag{3}$$

$$v_{s\ell}(\Delta T_{s\ell}) = \alpha(3k_B T_{s\ell}/m)^{1/2}\left(\exp\left(\beta\frac{L_m}{k_B T_m(p_s)}\frac{\Delta T_{s\ell}}{T_{s\ell}}\right) - 1\right) \tag{4}$$

$$L_m^{ne} = L_m(T_m(p_s)) + \Delta C_{ps}\Delta T_{s\ell} + \tfrac{\rho_s+\rho_\ell}{\rho_s-\rho_\ell}\tfrac{(u_s-u_\ell)^2}{2}, \quad L_m(T_m(p_s)) = L_{m,0} + \delta(T_m(p_s) - T_{m,0}),$$

$$\Delta C_{ps} = C_{ps} - C_{p\ell}, \quad \Delta T_{s\ell} = T_{s\ell} - T_m(p_s), \quad T_m(p_s) = T_{m,0} + \theta p_s,$$

where L_m^{ne} is nonequilibrium heat of melting, $L_{m,0}$, $T_{m,0}$ are equilibrium heat of melting and melting point, respectively, p_s is pressure on a solid surface. α, β, and δ are parameters determined from molecular modeling [59, 60]. For Al, $\alpha = 0.21$, $\beta = 5.28$, $\delta = 6.37$ J mol^{-1} K^{-1}, $\theta = 6.44 \times 10^{-3}$ K/atm.

2.1.3. *Model of heterogeneous evaporation* x = Γ_{lv}

Investigation of heterogeneous evaporation process began already in the century before last from experimental work: Hertz [61] and theoretical work: Knudsen [62] and continues at the present time, which is determined by its practical importance and not fully clarified features of nonequilibrium behavior of the substance during its evaporation. The mechanism of heterogeneous laser evaporation is realized in subcritical region of surface temperature T_{sur} and saturated vapor pressure $p_{sat} = p_{sat}(T_{sur})$, for which the following inequalities hold: $T_{sur} < T_{cr}$ and $p_{sat} < p_{cr}$. In the case $T_{sur} > T_{cr}$ or $p_{sat} > p_{cr}$, then a supercritical laser evaporation regime is realized, at which the state of the substance varies continuously [63].

Heterogeneous evaporation is characterized by a sharp phase boundary. Thin (several mean path lengths) nonequilibrium Knudsen layer (KL) is adjacent to the boundary surface. The nonequilibrium of the KL is determined by the flow of matter through the phase boundary. The total flow consists of flow of particles (atoms, molecules) leaving the condensed phase and flow of particles (atoms of molecules) returning from the evaporated matter as a result of collisions. These flows have different distribution functions, $f^{(+)}$ and $f^{(-)}$ respectively, and that leads to a strong nonequilibrium. The flow of injected particles satisfies the Maxwellian distribution function $f^{(+)}$ with parameters of the surface of the condensed medium. The return flow is determined by conditions of gas-dynamic expansion outside the KL. The behavior of particles inside a nonequilibrium Knudsen layer is described by Boltzmann equation, by solving of which one can determine the unknown parameters T_v and ρ_v that are boundary conditions for the continuum equations of gaseous medium. In the heterogeneous evaporation model, a special role is played by parameter M ($M = u_v/u_{sound}$ is Mach number) on the outer side of KL, which determines the degree of nonequilibrium of the phase transition.

In the phase equilibrium state, when the pressure of saturated vapor $p_{sat}(T_{sur})$ is equal to the external pressure p_v, the parameter $M = 0$. In the subsonic evaporation regime, when $M < 1$, the behavior of the interface depends on gas-dynamic perturbations in the vaporized matter flow. The maximum nonequilibrium is determined by the maximum value of mass flow, which is known to be achieved at $M = 1$, when flow of matter through the interface is maximal, and the recoil pressure at the heated surface is minimal. In this case, such evaporation regime can be realized when the behavior of condensed medium no longer depend on the external gas-dynamic problem, which greatly simplifies the description of evaporation process. However, the implementation area for such regime remains uncertain, because due to the nonlinearity of the gas dynamics equations, the transition line between $M < 1$ and $M \geq 1$ depends on desired solution and cannot be determined in advance [64]. Taking into account strong spatiotemporal diversity of processes in the nonequilibrium layer and continual media, KL is usually represented as a strong discontinuity in the gas-dynamic parameters. In such case, the kinetic processes in KL are not explicitly considered, and various phenomenological approaches [65–72] were used to determine boundary conditions on the outer side of the KL, which makes it possible without solving the kinetic problem, to determine the joining conditions under certain assumptions about the type of the nonequilibrium distribution function inside the discontinuity. One of the first works in which intensive evaporation was analyzed on the basis of phenomenological model is the work of Crout [65]. In this work, a nonequilibrium ellipsoidal Maxwellian particle distribution function was used written in an analytical form with the anisotropic in spatial directions longitudinal and perpendicular temperatures. Later papers used Mott-Smith approach [66], which was applied to the structure of the shock waves. With the help of this approach [67, 68] expressions for gas-dynamic parameters at $M = 1$ were obtained. Later in [69], this approach was extended to the entire range of evaporation $0 \leq M \leq 1$. In [70, 71], the approximation of the distribution function over the entire evaporation range $0 \leq M \leq 1$ was carried out by other more complicated expressions that satisfied the additional requirements that were formulated taking into account basic laws of gas dynamics. Following this laws, the used distribution functions must when $M \to 1$ to provide extreme values for all three complete flows of mass j_m, momentum j_i and energy j_e. The model [65] also satisfies these requirements. At the same time, in widely used model [69], the requirement of extrema of all flows at selected point $M = 1$ is not fulfilled. The calculations showed that total flows, j_m, j_i, j_e depending on M have extrema at $M = 0.88$, 1.18, 1.22, respectively. Nonfulfillment of the requirement of extremes indicates an unsuccessful choice of distribution function for the reverse flow $f^{(-)}$. The mathematical model of surface evaporation in the Knudsen layer approximation consists of three conservation laws and two additional relations from which the parameters on outer side of the Knudsen layer are determined: temperature T_v and density ρ_v. Velocity u_v and Mach number M are determined from the solution of the equations of gas dynamics:

$$\begin{aligned} x = \Gamma_{lv}(t): \qquad & j^m_{\ell v} = \rho_\ell(u_\ell - v_{\ell v}) = \rho_v(u_v - v_{\ell v}) \\ & j^i_{\ell v} = p_\ell + j^m_{\ell v}(u_\ell - v_{\ell v}) = p_v + j^m_{\ell v}(u_v - v_{\ell v}) \\ j^e_{\ell v} = -j^T_\ell + j^m_{\ell v}\left[H_\ell + \frac{(u_\ell - v_{\ell v})^2}{2}\right] - \sigma T^4 &= -j^T_v + j^m_{\ell v}\left[H_v + \frac{(u_v - v_{lv})^2}{2}\right] \end{aligned} \tag{5}$$

where $j^T_\ell = W_\ell = -\lambda(T_\ell)\frac{\partial T_\ell}{\partial x}$, $j^T_v = W_v = -\lambda(T_v)\frac{\partial T_v}{\partial x}$. With these expressions taken into account, energy conservation law can be represented as:

$$W_\ell - W_v = \rho_\ell v_{\ell v} L^{ne}_v + \sigma T^4, \tag{6}$$

where $L^{ne}_v = L^e_v(T_\ell) + C_{pv}(T_b - T_{\ell v}) + \frac{\rho_\ell + \rho_v}{\rho_\ell - \rho_v}\frac{(u_\ell - u_v)^2}{2}$ is the nonequilibrium heat of evaporation, σ is Stefan-Boltzmann law constant.

To determine parameters T_v, ρ_v, p_v, a modified Crout model was used [71].

$$T_v = \alpha_T(M)T_{sur}, \quad \rho_v = \rho_{sat}\alpha_\rho(M), \tag{7}$$

$$p_{sat} = p_b \exp\left[\frac{L_v(T)}{k_B}\left(\frac{1}{T_b} - \frac{1}{T_{sur}}\right)\right], \quad \alpha_T(M) = \frac{2\gamma M^2(m^2+0.5)^2}{(1+\gamma M^2)^2 m^2 t^2},$$

$$\alpha_\rho(M) = \frac{1}{\exp(-m^2) + \pi^{1/2}m(1+\mathrm{erf}(m))} \cdot \frac{(1+\gamma M^2)m^2}{\gamma M^2(m^2+0.5)^2},$$

the value m is determined from the equation $F(M)(m^2 + 0.5)^2 - m^2(m^2 + 1.5 + a) = 0$, where $F(M) = 1 + \frac{3\gamma M^2 - 1}{(\gamma M^2 - 1)^2}$, $a = 2t^2 - 0.5\pi^{1/2}mt - 1$, $t = \frac{2m}{\pi^{1/2}} + \frac{1+\mathrm{erf}(m)}{\exp(-m^2) + \pi^{1/2}m(1+\mathrm{erf}(m))}$.

ρ_{sat} and p_{sat} are the saturated vapor density and pressure, p_b is the pressure under normal conditions (1.01325×10^5 Pa), $u_{sound} = (\gamma RT)^{1/2}$ is the sound velocity, $M = u_v/u_{sound}$ is the Mach number, T_b is the boiling temperature, k_B is Boltzmann's constant. Taking into account p_{sat}, the value of ρ_{sat} was determined from the equation $p = p(\rho, T)$. For the laser radiation transfer equation on the target surface, the condition $G(\Gamma_{lv}) = (1 - R(T_{sur}))\,G_0\exp\left(-\left(\frac{t}{\tau}\right)^2\right)$ was used where $R(T_{sur})$ is the temperature dependence of reflectivity of target surface.

2.1.4. Model of surface condensation

When the inequality $p_{sat} < p$ is reached, the direction of phase transition changes, evaporation is replaced by condensation process on the target surface, to which the value $M < 0$ on outer side of the KL corresponds. Unlike evaporation, surface condensation can also occur in supersonic regime. According to [73] in subsonic condensation regime, gas dynamic quantities T_υ, M, and p_υ are related only by one two-parameter dependence $p_\upsilon/p_{sat} = F(T,M)$, where $T = T_\upsilon/T_{sur}$ and M are dimensionless parameters of temperature and velocity determined by the state of gas dynamic flow far from KL and are extrapolated from the gaseous medium to the discontinuity surface. The function $F(T,M)$ is determined in advance from the solution of Boltzmann equation, the results of which are then tabulated [74]. The values of the function $F(T,M)$ depend weakly on the parameter T and much stronger on the parameter M. Table values from [74] can be approximated by the expression [75]:

$$p_\upsilon/p_{sat} = F(T,\ M) \approx F(M) = 0.95 \times \exp(-2.42\ M) \quad (8)$$

When passing through the point $M = -1$ to supersonic regime of surface condensation, the boundary conditions change. In this case, all quantities on outer side of KL depend on the state of gas medium far from it and are extrapolated.

Thus, the description of the kinetics of heterogeneous evaporation is carried out by two one-parameter dependencies used as boundary conditions (for a known parameter M), and subsonic surface condensation is described by only one two-parameter (M,T) dependence.

2.1.5. Model of volumetric boiling of liquid phase heated by a laser pulse

The greatest difficulty in the continuum approach is the description of homogeneous mechanisms of phase transformations: melting crystallization and evaporation. Homogeneous mechanisms of phase transformations are characterized by nucleation of a new phase in a certain volume of superheated/supercooled matter. Representing them in continuum hydrodynamic models requires considerable additional efforts [57, 76] associated with formation of a cavity filled with vapor within a condensed medium.

The simplest scheme for simulating volumetric boiling in one-dimensional approximation can be represented by introducing, when certain criteria are satisfied into superheated liquid phase of artificial quasi-nuclei with thickness $h_i(t)$ bounded by moving planes of liquid-vapor interface $x_i(t) = \Gamma_{\ell\upsilon,i}(t)$, $x_{i+1}(t) = \Gamma_{\ell\upsilon,i+1}(t)$ where $i = 1, 2, \ldots$ is quasi-nuclei number. The criterion for nucleation beginning is a moment when superheating limit temperature $T_{\ell,\max}$ of liquid phase is reached and spatial coordinate. First quasi-nucleus of vapor phase with initial width $\Delta x_{\upsilon 1}(t) = \Gamma_{\ell\upsilon,2}(t) - \Gamma_{\ell\upsilon,1}(t) \sim 5$ nm is placed at this point. For a new region of vapor bounded by two flat surfaces, the initial conditions $T_{\upsilon,1} = T_{\ell,\max}$, $\rho_{\upsilon,i} = \rho_{sat}(T_{\ell,\max})$, $p_{\upsilon,1} = p_{sat}(T_{\ell,\max})$ were set. Under the influence of pressure difference between the inside quasi-nuclei and on the external irradiated surface of the liquid layer with the temperature T_{sur}, a rapidly expanding cavity filled with vapor is formed and blowout the liquid layer with thickness $\Delta x_{\ell 1}(t) = \Gamma_{\ell\upsilon,1}(t) - \Gamma_{\ell\upsilon}(t)$ in the direction of gaseous medium. The description of the processes in the expanding cavity and in the near-surface liquid layer separated from main target was carried out using gas-hydrodynamic system of Eq. (1). The model of heterogeneous evaporation (5)–(7) was used as boundary conditions on interphase planes $\Gamma_{\ell v,i}(t)$, $\Gamma_{\ell v,i+1}(t)$. Also, absorption and reflection of the intensity of laser radiation in the formed layer were taken into account. In case of repeated explosive boiling with the formation of next liquid fragment for each of them, as well as the main target, gaseous medium and formed vapor cavities, the solution algorithm remains unified.

The values of limit of superheat temperature depend on the rate of energy input and can be determined in advance from the molecular dynamics simulation. For Al, depending on the rate of heating, limit of superheat temperature of liquid phase T_{max} is in the range $T_{\max} \sim (0.89 \div 0.95) T_{cr}$ [77].

It should be noted that the use of gas dynamic equations inside the cavities which dimensions are small in the initial moments of time in comparison with the mean free path is a simplifying approximation that makes it easy to take into account the influence of expansion velocity of the cavity.

2.1.6. *Moving contact boundary*, $x = \Gamma_{vg}(t)$

On contact boundary and front of the shock wave, well-known standard relations are used [78]. At vapor-air interface, the boundary conditions were set in the form of equal values of velocity, pressure, and temperature:

$$x = \Gamma_{lv}(t): \quad u_v = u_g, \;\; P_v = P_g, \;\; T_v = T_g$$

2.1.7. *Moving shock wave*, $x = \Gamma_{sh,g}(t)$

A shock wave in air $\Gamma_{sh,g}(t)$ is a strong nonstationary discontinuity, on which three conservation laws are written in laboratory coordinate system [78]:

$$\begin{aligned} j^m_{sh,g} &= \rho_1 (u_1 - v_{sh,g}) = \rho_0 (u_0 - v_{sh,g}), \\ j^i_{sh,g} &= P_1 + \rho_1 (u_1 - v_{sh,g})^2 = P_0 + \rho_0 (u_0 - v_{sh,g})^2, \\ j^m_{sh,g} &\left(\varepsilon_1 + \frac{(u_1 - v_{sh,g})^2}{2} \right) - W_{T,1} = j^m_{sh,g} \left(\varepsilon_0 + \frac{(u_0 - v_{sh,g})^2}{2} \right) - W_{T,0} \end{aligned} \tag{9}$$

The indices 0 and 1 denote the values of the quantities on side of the background and the shock wave, respectively.

2.1.8. *Right moving boundary*, $x = \Gamma_g(t)$

The right boundary on the side of unperturbed gas is declared moving in order to improve economic effectiveness of the computational algorithm [79]. The speed of its motion is found from the differential equation of momentum.

2.2. Computational algorithm

The differential model (1)–(8) was approximated by a family of conservative finite-difference schemes [80, 81] written on computational grids with dynamic adaptation [79, 82, 83]. The method is based on the idea of transition to an arbitrary nonstationary coordinate system that allows calculations with an arbitrary number of discontinuous solutions, such as shock waves, propagating phase and temperature fronts, contact boundaries, and spalled fragments.

3. Temperature dependences of aluminum properties

In the mathematical modeling of the process of laser action on metals with an absorption coefficient of $\kappa_L \geq 10^7$ m^{-1}, the model of surface heating and evaporation is widely used. In this model, the maximum of the temperature profile T_{max} coincides with surface temperature T_{sur} from which evaporation occurs. At lower values of κ_L typical for dielectrics as well as for metals when irradiated with an electron beam or X-ray pulsed radiation or in the case of a metal-dielectric transition [84, 85], bulk heating of the substance and surface evaporation leads to the formation of temperature maximum below the surface of the substance. As a result, a volumetric explosive boiling of superheated liquid can occur in the region of the temperature maximum. The explosive boiling of superheated liquid is closely related to the concept of metal-dielectric phase transition. Zeldovich and Landau [86] denoted the possibility of a metal-dielectric phase transition for an expanded metal at subcritical temperatures. To perform the metal-dielectric transition, it is necessary that the energy acquired by metal atoms is higher than their binding energy in the crystal lattice and the distance between atoms is equal to the value causing a violation of their short-range order, which leads to the localization of the electrons on atoms [87].

With the advent of lasers capable of evaporating metals and producing plasma on their surfaces, it has become possible to observe laser-induced phase transitions metal-dielectric with the formation of transparency waves in the massive targets [88] and thin metallic films [89] in the subcritical temperature range. In other experimental and theoretical studies of the interaction of ns-laser pulses with an intensity of 10^7–10^8 W/cm^2 with metallic targets, the results of the appearance of metal-dielectric transitions accompanied by the formation of transparency waves are reported [53, 54, 85, 88–91].

Nevertheless, the physical mechanisms of interaction of laser radiation with metals taking into account metal-insulator transition have not yet been fully studied. This causes great difficulties in determining thermophysical and optical characteristics of metals in the vicinity of the critical region.

Figure 2 shows temperature dependences of thermophysical $\lambda(T)$, $L_v(T)$, $C_p(T)$ and optical $R(T)$, $\kappa_L(T)$ characteristics of aluminum. The curves were $L_v(T)$ and $C_p(T)$ obtained from molecular dynamics calculations, and $\lambda(T)$, $\kappa_L(T)$, and $R(T)$ were constructed on the basis of theoretical concepts [92–94] and reference data [95–97].

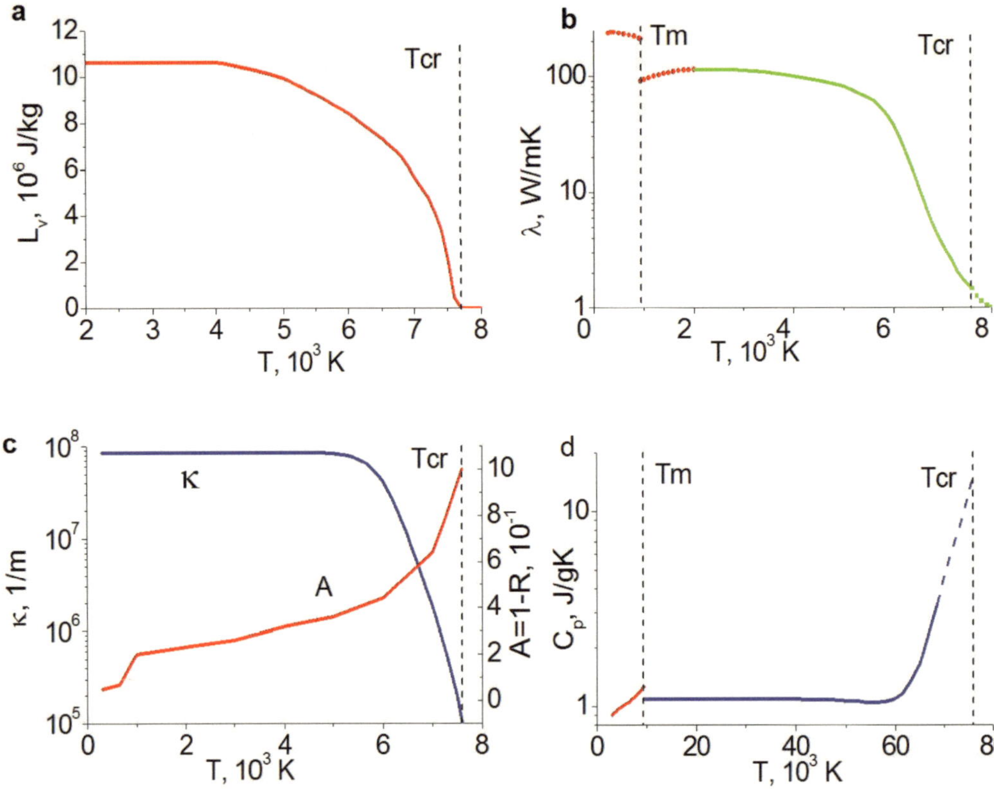

Figure 2. Temperature dependences of Al properties (a–d).

4. Results discussion

One of the purposes of this work is a detailed study of mechanism of explosive boiling in metals since explosive boiling is considered to be the most efficient thermal mechanism for the laser ablation of materials. At the same time, the difficulties associated with understanding of the mechanism of homogeneous phase transitions in metals occurring under the action of ns-laser pulses are known. In [98], based on analytical solution of thermal model, it was established that laser removal of material from a solid target, with a certain choice of irradiation parameters and material, is determined by the presence of near-surface temperature maximum. The validity of this position was confirmed by the results of numerical solution of thermal model for low-absorbing liquids irradiated by laser pulses [43] and nonmetallic solid materials [47]. For strongly absorbing media mainly metallic, calculations based on thermal model [44, 45, 47, 56] have shown that the magnitude of near-surface temperature maximum is several degrees. On this basis, overheating was excluded from consideration, up to the statement [56] that in metals, maximum temperature is always on the surface of the target, and sub-surface superheating is impossible. Explosive boiling was interpreted as a surface spitting of liquid phase when a critical temperature is reached on the surface [99]. This interpretation is not convincing, since it does not allow to determine even approximately the parameters of explosive boiling.

4.1. Mode of exposure

Let us consider ns-PLA process of aluminum target under the mode of exposure close to experimental conditions [53]. Al target with the thickness of 10^{-4} m is placed in the air under the normal conditions with pressure 1 bar and room temperature (T = 300 K). Laser pulse of Gaussian shape $G=(\beta/\pi)^{1/2}\ G_0 \exp(-\beta\ (t/\tau)^2)$ with full width at half maximum (FWHM) $\tau = 5 \times 10^{-9}$ s, with wavelength $\lambda_L = 1.06$ μm and fluence $F = 3.5$ J/cm^2, where $G_0 = 6.1 \times 10^8$ W/cm^2 is the peak intensity at $t = 0$, $-\infty < t < \infty$, $\beta = 4 \times \ln 2$ falls on the target surface from right to left. Air for the selected exposure mode is completely transparent for laser radiation, which is partially reflected from metal surface and partially absorbed by target material layer. The energy release of laser pulse has volumetric nature.

4.2. Evolution of processes

At the initial stage, time evolution of the processes at the surface of the target, in the target and in the gas medium near the target such as appearance of phase fronts (melting, evaporation), contact boundary, and shock wave and associated with formation of new phase media (liquid, vapor) occurs at the leading edge of laser pulse. Heterogeneous melting begins at the moment $t = -3.9 \times 10^{-9}$ s. The maximum velocity of melting front reaches a value $v_{sl} = 130$ ms^{-1}. As the interior of the target is heated, the melting front $\Gamma_{sl}(t)$ runs from its surface forming a new region of liquid phase. Further heating leads to appearance of heterogeneous evaporation front $\Gamma_{lv}(t)$ that runs inside the melt. A flow of the evaporated matter was formed at the surface of the melt pushing out air and creating another new phase—vapor. The new area occupied by vapor is limited on the one hand by the moving interface $\Gamma_{lv}(t)$ (evaporating surface) and on the other hand by the moving contact boundary vapor-air $\Gamma_{vg}(t)$. The surface evaporation process is controlled by the surface temperature $T_{sur}(t)$ and Mach number $M(t)$ on the outer side of KL ($0 < M(t) \le 1$). Evaporation begins at the moment $t = -2.3 \times 10^{-9}$ s when the saturated vapor pressure exceeds the pressure of external gas $p_{sat}(t) > p_g(t)$. The maximum velocity of evaporation front $v_{\ell v} = 88$ ms^{-1} is approximately 1.5 times smaller than velocity of melting front. The vaporized matter flow acting as a piston pushes out cold air, and performing a certain work warms up at peak G_0 intensity up to temperature $T_{vap} = 4.7 \times 10^3$ K. Under the pushing action of vapor flow, the compression of cold dense air which turns into a shock wave occurs, $t = -\ 2.1 \times 10^{-9}$ s. The spatial structure of erosion plume is shown in **Figure 3** by the main characteristics $T(x)$, $\rho(x)$, $u(x)$, $p(x)$ after the formation of shock wave.

By the moment $t = 0$, the shock wave propagates with velocity $v_{sh,g} = 2.9$ km s^{-1} and temperature $T_{sh,g} = 4.4 \times 10^3$ K, ahead of contact boundary moving with speed $v_{sh,g} = 2.5$ km s^{-1} toward the laser flow. The temperatures of vapor and air are in this case insufficient for initiation of ionization, and vapor-gas medium remains transparent for laser radiation.

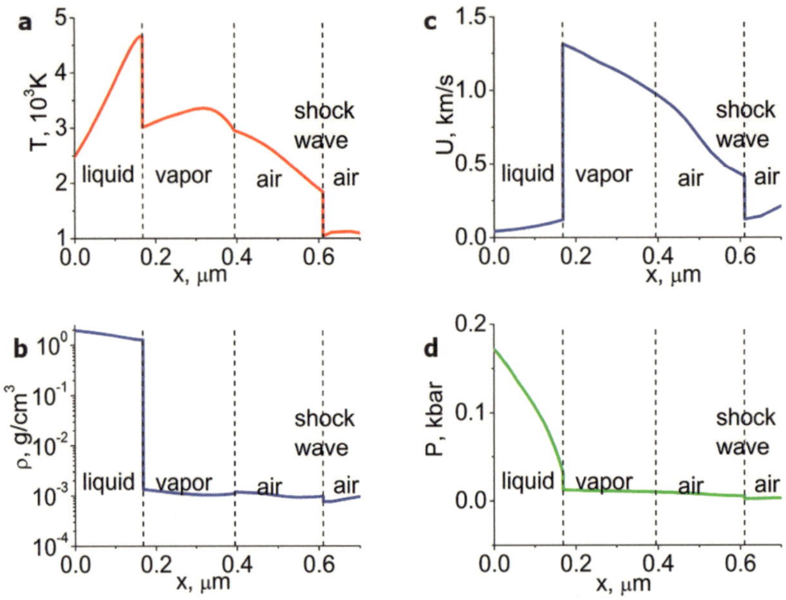

Figure 3. The spatial profiles of (a) $T(x)$, (b) $\rho(x)$, (c) $u(x)$, and (d) $p(x)$ in the target and the gas medium at the moment $t = -1.9$ ns.

4.3. Formation of near-surface temperature maximum

A consistent study of laser ablation is complicated by the fact that heterogeneous and homogeneous mechanisms of both melting and evaporation prove to be interrelated, and this interaction must be taken into account explicitly. Pulsed laser action on materials (including metals) has volumetric nature of energy release. Thus, when target is heated strongly in the vicinity of the critical point ($T_{cr} = 7600$ K, $\rho_{cr} = 0.47$ g cm^{-3}, $p_{cr} = 1.42$ kbar), thermophysical and optical properties of liquid phase change abruptly. Heat capacity C_p when approaching a critical point tends to infinity $C_p \to \infty$, thermal conductivity λl and absorption κ_L coefficients tend to zero, surface absorptivity $A = (1-R) \to 1$, **Figure 2a–c**. Under the joint influence of volumetric energy release of laser pulse and energy transfer by surface flow of evaporating substance, a near-surface maximum of temperature $T_{\ell,max}$ is formed in the depth of liquid phase, for which the following relations are satisfied: $T_{\ell,max} > T_{sur}$, $p_{\ell,max,sat} > p_{sur,sat}$.

Figure 4a shows a fragment of the spatial temperature distribution in near-surface layer of the target with overheating $T_{\ell,max} - T_{sur} = 170$ K at a depth of 30 nm from the irradiated surface at the moment immediately preceding the explosive boiling. For comparison, **Figure 4b** shows a fragment of temperature profile, calculated under the same conditions with help of molecular dynamics, which showed close results. The maximum was located at a depth of 70 nm, and the superheat value was 150 K. Thus, simulation results obtained by different methods indicate the presence of temperature inhomogeneity in metallic target (Al) caused by a decrease in the temperature of the irradiated surface by heterogeneous evaporation. Thus, the conditions for phase explosion were created with formation of a cavity in the region of maximum temperature,

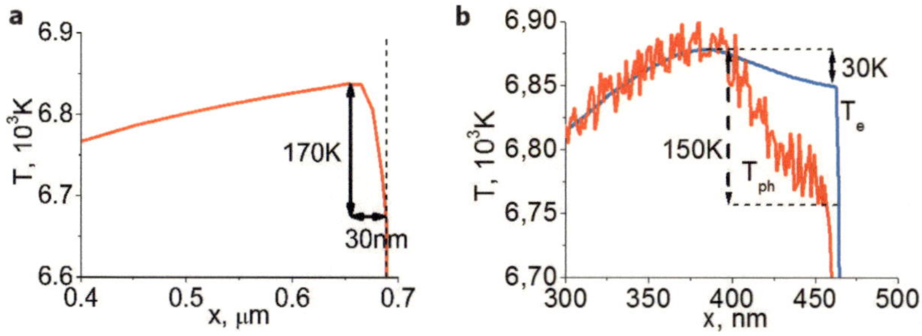

Figure 4. Spatial profiles of temperature maximum, (fragments) before explosive boiling: (a) continual model and (b) atomistic model.

where the deepest passing into the region of superheated metastable liquid critical point of liquid-vapor transition is achieved.

4.4. Explosive boiling

The process of explosive boiling begins at the backside front of the laser pulse. First boiling occurs at the moment t = +1.5 ns when the maximum of the temperature profile reaches the temperature of limiting superheating of liquid phase equal to $T_{\ell,max} = 0.9T_{cr} = 6840$ K. **Figure 5a–d** shows the spatial profiles of $T(x)$, $\rho(x)$, $u(x)$, $p(x)$. Arising cavity is the result of homogeneous nucleation in superheated liquid phase and is accompanied by sharp increase in pressure $p_{max} = p_{sat}(T_{\ell,max})$ which is more than two times greater than the recoil pressure on irradiated free surface $p_{sur} = 0.55p_{sat}(T_{sur})$. This pressure plays a decisive role in the expansion

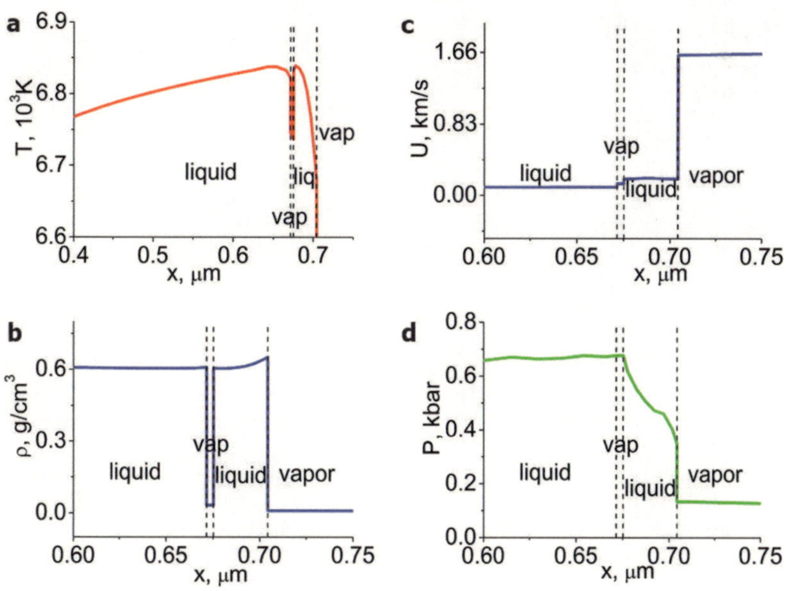

Figure 5. The spatial profiles of (a) $T(x)$, (b) $\rho(x)$, (c) $u(x)$, (d) $p(x)$ in the target and the gas medium at the moment $t = -1.7$ ns.

Fragment number, n	1	2	3	4	5
Moment of occurrence, ns	1.5	1.7	1.9	2.1	2.4
Fragment thickness d, nm	30	37	26	21	20

Table 1. Number, time of appearance and thickness of spalled fragments.

of cavity filled with vapor and in rapid growth of flyout velocity of thin $d_1 = 30$ nm fragment of liquid metal. Because of small thickness, flying fragment weakly absorbs laser radiation, the main part of which is released in the target increasing its temperature. Both surfaces of spalled fragment and surface of the target are subject to intense surface evaporation. Heterogeneous evaporation leads to decrease in temperature of the fragment and causes strong cooling of target surface, contributing to formation of next near-surface temperature maximum in which a new explosive boiling occurs when the temperature of limit overheating $T_{\ell,max} = 0.9T_{cr}$ of metastable liquid is reached. Repeated boiling occurred at a time $t = +1.7$ ns. The total number of explosive boilings reaches 5. The moments of occurrence and thickness of outgoing fragments are given in **Table 1**.

Figure 6a–d shows the state of all liquid fragments of explosive boiling at the time $t = +3.0$ ns. The resulting fragments eventually acquire a sufficiently high flyout velocity and under the influence of laser radiation absorption continue to evaporate until they disappear completely. The general trend in the evolution of all fragments is that over time the cavity size, flyout velocity and density of matter increase, and pressure in the cavity, thickness and temperature of the fragments decrease. So in the initially formed cavity the width increased from 5.0 nm to

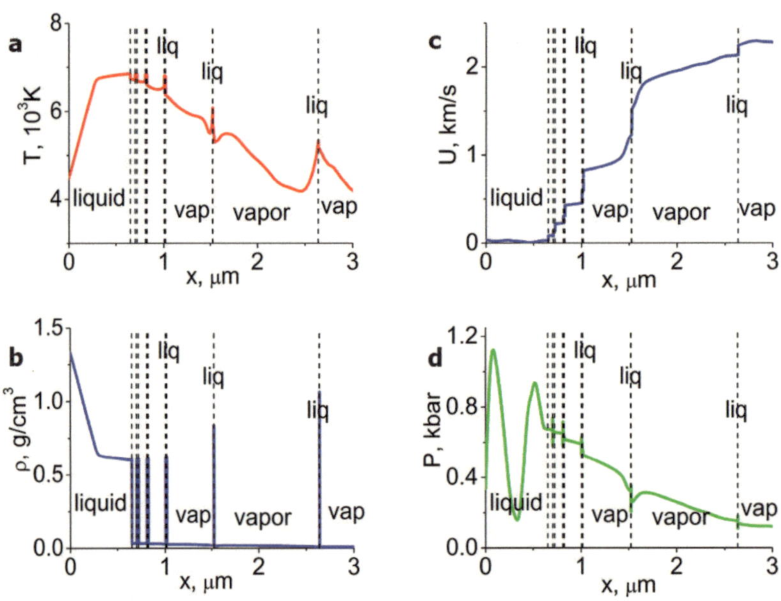

Figure 6. The spatial profiles of (a) $T(x)$, (b) $\rho(x)$, (c) $u(x)$, (d) $p(x)$ in the target and the gas medium after five explosive boiling at the moment $t = -3.0$ ns.

1.7 μm, the velocity increased up to 2.3 km/s and density due to cooling changed from 0.65 g/cm^3 to 1.3 g/cm^3. At the same time, the pressure in the cavity decreased to 200 bar, the thickness of fragment on the verge of extinction due to surface evaporation fell to h_ℓ = 1.9 nm, and the temperature decreased to T_ℓ = 5400 K.

The simulation results are in good qualitative agreement with the experimental data [53]. Quantitatively, the total depth due to release of superheated liquid phase (d_Σ = 37.0 nm) differs by about two times in excess. The total removal of matter by the mechanism of explosive boiling considerably exceeds the amount of evaporated matter. The trajectory of shock wave motion almost completely coincides with the experimental dependence [53] up to the moment t = 200 ns. From now, calculated curve is located above the experimental curve. As the fluence increases up to $F \sim 10$ J/cm^2, the target is heated faster and explosive boiling process begins to move to leading edge of the laser pulse. The frequency and amount of boiling thus increase to tens and hundreds of ejected liquid fragments, but the overall picture of processes does not undergo qualitative changes.

5. Conclusion

Within the framework of continuum approach, a multiphase, multifront hydrodynamic model and computational algorithm were constructed that were designed for modeling ns-PLA of metal targets embedded in gaseous media. The model contains temperature dependences of optical, thermal, and thermodynamic properties of metal target (Al). The temperature dependences of heat capacity $C_p(T)$ and thermal conductivity $\lambda(T)$ of aluminum take into account their singularity in the vicinity of critical point. In optical characteristics, the metal-insulator phase transition was taken into account accompanied by the formation of the transparency wave.

The computational algorithm is based on the method of dynamic adaptation, which ideally fit to the study of problems with heterogeneous phase transitions and allows the production of numerical solutions with an explicit allocation of unlimited number of interface boundaries in condensed medium, contact boundaries, and shock waves in a gaseous medium.

The calculations have shown that under the influence of temperature dependences of optical and thermophysical properties of Al the nature of laser heating of the target in high-temperature region changes substantially. The energy release has a volume nature, and irradiated surface is markedly cooled by the process of surface evaporation. An inhomogeneity arises in the spatial temperature profile—a near-surface maximum of temperature. When the limiting superheating temperature is reached, the conditions for explosive boiling were realized with formation of a cavity in the region of the maximum, where the maximum penetration into the region of superheated metastable liquid near the critical point of liquid-vapor transition is achieved. For Al in action modes under the consideration, the typical depths of temperature maximum are d = 20–70 nm with the values $\Delta T_{max} = T_{\ell,max} - T_{sur}$ = 150–170 K.

An approach was proposed to formulation in continuum models of homogeneous nucleation in a superheated liquid phase based on the introduction into hydrodynamic model of quasi-nuclei with thickness of 1–5 nm. Generation of quasi-nuclei is carried out by the criterion of maximum permissible overheating of initial region. The criterion of limit overheating is determined from molecular dynamics simulation.

Mathematical modeling using the developed technique allowed to obtain a sequence of five explosive boilings for the threshold fluence $F = 5.2$ J/cm^2 and 14 for $F = 6$ J/cm^2. In the first case, the depth of ablation due to explosive boiling was 37 nm, in the second—70 nm. Due to the surface evaporation, they were 17 nm and 12 nm, respectively. The obtained data make it possible to conclude that phase explosion is the main mechanism of material removal in nanosecond range of laser action with fluence $F \geq 5.2$ J/cm^2. The upper values of τ and F are limited by the processes of supercritical expansion and plasma formation and are subject to determination.

In metals, as in nonmetals, the transition from surface evaporation to volume removal of mass occurs in the region of temperature maximum near the critical temperature, where the influence of overheated metastable liquid phase determines the competition between surface evaporation and explosive boiling.

It should be noted that in explosive boiling regimes under investigation, cavity formation takes place in the region of positive values of pressure generated by homogeneous nucleation in superheated liquid phase, in contrast to the case of a series of spalles [57] arising in the region of negative pressures in unloading wave upon action to metal targets of ultrashort pico-femtosecond laser pulses.

Validation of obtained results was performed by comparison with experimental data and with calculations with atomistic models [77, 100].

Acknowledgements

The research was funded by the Russian Scientific Foundation (grant No 15-11-00032).

Author details

Vladimir I. Mazhukin[1,2]*

*Address all correspondence to: vim@modhef.ru

1 Keldysh Institute of Applied Mathematics, Russian Academy of Sciences, Moscow, Russia

2 National Research Nuclear University "MEPhI", Moscow, Russia

References

[1] Miller JC, editor. Laser Ablation. Principles and Applications. Berlin: Springer; 1994

[2] Phipps CR, editor. Laser Ablation and Its Applications. New York: Springer; 2007

[3] Bäuerle D. Laser Processing and Chemistry. Singapore: Springer; 2000

[4] Cheng J, Chang-Sheng Liu S, Shang D, Li W, Perrie G, Dearden K. Watkins. A review of ultrafast laser materials micromachining. Optics and Laser Technology. 2013;**46**:88-102

[5] Chrisey DB, Hubler GK, editors. Pulsed Laser Deposition of Thin Films. New York: John Wiley & Sons; 1994

[6] Eason R, editor. Pulsed Laser Deposition of Thin Films. Hoboken: John Wiley & Sons; 2007

[7] Singh JP, Thakur SN. Laser-Induced Breakdown Spectroscopy. Amsterdam: Elsevier; 2007

[8] Cremers DA, Radziemski LJ. Handbook of Laser-Induced Breakdown Spectroscopy. John Wiley & Sons; 2006

[9] Barcikowski S, Hahn AVK, Chichkov BN. Properties of nanoparticles generated during femtosecond laser machining in air and water. Applied Physics A: Materials Science & Processing. 2007;**87**:47-55

[10] Nicolodelli G, Lizarelli RFZ, Costa MM, Bagnato VS. Time progression of ultrashort laser ablation in a transparent material. Laser Physics. 2011;**21**(8):1420-1427

[11] Shukla P, Waugh DG, Lawrence J, Vilar R. Laser surface structuring of ceramics, metals and polymers for biomedical applications: A review. In: Vilar R, editor. Laser Surface Modification of Biomaterials. Techniques and Applications. Amsterdam, Boston, Gambridge: Elsevier; 2016. p. 281-299

[12] Al-Kattan A, Ryabchikov YV, Baati T, Chirvony V, Sánchez-Royo JF, Sentis M, Braguer D, Timoshenko VY, Anne Estève M, Kabashin AV. Ultrapure laser-synthesized Si nanoparticles with variable oxidation states for biomedical applications. Journal of Materials Chemistry B. 2016;**4**:7852-7858

[13] Zhang D, Gökce B, Barcikowski S. Laser synthesis and processing of colloids: Fundamentals and applications. Chemical Reviews. 2017:1-114

[14] Schmidt V, Belegratis MR, editors. Laser Technology in Biomimetics: Basics and Applications. Series: Biological and Medical Physics, Biomedical Engineering. Berlin, Heidelberg: Springer-Verlag; 2013

[15] Fadeeva E, Schlie-Wolter S, Chichkov BN, Paasche G, Lenarz T. 5—Structuring of biomaterial surfaces with ultrashort pulsed laser radiation. In: Vilar R, editor. Laser Surface Modification of Biomaterials, Techniques and Applications. No 111. 2016. p. 145-172

[16] Mazhukin VI, Nossov VV, Smurov I. Modeling of plasma-controlled evaporation and surface condensation of Al induced de 1.06 and 0.248 μm laser radiations. Journal of Applied Physics. 2007;**101**(2):24922-24937

[17] Hussein AE, Diwakar PK, Harilal SS, Hassanein A. The role of laser wavelength on plasma generation and expansion of ablation plumes in air. Journal of Applied Physics. 2013;**113**:143305

[18] Mazhukin VI, Mazhukin AV, Lobok MG. Comparison of nano- and femtosecond laser ablation of aluminium. Laser Physics. 2009;**19**(5):1169-1178

[19] Verhoff B, Harilal SS, Freeman JR, Diwakar PK, Hassanein A. Dynamics of femto- and nanosecond laser ablation plumes investigated using optical emission spectroscopy. Journal of Applied Physics. 2012;**112** 093303(1-9)

[20] Miotello A, Ossi PM, editors. Laser-Surface Interactions for New Materials Production. Tailoring Structure and Properties. Berlin, Heidelberg: Springer-Verlag; 2010 358 p

[21] Gladush GG, Smurov I. Physics of Laser Materials Processing. Theory and Experiment. Berlin, Heidelberg: Springer-Verlag; 2011 534 p

[22] Benavides O, Lebedeva O, Golikov V. Reflection of nanosecond Nd:YAG laser pulses in ablation of metals. Optics Express. 2011;**19**(22):21842-21848

[23] Li X, Yang Z, Jian W, Han J, Wei W, Jia S, Qiu A. The effect of target materials on colliding laser-produced plasmas. Journal of Applied Physics. 2016;**119** 133301(1-9)

[24] Farid N, Harilal SS, Ding H, Hassanein A. Emission features and expansion dynamics of nanosecond laser ablation plumes at different ambient pressures. Journal of Applied Physics. 2014;**115** 033107 (1–9)

[25] Mazhukin VI, Uglov AA, Chetverushkin BN. Low-temperature laser plasma near metal surfaces in high-pressure gases. Overview. Kvantovaya Elektronika. 1983;**10**(4):679-701

[26] Mazhukin V, Smurov I, Flamant G. Simulation of laser plasma dynamics: Influence of ambient pressure and intensity of laser radiation. Journal of Computational Physics. 1994;**112**(20):78-90

[27] Harilal SS, Miloshevsky GV, Diwakar PK, LaHaye NL, Hassanein A. Experimetal and computational study of complex shockwave dynamics in laser ablation plumes in argon atmosphere. Physics of Plasmas. 2012;**083504**(1–12):19

[28] Mazhukin VI, Samarskii AA. Mathematical modeling in the technology of laser treatments of materials. Review. Surveys on Mathematics for Industry. 1994;**4**(2):85-149

[29] Mazhukin VI, Nossov VV, Smurov I. Modeling of plasma-controlled surface evaporation and condensation of Al target under pulsed laser irradiation in the nanosecond regime. Applied Surface Science. 2007;**253**:7686-7691

[30] Autrique D, Clair G, L'Hermite D, Alexiades V, Bogaerts A, Rethfeld B. The role of mass removal mechanisms in the onset of ns-laser induced plasma formation. Journal of Applied Physics. 2013;**114**(2):023301

[31] Galasso G, Kaltenbacher M, Tomaselli A, Scarpa D. A unified model to determine the energy partitioning between target and plasma in nanosecond laser ablation of silicon. Journal of Applied Physics. 2015;**117** 123101 (1–12)

[32] Gleiter H. Nanostructured materials: Basic concepts and microstructure. Acta Metallurgica. 2000;**48**:1-29

[33] Meyers MA, Mishra A, Benson DJ. Mechanical properties of nanocrystalline materials. Progress in Materials Science. 2006;**51**:427-556

[34] Zhigilei LV, Lin Z, Ivano DS. Atomistic modeling of short pulse laser ablation of metals: Connections between melting, spallation, and phase explosion. Journal of Physical Chemistry C. 2009;**113**:11892-11906

[35] Wu C, Zhigilei LV. Microscopic mechanisms of laser spallation and ablation of metal targets from large-scale molecular dynamics simulations. Applied Physics A: Materials Science & Processing. 2014;**114**:11-32

[36] Upadhyay AK, Inogamov NA, Rethfeld B, Urbassek HM. Ablation by ultrashort laser pulses: Atomistic and thermodynamic analysis of the processes at the ablation threshold. Physical Review B. 2008;**78** 045437 (1-10)

[37] Mazhukin VI, Shapranov AV, Samokhin AA, Ivochkin AY. Mathematical modeling of non-equilibrium phase transition in rapidly heated thin liquid film. Mathematica Montisnigri. 2013;**27**:65-90

[38] Aghaei M, Mehrabian S, Tavassoli SN. Simulation of nanosecond pulsed laser ablation of copper samples: A focus on laser induced plasma radiation. Journal of Applied Physics. 2008;**104**(5):053303

[39] Clair G, L'Hermite D. 1D modelling of nanosecond laser ablation of copper samples in argon at P=1 atm with a wavelength of 532 nm. Journal of Applied Physics. 2011;**110** (8):083307

[40] Chen Z, Bogaerts A. Laser ablation of Cu and plume expansion into 1 atm ambient gas. Journal of Applied Physics. 2005;**97**:063305 (1-12). DOI: 10.1063/1.1863419

[41] Autrique D, Alexiades V. Comment on "laser ablation of Cu and plume expansion into 1 atm ambient gas" [J. Appl. Phys. 97, 063305 (2005)]. Journal of Applied Physics. 2014; **115**:166101 (1-4). DOI: 10.1063/1.4872325

[42] Moscicki T, Hoffman J, Chrzanowska J. The absorption and radiation of a tungsten plasma plume during nanosecond laser ablation. Physics of Plasmas. 2015;**22**:103303 (1-7). DOI: 10.1063/1.4933214

[43] Kozlov BM, Samokhin AA, Uspenskii AB. O chislennom analize pulsiruiushchego rezhima ispareniia kondensirovannogo veshchestva pod deistviem lazernogo izlucheniia. Kvantovaia elektronika. 1975;**2**(9):2061-2063

[44] Miotello A, Kelly R. Critical assessment of thermal models for laser sputtering at high fluences. Applied Physics Letters. 1995;**67**(24):3535-3537

[45] Mazzi A, Miotello A. Simulation of phase explosion in the nanosecond laser ablation of aluminum. Journal of Colloid and Interface Science. 2016;**489**:1-5

[46] Yoo JH, Jeong SH, Mao XL, Greif R, Russo RE. Evidence for phase-explosion and generation of large particles during high power nanosecond laser ablation of silicon. Applied Physics Letters. 2000;**76**:783

[47] Bulgakova N, Bulgakov A. Pulsed laser ablation of solids: Transition from normal vaporization to phase explosion. Applied Physics A. 2001;**73**:199-208

[48] Xu X. Phase explosion and its time lag in nanosecond laser ablation. Applied Surface Science. 2002;**197-198**:61-66

[49] Andreev SN, Kartashov IN, Samokhin AA. Modeling of volumetric boiling during laser evaporation of absorbing condensed media. Kratkie soobshcheniya po fizike FIAN. 2006;**6**:10-21

[50] Porneala C, Willis DA. Observation of nanosecond laser-induced phase explosion in aluminum. Applied Physics Letters. 2006;**89** 211121(1-3)

[51] Yoh JJ, Lee H, Choi J, Lee KC, Kim KH. Ablation-induced explosion of metal using a high-power Nd:YAG laser. Journal of Applied Physics. 2008;**103**:043511

[52] Gragossian A, Tavassoli SH, Shokri B. Laser ablation of aluminum from normal evaporation to phase explosion. Journal of Applied Physics. 2009;**105** 103304 (1-7)

[53] Porneala C, Willis DA. Time-resolved dynamics of nanosecond laser-induced phase explosion. Journal of Physics D: Applied Physics. 2009;**42** 155503 (1-7)

[54] Fishburn JM, Withford MJ, Coutts DW, Piper JA. Study of the fluence dependent interplay between laser induced material removal mechanisms in metals: Vaporization, melt displacement and melt ejection. Applied Surface Science. 2006;**252**:5182-5188

[55] Lutey AHA. Modeling of thin-film single and multilayer nanosecond pulsed laser processing. Journal of Manufacturing Science and Engineering. 2013;**135** 061003 (1-8)

[56] Jiang MQ, Wei YP, Wilde G, Dai LH. Explosive boiling of a metallic glass superheated by nanosecond pulse laser ablation. Applied Physics Letters. 2015;**106**(2) 021904(1-6)

[57] Mazhukin VI, Demin MM, Shapranov AV. High-speed laser ablation of metal with pico- and subpicosecond pulses. Applied Surface Science. 2014;**302**:6-10

[58] Jackson KA. The interface kinetics of crystal growth processes. Interface Science. 2002;**10**:159-169

[59] Mazhukin VI, Shapranov AV, Mazhukin AV, Koroleva ON. Mathematical formulation of a kinetic version of Stefan problem for heterogeneous melting/crystallization of metals. Mathematica Montisnigri. 2016;**36**:58-77

[60] Mazhukin VI, Shapranov AV, Perezhigin VE. Matematicheskoe modelirovanie teplofizicheskikh svoistv, protcessov nagreva i plavleniia metallov metodom molekuliarnoi dinamiki. Mathematica Montisnigri. 2012;**24**:47-65

[61] Hertz H. Uber die verdunstung der flussigkeiten, in besondere des quecksilbers, im luftleeren raume. Annalen der Physik und Chemie. 1882;**17**:177-190

[62] Knudsen M. Maximum rate of vaporization of mercury. Annals of Physics. 1915;**47**:697-705

[63] Afanasev YV, Krokhin ON. Isparenie veshchestva pod deistviem izlucheniia lazera. JETF. 1967;**52**(4):966-975

[64] Mazhukin VI, Samokhin AA. Kinetics of a phase transition during laser evaporation of a metal. Soviet Journal of Quantum Electronics. 1984;**14**(12):1608-1611

[65] Crout PD. An application of kinetic theory to the problems of evaporation and sublimation of monoatomic gases. Journal of Mathematical Physics. 1936;**15**:1-54

[66] Mott-Smith HM. The solution of the Boltzmann equation for a shock wave. Physics Review. 1951;**82**:885

[67] Anisimov SI. Vaporization of metal absorbing laser radiation. Soviet Physics—JETP. 1963;**27**(1):182-183

[68] Ytrehus T. Theory and experiments on gas kinetics in evaporation. In: Potter JL, editor. Rarefied Gas Dynamics. New York: AIAA; 1977. p. 1197-1212

[69] Knight CJ. Theoretical modeling of rapid surface vaporization with back pressure. AIAA Journal. 1979;**17**(5):81-86

[70] Mazhukni VI, Prudkovski PA, Samokhin AA. O gazodinamicheskikh granichnykh usloviiakh na fronte ispareniia. Matematicheskoe modelirovanie. 1993;**5**(6):3-9

[71] Mazhukin VI, Samokhin AA. Boundary conditions for gas-dynamical modeling of evaporation processes. Mathematica Montisnigri. 2012;**24**:8-17

[72] Avdeev AA, Zudin YB. Kinetic analysis of intensive evaporation (method of reverse balances). High Temperature. 2012;**50**(4):527-535

[73] Sone Y, Takata S. Notes on the boundary conditions for fluid-dynamic equations on the interface of a gas and its condensed phase. Physics of Fluids. 2001;**13**:324-334

[74] Aoki K, Sone Y, Yamada T. Numerical analysis of gas flows condensing on its plane condensed phase on the basis of kinetic theory. Physics of Fluids A. 1990;**2**(10):1867-1878

[75] Mazhukin VI, Nossov VV, Smurov I. Analysis of laser-induced evaporation of Al target under conditions of vapour plasma formation. Thin Solid Films. 2004;**453 – 454**:353-361

[76] Andreev SN, Mazhukni VI, Samokhin AA, Demin MM. O modelirovanii spinodal'nogo raspada peregretoi zhidkosti. Kratkie soobshcheniia po fizike. 2006;**7**:50-54

[77] Mazhukin VI, Shapranov AV, Demin MM, Samokhin AA, Zubko AE. Molecular dynamics modeling of nanosecond laser ablation: Subcritical regime. Mathematica Montisnigri. 2016;**37**:24-42

[78] Zeldovich YB, Raizer YP. Physics of Shock Waves and High Temperature Hydrodynamics Phenomena I. New York: Academic; 1967

[79] Mazhukin VI, Smurov I, Shapranov AV, Demin MM. The method of constructing dynamically adapting grids for problems of unstable laminar combustion. Numerical Heat Transfer: Part B: Fundamentals. 2003;**44**(4):387-415

[80] Samarskii AA, Matus PP, Mazhukin VI, Mozolevski IE. Monotone difference schemes for equations with mixed derivatives. Computers & Mathematics with Applications. 2002;**44**(3–4):501-510

[81] Mazhukin VI, Malaphei DA, Matus PP, Samarskii AA. Difference schemes on irregular grids for equations of mathematical physics with variable coefficients. Computational Mathematics and Mathematical Physics. 2001;**41**(3):379-391

[82] Mazhukin AV, Mazhukin VI. Dynamic adaptation for parabolic equations. Computational Mathematics and Mathematical Physics. 2007;**47**(11):1833-1855

[83] Breslavskii PV, Mazhukin VI. Dynamic adaptation method in gas dynamic simulations with nonlinear heat conduction. Computational Mathematics and Mathematical Physics. 2008;**48**(11):2102-2115

[84] Yoo JH, Jeong SH, Greif R, Russo RE. Explosive change in crater properties during high power nanosecond laser ablation of silicon. Journal of Applied Physics. 2000;**88**:1638-1649

[85] Andreev SN, Mazhukin VI, Nikiforova NM, Samokhin AA. On possible manifestations of the induced transparency during laser evaporation of metals. Quantum Electronics. 2003;**33**:771-776

[86] Zeldovich YB, Landau LD. On the relation between the liquid and the gaseous states of metals. Zhurnal Eksperimental'noi i Teoreticheskoi Fiziki. 1944;**14**:32-36

[87] Mott NF. Metal–Insulator Transitions. London: Taylor & Fransis; 1974 285 p

[88] Batanov VA, Bunkin FV, Prokhorov AM, Fedorov VB. Evaporation of metallic targets caused by intense optical radiation. Soviet Physics—JETP. 1973;**36**:311-322

[89] Senatsky YV, Bykovsky NE, Pershin SM, Samokhin AA. Transmittance in a thin aluminum layer at nanosecond pulsed laser ablation. Laser and Particle Beams. 2017:1-5

[90] Mazhukin VI, Samokhin AA, Demin MM, Shapranov AV. Explosive boiling of metals upon irradiation by a nanosecond laser pulse. Quantum Electronics. 2014;**44**:283-285

[91] Mazhukin VI, Samokhin AA, Demin MM, Shapranov AV. Modeling of nanosecond laser vaporization and explosive boiling of metals. Mathematica Montisnigri. 2014;**29**:68-90

[92] Wu B, Shin YC. Absorption coefficient of aluminum near the critical point and the consequences on high-power nanosecond laser ablation. Applied Physics Letters. 2006;**89** 111902(1-3)

[93] Mazhukin VI, Mazhukin AV, Koroleva ON. Optical properties of electron Fermi-gas of metals at arbitrary temperature and frequency. Laser Physics. 2009;**19**(5):1179-1186

[94] Marla D, Bhandarkar UV, Joshi SS. Models for predicting temperature dependence of material properties of aluminum. Journal of Physics D: Applied Physics. 2014;**47** 105306 (1-12)

[95] Borchers VH, Schmidt E, editors. Landolt – Börnstein: Numerical Data and Functional Relatioships in Science and Technology. Vol. 6. Berlin-Gőttingen-Heidelberg-New York: Springer — Verlag; 1964

[96] Smithells CJ. In: Gale WF, Totemeier TC, editors. Smithells Metals Reference Book. 8th ed. Butterworth-Heinemann; 2003 2080 p

[97] Mazhukin VI, Nossov VV, Smurov I. Modeling of plasma-controlled evaporation and surface condensation of Al induced by 1.06 and 0.248μm laser radiations. Journal of Applied Physics. 2007;**101** 024922 (1–16)

[98] Dabby F, Paek UC. High-intensity laser-induced vaporization and explosion of solid material. IEEE Journal of Quantum Electronics. 1972;**8**(2):106-111

[99] Kelly R, Miotello A. Comments on explosive mechanisms of laser sputtering. Applied Surface Science. 1996;(96–98):205-215

[100] Mazhukin VI, Shapranov AV, Demin MM, Samokhin AA, Zubko AE. Molecular dynamics modeling of nanosecond laser ablation: Transcritical regime. Mathematica Montisnigri. 2017;**37**:78-88

Chapter 3

Protected Laser Evaporation/Ablation and Deposition of Organic/Biological Materials: Thin Films Deposition for Nano-biomedical Applications

Gianina-Florentina Popescu-Pelin,
Carmen-Georgeta Ristoscu, Maria Badiceanu and
Ion N. Mihailescu

Additional information is available at the end of the chapter

http://dx.doi.org/10.5772/intechopen.70615

Abstract

This chapter reviews the laser ablation of delicate organic/biological substances by matrix-assisted pulsed laser evaporation (MAPLE). It is shown that direct ablation in this case is possible but sometimes not workable at all in adverse conditions. The considered solution is the protection by a prevalent dissolving/suspending component that can allow for a "shielded" ablation by the frozen solvent followed by its gradual evaporation by melting, evaporation and evacuation by pumping system. We extend the study to the case of non-UV absorbing solvents, e.g., water, when the primary interaction between laser and solute ignites evaporation process at a lower ablation threshold due to reduced pressure inside irradiation chamber. We called this case as "generalized" MAPLE interaction. Relevant examples are provided and critically analyzed in view of potential applications for nanobiomedicine, biosensors, advanced implants and chemical technologies.

Keywords: laser interaction and ablation mechanisms of organic/biological compounds, thin films deposition, nanobiomedicine, protected ablation, functional organic/biological layers

1. Introduction

Laser ablation of "delicate," organic and/or biological materials is reviewed. Particular attention was recently paid to this field stimulated by the progress of laser sources, the advance of "smart targets" and new applications in key technological areas like chemistry environment, biology, and medicine at micro- and nanoscale. The ablation of organic/biological materials was initiated and promoted under "protection" and investigated in various stages after generation, during propagation in vacuum or different ambiances till the final deposition of expulsed material on substrates of interest. Appropriate models were developed for describing the coupling of radiation with the organic/biological materials, bare or protected by a matrix, while complementary techniques were used for obtaining exhaustive information about the ablated substance in plasma plume, liquid, and final solid states. The progress was possible on this basis of a new generation of composite thin films for applications in drug delivery systems, biosensing, and advanced coatings for metallic implants.

The "protected" ablation and transfer of organic and biological substances can be obtained by multipulse laser irradiation. In this case, the "delicate" solute substance is dissolved in more solvents while the obtained mixture is frozen. The resulting "icy" target is then submitted to the laser ablation. The solute is ablated and transferred under the protection of a frozen solvent layer. During transfer, the icy layer is gradually melted, evaporated, and evacuated by the vacuum system whereas the solute reaches the final destination with the minimum or no perturbation or damage and is deposited in the form of a thin film. The technique was invented in Naval Research Laboratory, Washington D.C. Refs. [1, 2] describe it as matrix-assisted pulsed laser evaporation (MAPLE) and is extensively applied since 2000 for the ablation and deposition of a large class of compounds with application in many top field domains, like nanobiomedicine, photocatalysis, synthesis of hard layers, and so on [3]. The choice of an appropriate solvent plays a key role in MAPLE. It is generally accepted now that the main requirements for these selections are: laser fluence must have proper values, lower than in pulsed laser deposition(PLD); incident laser energy must be the majority absorbed by solvent molecules and not by organic molecules of the base material (0.5–10) wt%; frozen solvent must be characterized by a high absorption at working laser wavelength; solvent has to be selected so that the organic material presents a good solubility; solvent has to present a high freezing point; and solvent must not produce chemical reaction under laser radiation exposure. For illustration, we give in **Figure 1** a general scheme of a MAPLE setup and the involved fundamental physical-chemical processes in ablation of the frozen target.

The main difference between MAPLE and "classical" pulsed laser deposition (PLD) [4, 5] consists of target preparation and the laser interaction (ablation) mechanisms. This makes the ablation process in case of MAPLE substantially different to that in PLD. Fundamental mechanism and processes in MAPLE ablation were studied in [6, 7]. The ablation of organic/biological materials is followed by time-resolved plume imaging [8] in many cases with spatial and/or temporal resolution. The expulsed substance was characterized by combined techniques like time program desorption mass spectrometry and atomic spectroscopy [9, 10].

Nevertheless, the selection of the perfect solvent is not easy and sometimes impossible because of the limits introduced by very toxic characters of various solvents, available quantities and prices, and good mixing between a solvent and solute, as being the most important.

Therefore, we refer in Section 2 to a simple and cheap solution of using solvents by hand when all the aforementioned criteria are properly applied.

http://dx.doi.org/10.5772/intechopen.70615

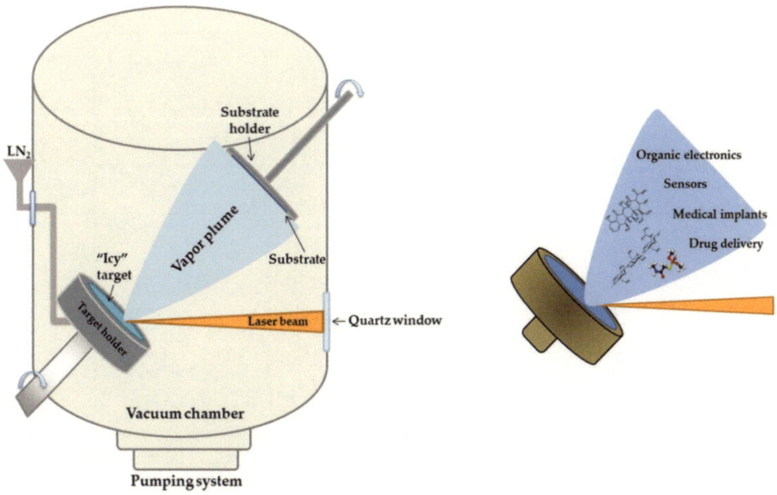

Figure 1. Schematic of MAPLE setup and interaction mechanism.

Selected information are introduced about ablation and deposition of polysaccharides (triacetate-pullulan), enzyme (urease), proteins (fibro- and vitro-nectines), and biopolymers (papain, lysozyme, poly(lactic-co-glycolic acid). All structures were studied by physical-chemical methods and assessed biologically by *in vitro* and more recently *in vivo* studies. Recent literature results are mentioned and evaluated critically.

2. "Generalized" MAPLE mechanism

Smausz *et al.* [11] synthesized urease thin films by MAPLE and PLD from frozen water solutions of urease (1–10 wt%) and pure urease pellets. The KrF* excimer laser fluence was varied between 300 and 2200 mJ/cm^2. The FT-IR spectra of the PLD films at the same fluence range evidenced no significant difference as compared to the MAPLE ones (**Figure 2**). This can be considered an unexpected result, because the direct irradiation, and in particular higher fluence, is predicted to severely affect the molecular structure of the biomaterial. Measurements of absorption coefficient indicated that the idea of absorbing matrix does not work when using water solvent, that is, the relatively high absorptivity of matrix is not a general requirement for a successful MAPLE experiment. Accordingly, the laser energy is absorbed by the organic molecules and the heat is transferred to the surrounding solvent, while the matrix protects the delicate organic material from overheating and consequently thermal decomposition.

Different direct measurements or numerical simulations inferred absorption coefficients below 1 m^{-1} around 250 nm wavelength [12]. For the concrete case of a 1 wt% frozen urease solution [11], this is about two orders of magnitude lower, indicating that the photons are generally absorbed by the urease molecules instead of the water solvent. It has to be mentioned that pure ice could not be ablated in vacuum chamber at 1 J/cm^2 fluence. This is,

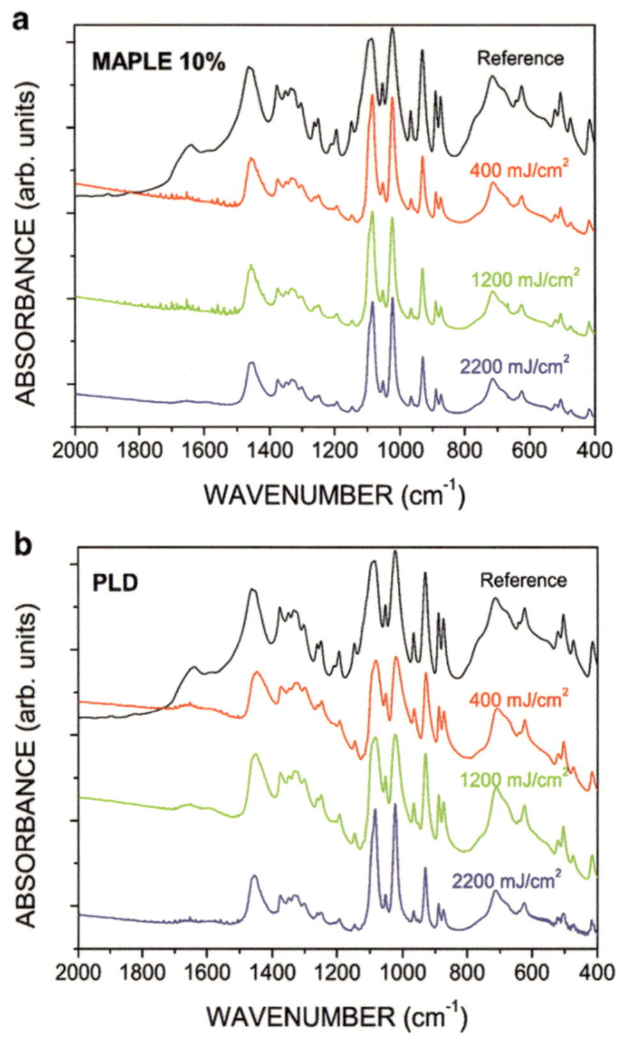

Figure 2. FT-IR spectra of urease thin films deposited by 10% MAPLE and conventional PLD at different fluences (reproduced with permission from [11]).

however, in contradiction with the principle of "classical" MAPLE based on an absorptive matrix. In our opinion [11], the laser energy is absorbed by the organic molecules (and/or molecule cluster) (**Figure 3a** and **b**), leading to an increase in their temperature, thus melting and heating the water in their close vicinity (**Figure 3c**). Because of the working pressure in the reaction chamber, the molten 0°C water starts boiling, the upper ~micrometer layer evaporates from the surface (**Figure 3d**) carrying away the urease molecules, while the deeper layers refroze. Accordingly, the ablation of the delicate material can be achieved at lower temperature (well below the denaturation threshold of 90°C) than in the case of bulk material ablation. Similar ablation process was reported in [13] for aqueous solutions of absorptive materials.

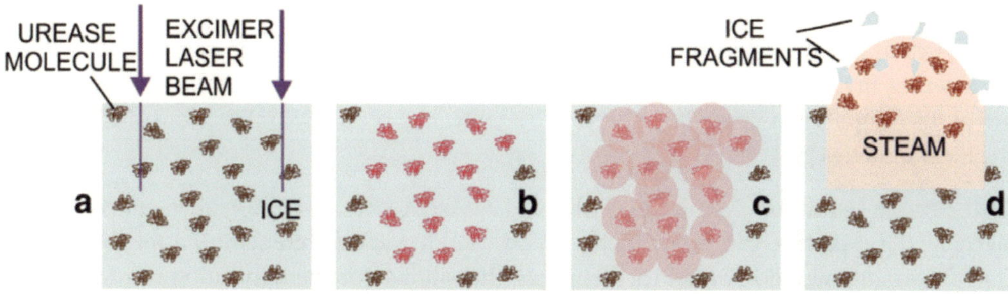

Figure 3. Scheme of the ablation process of the frozen water solution of urease: the laser energy is absorbed by the urease molecules (a) increasing their temperature (b). Part of the heat is transferred to the surrounding ice medium (c), the upper layer evaporates and the deeper layers refreeze (d) (reproduced with permission from [11]).

Experiments using water as MAPLE matrix were performed at 355 nm [14]. In this case, the absorption of ice in UV-visible range is close to the local minimum. This confirms that successful MAPLE deposition can be accomplished without the principal absorption of matrix material, e.g., water.

3. Relevant examples

3.1. Polysaccharides

Polysaccharides, described as complex molecules, are an interesting class of materials due to their biological and chemical properties such as biodegradability, nontoxicity, biocompatibility, nonimmunogenicity, and increased chemical reactivity [15–17]. Moreover, most polysaccharides are of natural origin (plants, animals, and microorganism), and depending on the sources, they can vary with respect to the molecular weight and structure [18, 19]. The presence of glycosaminoglycans (part of the extracellular matrix) in the composition of natural polysaccharides is an important feature, which proved to increase the wound healing process by binding to proteins at hierarchical peculiarity [17, 20, 21]. Since the biological activity of polysaccharides is dependent on their properties, the further advancement of polysaccharide-based nanomedicine, which is a current direction of interest, proposes the development of alternative methods to produce polysaccharides with reliable features [17, 19].

Tissue engineering and drug delivery are also two directions of permanent interest for the medical field, and so a variety of polysaccharides have been used in order to bring new solutions to the encountered issues [17]. In this respect, there were reported in the literature the benefits of alginate [22–24], gellan [25, 26], dextran [27–29], hyaluronic acid [30], chitosan [31, 32], and pullulan [29, 33–35] for specific applications [17].

3.1.1. Pullulan

Pullulan, with the molecular formula $(C_6H_{10}O_5)_n$, is a neutrally charged polysaccharide, which is soluble in water and produced by yeast-like fungus *Aureobasidium pullulans* [16, 33]. Due to its unique structural features, excellent mechanical properties and biocompatibility, and to the high hydration capacity, the native pullulan and its derivatives possess various biomedicinal and pharmaceutical applications [33, 36]. The application of pullulan in tissue engineering involves the surface modification practices. Consequently, the surface properties of this biopolymer could be enhanced by replacing, on its hydroxyl groups, the desired chemical moieties [33].

Furthermore, thin films of pullulan, which are biodegradable, biocompatible, and with good mechanical properties, can be synthetized and used in various biomedical applications [33]. One processing method is MAPLE technique that allows for the deposition of high-quality pullulan films, e.g., triacetate-pullulan, cinnamate-pullulan and tosylate-pullulan [35, 37, 38]. The laser processing of such organic materials in the form of thin films requires the preservation of pullulan molecular structure and functionality [10]. Cristescu *et al.* synthetized MAPLE thin films of pullulan and its derivatives from a frozen target of 2 wt% biopolymer in dimethyl sulfoxide, under the irradiation with a KrF* excimer laser source [35, 37].

Based on the biofunctionality of this biopolymer, Bulman *et al.* proved that the addition of pullulan enhanced mesenchymal stem cells (MSCs) retention at a diseased cartilage surface, increasing MSC therapeutic efficiency and acting as a cellular adhesive [39]. In 2016, Atila *et al.* reported for the first time the production of pullulan and cellulose acetate scaffolds (in various combinations) by wet electrospinning technique, in order to fulfill the requirements of an engineered-tissue construct [40]. As for the use of pullulan in drug delivery systems, a relatively new study revealed the synthesis and efficiency of a system based on pullulan and poly(β-amino)ester for gene co-delivery (hepatoma-targeted) and chemotherapy agent [41].

These are only a small part of pullulan applications that demonstrate its significance for the present and future research directions, offering a wide field of activity due to its versatile composition.

3.2. Proteins

Proteins are macromolecules distinguished from polysaccharides by their content of approximately 20 amino acid monomers and can be found in all biological systems, from inferior prokaryotes to complex eukaryotes [42–44]. According to the chemical properties, amino acids are classified as non-polar aliphatic (hydrophobic), non-polar aromatic (hydrophobic, except tyrosine), polar uncharged (hydrophilic), polar negatively charged at pH 7, polar positively charged at pH 7, and sulfur-containing (maintain the structure of the protein) groups [45]. The attachment of each amino acid to the central carbon by a different side group leads to the unique character of proteins [43].

Furthermore, proteins comprise a significant number of reactive groups that ensure flexibility by their chemical modification [42]. This statement is strengthened by the fact that proteins have multiple sites for chemical interaction, which can allow for the improvement and tailoring of their properties [46]. Proteins are known to occur essentially in aqueous or membrane environments, are insoluble in non-polar solvents, and cover a wide range of polymeric compounds [43, 44]. Thus, van der Waals, hydrogen bonding, electrostatic, hydrophobic, and disulfide cross-link interactions between the amino acid units are connections responsible for the structural modifications of proteins along the polymeric chain [43].

The specific requirements of various applications can be fulfilled due to the possible adjustment of the properties of proteins [42]. One of the most discussed and studied biomedical applications of proteins are diagnostic imaging [47–49], therapeutic delivery [47, 50, 51], and tissue engineering [52–54].

The progress on the processing of different protein materials (e.g., silk, gelatin, collagen, casein, keratin, etc.) into coatings with applicability in tissue engineering, cell adhesion, implants proved to be a key factor for the improvement of the medical field [43, 55, 56]. The presence of the different side groups attached to the central carbon in the protein structure is an advantage for film fabrication or even for its improvement (stability) because proteins can be submitted to chemical and/or mechanical modifications [42, 46]. Furthermore, protein-based films or coatings are known to be biodegradable, being thus a source of nitrogen (proteins contain about 16% nitrogen by mass) which contributes as fertilizers [44, 46].

As a component, proteins of the extracellular matrix (ECM), which is present within all tissues and organs, fibronectin (FN), vitronectin (VN), and collagen I (Col1) proved their important roles in wound-healing processes [57].

Our own interest was focused on the study of FN and VN in which *in vitro* have been shown to increase the attachment of specific cells for a designed substrate [58].

3.2.1. *Fibronectin*

Fibronectin is one of the most important and intensively studied ECM proteins, exists as a dimer, induces mineralization, and is a soluble circulating protein in body fluids (like plasma) [59–61]. At physiological pH (7.4), FN proved to be negatively charged due to its acidic isoelectric point (pI = 5.5–6.0) [59]. From the biological point of view, FN plays a key role in the adhesion, spreading, migration, differentiation, and proliferation of various cells and support the accumulation of multiple growth factors (GF) [59, 60]. In addition, FN can enhance the GF growth-promoting function [60]. Due to its stimulus in cell attachment and migration processes, FN, also known as a multifunctional extracellular glycoprotein, is extensively studied and used as a coating in tissue engineering [62].

In 2013, the fabrication of functional FN patterns onto Ti substrates by using the laser direct write (LDW) technique was described by Grigorescu *et al.* [63]. In this study, the authors validated the suitability of LDW technique for FN molecules' transfer, preserving the compositional protein features and functionality [63]. The integrity preservation of the transferred FN was evaluated by ATR–FTIR measurements, and one could observe no structural modifications between FN dropcast and the LDW ribbon (**Figure 4**). The characteristic amide I and II bands are clearly visible and correspond well to the similar bands of the positive control consisting of a FN solution dropcasted onto a Si substrate (**Figure 4**—curve 5). The spectrum presented in **Figure 4**—curve 2 corresponds to a negative control of a heat-denatured protein sample. One can clearly notice that the two proteic bands are no longer visible in this case.

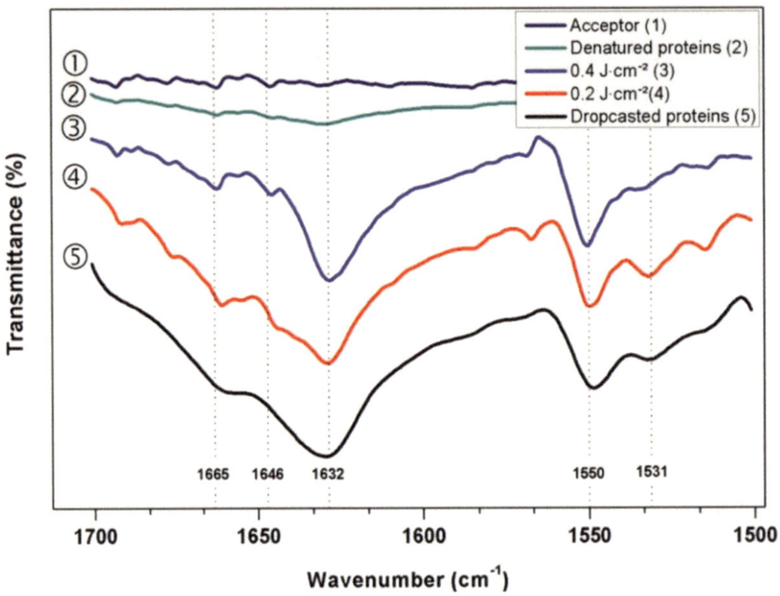

Figure 4. Representative ATR-FTIR spectra of Si acceptor (1), a negative control of heat-denatured proteins (2), proteins transferred at 0.4 J/cm^{-2} (3), proteins transferred at 0.2 J/cm^{-2} (4), and a positive control consisting of dropcasted proteins (5).

This implies that proteins exposed at a temperature of 100°C for 1 min lose their secondary structure. However, any heat damage susceptible to occur is strictly localized and does not influence the secondary structure of the proteins in a detectable manner, which is majoritarily preserved after the transfer.

In addition, Western blot assays were conducted onto a control-purified FN solution and that of protein desorbed from the LDW acceptor. One can observe similar molecular weights between the tested solutions (**Figure 5**) [63].

In vitro tests performed with MC3T3-E1 preosteoblast cells validated the transfer of secondary structure and functionality. Particular attention was paid to the presence and availability of the cell binding domain within the FN molecules, as evidenced by immunostaining with specific monoclonal antibodies followed by microscopic observations. MC3T3 osteoblasts preferentially attached onto the FN features (**Figure 6a** and **b**). The spreading of cells close to the FN features was slightly enhanced in comparison with cells onto the bare substrate. The cellular morphology was generally influenced by the presence of the FN spots. Thus, the cells situated in the close vicinity of FN exhibited an asymmetric aspect with elongated cytoplasmic extensions, whose tip-adhesion site ended in most cases on a FN region. Cells situated completely onto the FN features present a stronger polarity, with longer filopodia, which suggest an improved fixation on the substrate. A characteristic representation of such cellular behavior is depicted in **Figure 6e** and **f**, wherefrom one can observe the actin filaments (**Figure 6f**) oriented along the FN formations (**Figure 6e**). In all cases, cell nuclei did not exhibit significant morphological modifications such as fragmentations or condensation, which could suggest

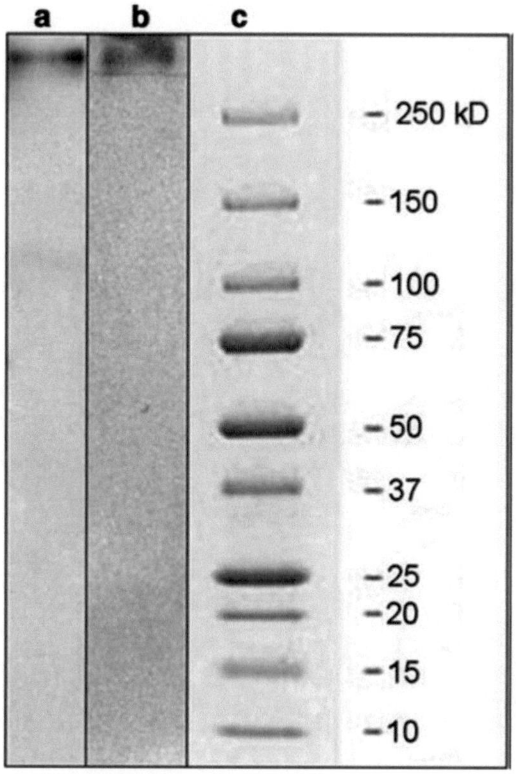

Figure 5. Western blot profiles obtained for the control-purified FN solution (a) and that of protein desorbed from the LDW acceptor (b) compared with the BioradAllBlue calibration molecular weight reference proteins (c) (reproduced with permission from [63]).

cellular apoptosis. The *in vitro* tests revealed the full functionality of the transferred protein features in both deposition regimes, 0.2 or 0.4 J/cm^2. However, it was observed that slight differences concerning the local uniformity, as well as the aspect repeatability of the protein features, do exist. For obtaining uniform, highly repetitive features, lower fluences are definitely desirable.

The biological response of the transferred FN was also assessed by fluorescence microscopic measurements after the seeding of Swiss 3 T3 fibroblasts cells onto the tested samples [63]. The shape of the imprinted FN features strongly influenced the morphology of the adherent Swiss 3 T3 fibroblasts cells. A weaker attachment and a round shape of the attached cells onto the Ti control was observed, whereas for the transferred FN patterns, they were clearly spread, actin filaments being organized according to the FN spots aspect [63]. These results are in good agreement with those obtained in case of preosteoblast MC3T3 cells.

The biochemical characterization of the FN structures open new perspectives on the fabrication of complex biomaterials including proteins (with high molecular weight) for use in biomedical applications (e.g., drug delivery, regenerative tissue engineering) [63].

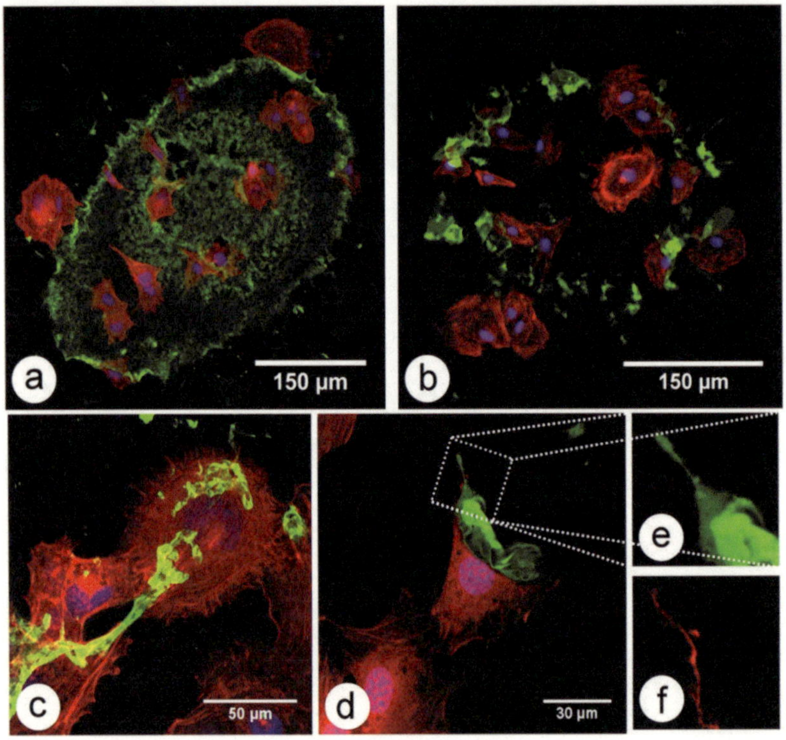

Figure 6. Characteristic microscopic aspect of MC3T3 cells in the neighboring region to laser transferred fibronectin features. (a) Osteoblasts attached to the FN features obtained at 0.2 J/cm². (b) Osteoblasts attached to the FN features transferred at 0.4 J/cm². (c) Details of cytoskeleton actin fibers oriented along the FN subfeatures obtained at 0.4 J/cm². (d) Specific modified cell morphology in the close vicinity of FN (0.4 J/cm²), where a certain bending of the protein subfeature can be deduced as a result of the cellular activity. (e and f) Detailed image of a cytoplasmic extension (filopodia) placed along a FN subfeature. Confocal scanning microscopy was conducted using TRITC-coupled phalloidin for the F-actin, DAPI for the DNA, and FITC-coupled antibodies for the FN.

MAPLE technique was also applied in the synthesis of fibronectin coatings. A recent report of Sima *et al.* describe the fabrication, in two steps, of a new hybrid biomimetic thin structure (inorganic-organic bilayer) using two different laser techniques (PLD for hydroxyapatite (HA) and MAPLE for FN deposition, respectively) [54]. The authors quantified by temperature-programmed desorption mass spectrometry the deposited FN over the HA layer and verified its biological efficiency by multiple cellular studies [54].

In a recent study published in 2015, Agarwal *et al.* reported the enhanced implant-bone integration in healthy and osteoporotic rats by applying FN coatings obtained by immersion of clinical grade stainless steel implants in purified solutions of FN for 30 min [64]. Following the same direction on integration of biomedical implants with tissue, other research groups attained the idea of producing an interface able to mimic the ECM and improve cell behavior [59]. Gand *et al.* proposed the synthesis of a layer-by-layer film by direct incorporation of FN (as anionic polymer) into poly(L-lysine) (PLL) (positively charged) in order to obtain an FN-enriched structure, capable to mimic the ECM [59].

From reported data, one can notice the fast development on FN processing, thus the principal objective being the chemical preservation and the improvement of the biological response.

3.2.2. Vitronectin

A similar and important constituent of the interstitial ECM, vitronectin, is readily adsorbed at the interface of biomaterials and is known for its adhesive functions during development and angiogenesis [65, 66]. VN is the main glycoprotein adsorbed from the serum onto synthetic polymers and was identified as an essential adhesion and spreading mediator in many cells [65]. Given its multifunctional physiological activity, vitronectin is also described as S protein, epibolin, or "serum spreading factor" [65–67].

In order to prevent fibrosis and to support migration of adjacent cells, VN have the capability to trigger the enzymatic degradation of provisional ECM, influencing thus the fate of implants. In addition, it encompasses an approach for improving the endothelization of implants [66]. As reported [68], VN and FN are vital for the *in vitro* adhesion of osteoblast cells.

In 2009, Whitlow *et al.* reported the cover of a platinum aneurysm coil with gold (Au) or VN by using an electrostatic self-assembly process trying to evaluate the degree of neointimal change associated with the applied ultrathin layer [69]. The neointimal response revealed the improved reaction in case of Au and VN coil groups as compared with the uncoated platinum group [69].

Later, in 2011, Sima *et al.* reported the transfer and immobilization of VN by MAPLE technique on a collector coated with an HA layer (by PLD). In this study, the biological response was evaluated by using human osteoprogenitor (HOP) cells, and the obtained results revealed the improved adherence, spreading, and growth in respect with HA coatings [70].

The vitronectin multifunctionality makes it an attractive biopolymer for the tissue engineering when used as a surface coating and moreover, opens new directions for research.

3.3. Enzymes

Enzymes are proteins that consist of one or more polypeptide chains and have an active site, thus determining their specificity and flexibility [71, 72]. Enzymes are also known as biocatalysts that accelerate a chemical reaction without interfering their equilibrium [71, 73]. Enzymes are recognized for their high catalytic activity and excellent selectivity for the targeted substrate, being thus described as optimal biorecognition molecules [74]. As biosensor components, enzymes are considered the shortest lived elements because they progressively lose activity [74].

For a long time, enzymes have been used in microbial processes, helping (by fermentation process) in the preparation of cheeses, wines, and other milk products [71]. In course of time, the interest of some researchers leads to the use of enzyme in the medical field, in the diagnosis, and in the treatment of various diseases [75]. In this respect, they try to study, understand, and obtain the basic information on toxicology, immunological reactions, and chemical stability of the organism *in vivo* [76].

Various deposition techniques could be applied in order to immobilize different enzymes onto a solid holder. This task is not an easy one due to the fact that one can deal with the transfer of complex molecules [77]. A possible approach for enzyme immobilization could be MAPLE technique recognized for its capability to transfer soft materials, avoiding the changes of the chemical and biological properties.

3.3.1. *Papain*

Papain, a proteolytic enzyme, found naturally in papaya is of key importance in various vital biological processes [78]. Due to its proteolytic and elastolytic properties, papain exhibits antibacterial, antifungal, and anti-inflammatory activities [79].

The use of papain as a coating for surface improvement in different medical devices is still at the beginning. In 2013, Motoc *et al.* [80] reported the deposition of a simple HA (Ti/HA or papain Ti/papain) and a bilayer (consisting of HA and papain synthetized by PLD and MAPLE, respectively: Ti/HA/papain). The authors tried to functionalize the titanium surface in two steps, enhancing the bioactivity (HA layer) and activating the localized drug with antimicrobial and anti-inflammatory effects (papain layer) [80]. In order to evaluate the stoichiometric transfer of papain, the used powders and the obtained coatings were analyzed by FTIR. The spectra revealed the presence of the proteins characteristic bands (**Figure** 7) [80].

Moreover, the osteoblasts-like SaOs2 human cells were used in order to evaluate the biological performances of papain coatings, by adhesion (immunofluorescence microscopy: **Figure 8**) and proliferation tests (MTS assay: data not shown here) [80].

Taking into account the positive results, one can consider the papain-based coatings as potential implant material with good antimicrobial properties and improved integration.

3.3.2. *Lysozyme*

Lysozyme, a small and stable lytic enzyme, is found in nature, being present not only in almost all secretions, body fluids (tears, saliva, and sweat), and tissues (nasal cavity) of the living organisms, but also in some plants, bacteria, or egg white [81, 82]. It has a specific hydrolytic activity against the cell walls of liable bacteria [81, 83]. Thus, lysozyme increases the permeability of Gram-positive bacteria and causes the burst of cells [82, 84]. Oppositely, the antimicrobial efficiency of lysozyme against the Gram-negative bacteria is limited and even lacks toward eukaryotic cell walls [82]. Additionally, there are studies that reveal the potential of lysozyme to inhibit tumor formation and growth (anticancer agent) [85, 86].

Due to its multifunctionality and recognized antimicrobial efficiency, lysozyme is used, more and more, in the biomedical field. A special attention was paid to the development of medical devices by surface functionalization with lysozyme or composites based on it, which can also act as drug delivery systems.

In this respect, Visan *et al.* reported the deposition of composite coatings based on lysozyme embedded in a mixture of polyethylene glycol (PEG) and polycaprolactone (PCL) by both MAPLE and dip-coating techniques. The authors demonstrated the antimicrobial efficiency

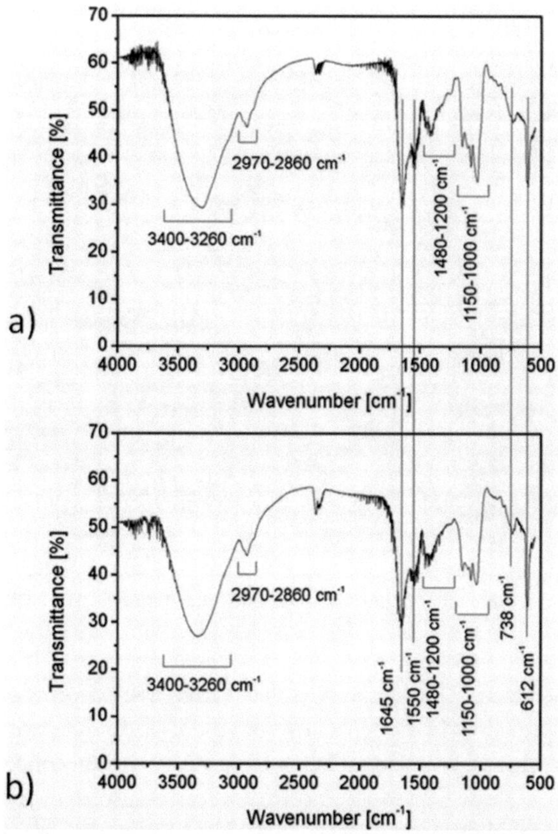

Figure 7. FTIR micrographs of the papain dropcast (a) and of the thin films obtained by MAPLE (b) for 5 wt% papain in distilled water and a laser fluence of 0.75 J/cm^2 (reproduced with permission from [79]).

of PCL:PEG-lysozyme composite coatings, which was in good agreement with the viability results with osteoblasts and MSC cells [87].

The incorporation of lysozyme into polymeric matrices was also studied by Grumezescu *et al.* This time, poly(3-hydroxybutyric acid-co-3-hydroxyvaleric acid) (P(3HB-3 HV)) and PEG biopolymers were used as core shells for lysozyme, and the composite material was deposited by MAPLE in the form of thin films. Both P(3HB-3 HV)/PEG/Lys and P(3HB-3 HV)/Lys coatings proved a significant antimicrobial activity and an improved biocompatibility [88].

Such complex coatings based on lysozyme are promising for the nanobiomedical field due to their potential use as bone implants.

3.3.3. *Urease*

Urease, a non-redox metalloenzyme, known also as nickel-dependent enzyme, induces the hydrolysis of urea into ammonium and carbon dioxide [89–91]. Urease can also provide a defense mechanism against pathogens by controlling the nitrogen content in the biological

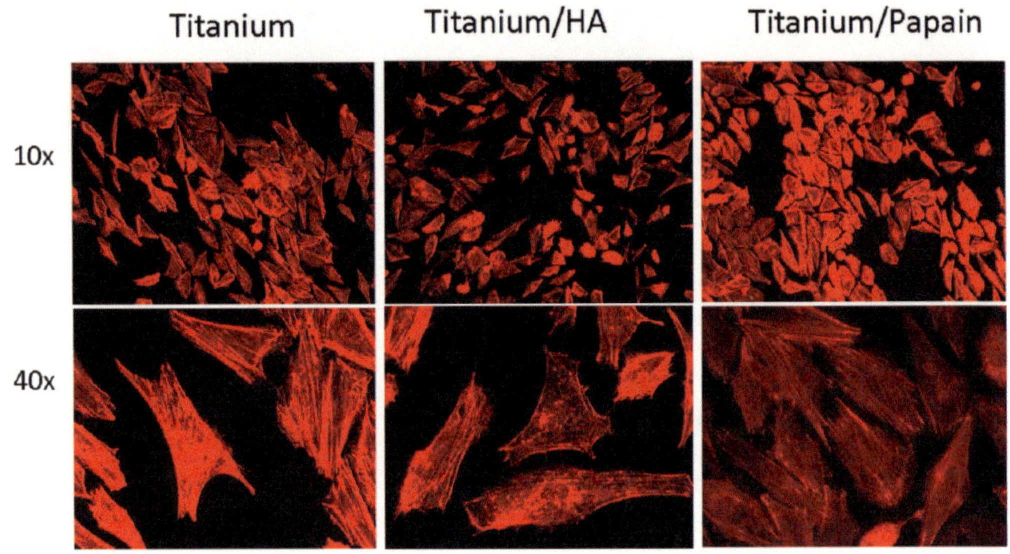

Figure 8. Fluorescence microscopic images of human SaOs2 cells grown on bare Ti, Ti/HA, and Ti/papain for 5 wt% papain in distilled water and a laser fluence of 0.75 J/cm^2 (reproduced with permission from [79]).

environments [92]. In addition, urease can be isolated from a selection of organisms, comprising bacteria, fungi, and plants [91]. Urease is a key enzyme used to determine the amount of urea in biological solutions (blood), where urea being toxic above certain concentrations [93, 94]. The removal of urea from waste water, food, and fruit juices is also relevant for domains such as environmental analyses and the food industry [94]. The presence of urea can be determined by electrochemical or optical methods, on the basis of the formed ammonia [90].

The use of urease in applications such as clinical diagnosis, environmental analysis and detection of food adulteration requires the maintenance of the enzyme stability, functionality and activity as close as possible to its natural state [95].

The use of urease in biosensing applications was and still is an area of interest for a relatively large number of scientists. In 2010, György *et al.* reported the immobilization of urease (from "icy" targets with different urease concentrations) in the form of thin films by using MAPLE technique. The authors proved the stoichiometric transfer of urease, and further, the enzymatic activity was preserved. The kinetic and enzymatic activity of the immobilized urease was evaluated by using the Worthington assay method (**Figure 9**) [90]. This study proved that the thin film of urease could be active in urea disintegration if it is obtained under optimum deposition conditions [90].

Furthermore, the key factor in the achievement of a high-sensitive biosensing unit could be considered the optimum combination between biomolecules and nanomaterials [96]. In this respect, Siqueira *et al.* [96] studied the electrochemical properties of urea-based electrolyte-insulator-semiconductor (EIS) sensor for urea detection. The authors described the synthesis of LbL films for urease immobilization on capacitive EIS chips by using polyamidoamine

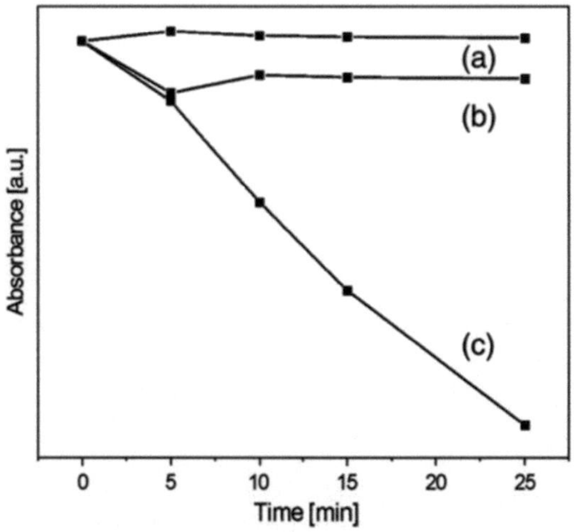

Figure 9. The kinetic assay of urease thin film for (a) 1, (b) 3.5, and (c) 10 wt% concentrations in the frozen composite targets deposited at 0.4 J/cm^2 laser fluence (reproduced with permission from [89]).

(PAMAM)/carbon nanotubes (CNT) as stabilizing matrices. The achievement of PAMAM/CNT LbL films for urease immobilization opens new directions of research in the field of biosensors, with the possibility of extension to other enzymes [96].

4. Conclusions

Direct laser ablation of organic/biological materials was considered for a long time inaccessible because of risks of decomposition and irreversible damages. The recent progress of "soft" ablation laser techniques makes possible the safe expulsion and transfer of materials, from target to substrate for synthesis of structures of various bio-, nano-, and more recently meta-materials. This opened the access toward pulsed laser technologies utilization for the ablation of "delicate" simple and composite materials. Systematic complementary investigations demonstrated that, under proper irradiation conditions and "special protection" as ensured by cryogenic utilization, the preservation is possible for basic material composition, structure, morphology, and more likely functionality. Simultaneous or subsequent ablation of organic/biological materials was reached for new top applications in technology and in particular in nanobiomedicine. The chapter is based on recent original results of the authors and a selection of the relevant existing new data from specialized literature.

Acknowledgements

This work was supported by the contracts NATO G4890 and 43 NATO SPS.

Author details

Gianina-Florentina Popescu-Pelin[1,2], Carmen-Georgeta Ristoscu[1], Maria Badiceanu[1,2] and Ion N. Mihailescu[1*]

*Address all correspondence to: ion.mihailescu@inflpr.ro

1 National Institute for Lasers, Plasma and Radiation Physics, Magurele-Ilfov, Romania

2 Faculty of Physics, University of Bucharest, Magurele-Ilfov, Romania

References

[1] McGill RA, Chrisey DB. Method of producing a film coating by matrix assisted pulsed laser deposition. Google Patents; 2000

[2] Chrisey D, Pique A, McGill R, Horwitz J, Ringeisen B, Bubb D, et al. Laser deposition of polymer and biomaterial films. Chemical Reviews. 2003;**103**(2):553-576

[3] Jelinek M, Kocourek T, Remsa J, Cristescu R, Mihailescu IN, Chrisey DB. MAPLE applications in studying organic thin films. Laser Physics. 2007;**17**(2):66-70(5)

[4] Ristoscu C, Mihailescu IN. Biomimetic coatings by pulsed laser deposition, chapter 7. In: Belegratis M, Schmidt V, editors. Laser Technology in Biomimetics: Basics and Applications, Springer—Biological and Medical Physics, Biomedical Engineering. Berlin Heidelberg: Springer-Verlag; 2013. p. 163-191 ISBN 978-3-642-41340-7

[5] Nelea V, Jelinek M, Mihailescu IN. Biomaterials: New issues and breakthroughs for biomedical applications, chapter 18. In: . Pulsed Laser Deposition of Thin Films: Applications-Lead Growth of Functional Materials. Wiley & Sons; Hoboken, New Jersey; 2007. p. 421-459

[6] Leveugle E, Zhigilei LV. Molecular dynamics simulation study of the ejection and transport of polymer molecules in matrix-assisted pulsed laser evaporation. Journal of Applied Physics. 2007;**102**(7):074914

[7] Zhigilei LV, Lin Z, Ivanov DS, Leveugle E, Duff WH, Thomas D, Sevilla C, Stephen JG. Atomic/molecular-level simulations of laser–materials interactions, chapter 3. In: Miotello A, Ossi PM, editors. Laser-Surface Interactions for New Materials Production Tailoring Structure and Properties, Series: Springer Series in Materials Science. Vol. 130. Springer Heidelberg Dordrecht London New York; 2010. p. 43-79

[8] Bubb DM, Johnson SL, Collins B, Haglund RF. Thermal confinement and temperature-dependent absorption in resonant infrared ablation of frozen liquid targets. The Journal of Physical Chemistry. 2010;**114**:5611

[9] Itina TE, Zhigilei LV, Garrison BJ. Matrix-assisted pulsed laser evaporation of polymeric materials: A molecular dynamics study. Nuclear Instruments and Methods in Physics Research Section B: Beam Interactions with Materials and Atoms. 2001;**180**:238-244

[10] Caricato AP. MAPLE and MALDI: Theory and Experiments, Chapter 12. In: Castillejo M, Ossi PM, Zhigilei L, editors. Lasers in Materials Science. Springer Cham Heidelberg New York Dordrecht London; 2014. p. 295-323

[11] Smausz T, Megyeri G, Kékesi R, Vass C, György E, Sima F, Mihailescu IN, Hopp B. Comparative study on pulsed laser deposition and matrix assisted pulsed laser evaporation of urease thin films. Thin Solid Films. 2009;**517**:4299-4302

[12] Warren SG, Brandt RE, Grenfell TC. Visible and near-ultraviolet absorption spectrum of ice from transmission of solar radiation into snow. Applied Optics. 2006;**45**:5320-5334

[13] Esenaliev RO, Karabutov AA, Podymova NB, Letokhov VS. Laser ablation of aqueous solutions with spatially homogeneous and heterogeneous absorption. Applied Physics B—Springer. 1994;**59**:73-81

[14] Purice A, Schou J, Kingshott P, Dinescu M. Production of active lysozyme films by matrix assisted pulsed laser evaporation at 355 nm. Chemical Physics Letters. 2007;**435**:350-353

[15] Ikuma K, Madden AS, Decho AW, Lau BLT. Deposition of nanoparticles onto polysaccharide coated surfaces: Implications for nanoparticle–biofilm interactions. Environmental Science: Nanomaterials. 2014;**1**:117-122

[16] Uthaman S, Lee SJ, Cherukula K, Cho C-S, Park I-K. Polysaccharide-coated magnetic nanoparticles for imaging and gene therapy. BioMed Research International. 2015;**14**: Article ID 959175

[17] Donald C. Aduba Jr, Hu Yang. Polysaccharide fabrication platforms and biocompatibility assessment as candidate wound dressing materials. Bioengineering. 2017;**4**:1. DOI: 10.3390/bioengineering4010001.

[18] Samal PK, Dangi JS. Isolation, preliminary characterization and hepatoprotective activity of polysaccharides from *Tamarindus indica* L. Carbohydrate Polymers. 2014;**102**:1-7

[19] Doh K-O, Yoon Yeo. Application of polysaccharides for surface modification of nanomedicines. Therapeutic Delivery. 2012;**3**(12):1447-1456

[20] Senni K, Pereira J, Gueniche F, Delbarre-Ladrat C, Sinquin C, Ratiskol J, Godeau G, Fischer A-M. Dominique Helley and Sylvia Colliec-Jouault marine polysaccharides: A source of bioactive molecules for cell therapy and tissue engineering. Marine Drugs. 2011;**9**:1664-1681. DOI: 10.3390/md9091664

[21] Lohmann N, Schirmer L, Atallah P, ElkeWandel RAF, Werner C, Simon JC, Franz S, Freudenberg U. Glycosaminoglycan-based hydrogels capture inflammatory chemokines and rescue defective wound healing in mice. Science Translational Medicine. 2017;**9**:eaai9044

[22] Sun J, Tan H. Alginate-based biomaterials for regenerative medicine applications. Materials. 2013;**6**:1285-1309

[23] Wang Y, Wang X, Shi J, Zhu R, Zhang J, Zhang Z, Ma D, Hou Y, Lin F, Yang J, Mizuno M. A biomimetic silk fibroin/sodium alginate composite scaffold for soft tissue engineering. Scientific Reports. 2016;**6**: Article number 39477

[24] Lu L, Qi Y, Tian J, Zhou C. Alginate/hydroxyapatite hydrogel as biodegradable in situ forming scaffold. In: Peng Y, Weng X, editors. 7th Asian-Pacific Conference on Medical and Biological Engineering. IFMBE Proceedings. Vol. 19. Berlin, Heidelberg: Springer; 2008

[25] Oliveira JT, Martins L, Picciochi R, Malafaya PB, Sousa RA, Neves NM, Mano JF, Reis RL. Gellan gum: A new biomaterial for cartilage tissue engineering applications. Journal of Biomedical Materials Research Part A. 2009;852-863

[26] Cengiz IF, Miguel J, Oliveira, Reis RL. Tissue engineering and regenerative medicine strategies for the treatment of osteochondral lesions, chapter 2. In: Ratib O, Choi H, editors. 3D Multiscale Physiological Human, Nadia Magnenat-Thalmann. London: Springer-Verlag; 2015. p. 25-47

[27] Sun G, Mao JJ. Engineering dextran-based scaffolds for drug delivery and tissue repair. Nanomedicine (Lond). 2012;**7**(11):1771-1784. DOI: 10.2217/nnm.12.149

[28] Liu J, Qi C, Tao K, Zhang J, Zhang J, Luming X, Jiang X, Zhang Y, Huang L, Li Q, HongjianXie JG, Shuai X, Wang G, Wang Z, Wang L. Sericin/dextran injectable hydrogel as an optically trackable drug delivery system for malignant melanoma treatment. ACS Applied Materials & Interfaces. 2016;**8**(10):6411-6422

[29] Schlaubitz S, Mohammed Derkaoui S, Marosa L, Miraux S, Renard M, Catros S, Visage C, Letourneur D, Amédée J, Fricain J-C. Pullulan/dextran/nhamacroporous composite beads for bone repair in a femoral condyle defect in rats. PLoS ONE. 2014;**9**(10):e110251. DOI: 10.1371/journal.pone.0110251

[30] Maurice NC. Hyaluronic Acid for Biomedical and Pharmaceutical Applications. Smithers Rapra Technology, eBook, 238 pages; 2014

[31] Visan A, Stan GE, Ristoscu C, Popescu-Pelin G, Sopronyi M, Besleaga C, Luculescu C, Chifiriuc MC, Hussien MD, Marsan O, Kergourlay E, Grossin D, Brouillet F, Mihailescu IN. Combinatorial MAPLE deposition of antimicrobial orthopedic maps fabricated from chitosan and biomimetic apatite powders. International Journal of Pharmaceutics. 2016;**511**(1):505-515

[32] Riva R, Ragelle H, des Rieux A, Duhem N, Jérôme C, Préat V. Production chitosan derivatives. In: Jayakumar R, Prabaharan M, Muzzarelli RAA, editors. Drug Delivery and Tissue Engineering, Chapter 2 in Chitosan for Biomaterials II. Advances in Polymer Science. Berlin Heidelberg: Springer Verlag; 2014. p. 19-44. DOI: 10.1007/12_2011_137.sau

[33] Singh RS, Kaur N, Rana V, Kennedy JF. Pullulan: A novel molecule for biomedical applications. Carbohydrate Polymers. 2017;**171**:102-121

[34] Wong VW, Rustad KC, Galvez MG, Neofytou E, Glotzbach JP, Januszyk M, Major MR, Sorkin M, Longaker MT, Rajadas J, Geoffrey C. Gurtner "engineered pullulan–collagen composite dermal hydrogels improve early cutaneous wound healing". Tissue Engineering: Part A. 2011;**17**(5-6):631-644

[35] Cristescu R, Dorcioman G, Ristoscu C, Axente E, Grigorescu S, Moldovan A, Mihailescu IN, Kocourek T, Jelinek M, Albulescu M, Buruiana T, Mihaiescu D, Stamatin I, Chrisey DB. Matrix assisted pulsed laser evaporation processing of triacetate-pullulan polysaccharide thin films for drug delivery systems. Applied Surface Science. 2006;**252**:4647-4651

[36] Ponnusami V, Gunasekar V. Production of pullulan by microbial fermentation. In: Ramawat KG, Mérillon J-M, editors. Polysaccharides: Bioactivity and Biotechnology. Springer Cham Heidelberg New York Dordrecht London; 2014. p. 1-13. DOI: 10.1007/978-3-319-03751-6_58-1

[37] Cristescu R, Stamatin I, Mihaiescu DE, Ghica C, Albulescu M, Mihailescu IN, Chrisey DB. Pulsed laser deposition of biocompatible polymers: A comparative study in case of pullulan. Thin Solid Films. 2004;**453-454**:262-268

[38] Jelinek M, Cristescu R, Axente E, Kocourek T, Dybal J, Remsa J, Plestil J, Mihaiescu D, Albulescu M, Buruiana T, Stamatin I, Mihailescu IN, Chrisey DB. Matrix assisted pulsed laser evaporation of cinnamate-pullulan and tosylate-pullulan polysaccharide derivative thin films for pharmaceutical applications. Applied Surface Science. 2007;**253**:7755-7760

[39] Sarah E, Bulman, Coleman CM, Mary Murphy J, Medcalf N, Ryan AE, Barry F. Pullulan: A new cytoadhesive for cell-mediated cartilage repair. Stem Cell Research & Therapy. 2015;**6**:34. DOI: 10.1186/s13287-015-0011-7.

[40] Atila D, Keskin D, Tezcaner A. Crosslinked pullulan/cellulose acetate fibrous scaffolds for bone tissue engineering. Materials Science and Engineering C. 2016;**69**:1103-1115

[41] Liu Y, Wang Y, Zhang C, Zhou P, Yang L, An T, Sun D, Zhang N, Wang Y. Core–shell nanoparticles based on pullulan and poly(β-amino) ester for hepatoma-targeted codelivery of gene and chemotherapy agent. ACS Applied Materials & Interfaces. 2014;**6**(21):18712-18720

[42] Catherine Guérin, Jean-Luc Audic. Egg-protein-based films and coatings, chapter 31 In: Rainer Huopalahti, Rosina López-Fandiño, Marc Anton, Rüdiger Schade editors. Bioactive Egg Compound. Berlin Heidelberg: Springer-Verlag; 2007. pp. 265-273

[43] John MK. Proteins as raw materials for films and coatings: Definitions, current status, and opportunities, chapter 1. In: Gennadios A, editor. Protein-based Films and Coatings. Boca Raton-London-New York-Washington D.C.: CRC Press LLC; 2002

[44] Cozzone AJ. Proteins: Fundamental chemical properties, encyclopedia of life sciences. Macmillan Publishers Ltd, Nature Publishing Group; 2002. p. 10

[45] Revest P. Biochemistry and cell biology, chapter 2. In: Naish J, Revest P, Court DS, editors. Medical Sciences. Saunders: Elsevier; 2009. p. 15-88

[46] Dangaran K, Tomasula PM, Qi P. Structure and function of protein-based edible films and coatings, chapter 2. In: Huber KC, Embuscado ME, editors. Edible Films and Coatings for Food Applications. New York: Springer; 2009. p. 25-56. DOI: 10.1007/978-0-387-92824-1_2

[47] Gagner JE, Kim W, Chaikof EL. Designing protein-based biomaterials for medical applications. Acta Biomaterialia. 2014;**10**:1542-1557

[48] McCann TE, Kosaka N, Choyke PL, Kobayashi H. The use of fluorescent proteins for developing cancer-specific target imaging probes, chapter 13. In: Hoffman RM, editor. In Vivo Cellular Imaging Using Fluorescent Proteins. Vol. 872. New York: Humana Press Springer Science+Business Media, eBook; 2012. p. 191-204. DOI: 10.1007/978-1-61779-797-2_13

[49] Löfblom J, Feldwisch J, Tolmachev V, Carlsson J, Ståhl S, Frejd FY. Affibody molecules: Engineered proteins for therapeutic, diagnostic and biotechnological applications. FEBS Letters. 2010;**584**:2670-2680

[50] Rehman K, Hamid Akash MS, Akhtar B, Tariq M, Mahmood A, Ibrahim M. Delivery of therapeutic proteins: Challenges and strategies. Current Drug Targets. 2016;**17**(10):1172-1188

[51] Yin L, Yuvienco C, Montclare JK. Protein based therapeutic delivery agents: contemporary developments and challenges. Biomaterials. 2017;**134**:91-116

[52] Gomes S, Leonor IB, Mano JF, Reis RL, Kaplan DL. Natural and genetically engineered proteins for tissue engineering. Progress in Polymer Science. 2012;**37**(1):1-17

[53] Sayin E, TürkerBaran E. Vasif Hasirci Protein-based materials in load-bearing tissue-engineering applications. Regenerative Medicine. 2014;**9**(5):687-701

[54] Sima F, Davidson PM, Dentzer J, Gadiou R, Pauthe E, Gallet O, Mihailescu IN, Anselme K. Inorganic–organic thin implant coatings deposited by lasers. ACS Applied Materials & Interfaces. 2015;**7**:911–920

[55] Borkner CB, Elsner MB, Scheibel T. Coatings and films made of silk proteins. ACS Applied Materials & Interfaces. 2014;**6**(18):15611-15625

[56] Wang H-J, Di L, Ren Q-S, Wang J-Y. Applications and degradation of proteins used as tissue engineering materials. Materials. 2009;**2**:613-635

[57] Frantz C, Stewart KM, Weaver VM. The extracellular matrix at a glance. Journal of Cell Science. 2010;**123**:4195-4200

[58] Li M, Mills DK, Cui T, McShane MJ. Cellular response to gelatin- and fibronectin-coated multilayer polyelectrolyte nanofilms. IEEE Transactions on Nanobioscience. 2005;**4**(2):170-179

[59] Gand A, Tabuteau M, Chat C, Ladam G, Atmani H, Tassel PRV, Pauthe E. Fibronectin-based multilayer thin films. Colloids and Surfaces B: Biointerfaces. 2017;**156**:313-319

[60] Andreeva NV, Leonova OG, Popenko VI, Belyavsky AV. Controlled formaldehyde fixation of fibronectin layers for expansion of mesenchymal stem cells. Analytical Biochemistry. 2016;**514**:38-41

[61] Sangkert S, Kamonmattayakul S, Chai WL, Meesane J. A biofunctional-modified silk fibroin scaffold with mimic reconstructed extracellular matrix of decellularized pulp/collagen/fibronectin for bone tissue engineering in alveolar bone resorption. Materials Letters. 2016;**166**:30-34

[62] Li G, Yang P, Huang N. Layer-by-layer construction of the heparin/fibronectin coatings on titanium surface: Stability and functionality. Physics Procedia. 2011;**18**:112-121

[63] Grigorescu S, Hindie M, Axente E, Carreiras F, Anselme K, Werckmann J, Mihailescu IN, Gallet O. Fabrication of functional fibronectin patterns by nanosecond excimer laser direct write for tissue engineering applications. Journal of Materials Science Materials in Medicine. 2013;**24**:1809-1821

[64] Agarwal R, Gonzalez-García C, Torstrick B, Guldberg RE, Salmeron-Sanchez M, García AJ. Simple coating with fibronectin fragment enhances stainless steel screw osseointegration in healthy and osteoporotic rats. Biomaterials. 2015;**63**:137-145

[65] Schleicher I, Parker A, Leavesley D, Crawford R, Upton Z, Xiao Y. Surface modification by complexes of vitronectin and growth factors for serum-free culture of human osteoblasts. Tissue Engineering. 2005;**11**(11/12):1688-1698

[66] Toromanov G, Gugutkov D, Gustavsson J, Planell J, Salmeron-Sanchez M, Altankov G. Dynamic behavior of vitronectin at the cell–material interface. ACS Biomaterials Science Engineering. 2015;**1**(10):927-934

[67] Yamada KM. Fibronectin and other cell interactive glycoproteins, chapter 4. In: Hay ED, editor. Cell Biology of Extracellular Matrix. 2nd ed. USA: Springer; 1991. p. 111-146

[68] McConigle J, Thomas J. Webster functional tests for biocompatability, chapter 20. In: Glocker D, Ranade S, editors. Medical Coatings and Deposition Technologies. Wiley Scrivener Publishing, eBook; 2016. p. 671-702

[69] Whitlow CT, Geer CP, Mattern CWT, Mussat-Whitlow BJ, Yazdani SK, Berry JL, Lalli JH, Claus RO, Challa VR, Morris PP. Endovascular histologic effects of ultrathin gold- or vitronectin-coated platinum aneurysm coils in a rodent arterial occlusion model: A preliminary investigation. AJNR American Journal of Neuroradiology. 2009;**30**:85-90

[70] F. Sima, P. Davidson, E. Pauthe, O. Gallet, K. Anselme,·IN. Mihailescu. Thin films of vitronectin transferred by MAPLE, Applied Physics A 105:611-617 (2011).

[71] Mäntsälä P, Niemi J. Enzymes: The biological catalysts of life, chapter 1. In: Hanninen OOP, Atalay M, editors. Physiology and Maintenance: Enzymes: The Biological Catalysts of Life, Nutrition and Digestion. Vol. 2. EOLSS Publisher Co. Ltd., Oxford, United Kingdom; 2009. p. 1-21

[72] Nature of Biology 2. Enzymes and biochemical pathways [Internet]. 2016. Available from: http://www.wiley.com/legacy/Australia/PageProofs/c03EnzymesAndBiochemicalPathways_web.pdf [Accessed: 2017-07-30]

[73] Pratul KA. Enzymes: An integrated view of structure, dynamics and function. Microbial Cell Factories. 2006;**5**:2-12. DOI: 10.1186/1475-2859-5-2

[74] Putzbach W, Ronkainen NJ. Immobilization techniques in the fabrication of nanomaterial-based electrochemical biosensors: A review. Sensors (Basel). 2013;**13**(4):4811-4840

[75] Leitgeb M, Knez Ž, Vasić K. Micro- and nanocarriers for immobilization of enzymes, chapter 2. In: Stanciu SG, editor. Micro and Nanotechnologies for Biotechnology. Croatia: InTech; 2016. p. 21-58

[76] Costa SA, Helena SA, Reis RL. Enzyme immobilization in biodegradable polymers for biomedical applications, chapter 17. In: Reis RL, San Román J, editors. Biodegradable Systems in Tissue Engineering and Regenerative Medicine. CRC Press, eBook; 2004. p. 301-323

[77] Popescu C, Roqueta J, Pérez delPino A, Moussaoui M, Nogués MV, György E. Processing and immobilization of enzyme Ribonuclease A through laser irradiation. Journal of Materials Research. 2011;**26**(6):815-821

[78] Amri E, Mamboya F. Papain, a plant enzyme of biological importance: A review. American Journal of Biochemistry and Biotechnology. 2012;**8**(2):99-104

[79] Manohar CM, Prabhawathi V, Sivakumar PM, Doble M. Design of a papain immobilized antimicrobial food package with curcumin as a crosslinker. PLoS One. 2015;**10**(4):e0121665

[80] Motoc MM, Axente E, Popescu C, Sima LE, Petrescu SM, Mihailescu IN, Gyorgy E. Active protein and calcium hydroxyapatite bilayers grown by laser techniques for therapeutic applications. Journal of Biomedical Materials Research A. 2013;**101A**(9):2706-2711

[81] Park S-I, Daeschel MA, Zhao Y. Functional properties of antimicrobial lysozyme–chitosan composite films. Journal of Food Science. 2004;**69**(8):M215-M221

[82] Liburdi K, Benucci I, Esti M. Lysozyme in wine: An overview of current and future applications. Comprehensive Reviews in Food Science and Food Safety. 2014;**13**:1062-1073

[83] Dzhavakhiya VG, Shcherbakova LA. Creation of disease-resistant plants by gene engineering, chapter 16. In: Dyakov YT, Dzhavakhiya VG, Korpela T, editors. Comprehensive and Molecular Phytopathology part 3: Practical Use of Molecular Studies. Elsevier, eBook; 2007. p. 439-466

[84] Oliver WT, Wells JE. Lysozyme as an alternative to growth promoting antibiotics in swine production. Journal of Animal Science and Biotechnology. 2015;**6**(1):35

[85] Pacor S, Giacomello E, Bergamo A, Clerici K, Zacchigna M, Boccu E, Sava G. Antimetastatic action and lymphocyte activation by the modified lysozyme mPEG-Lyso in mice with MCa mammary carcinoma. Anticancer Research. 1996;**16**:2559-2564

[86] Mahanta S, Paul S, Srivastava A, Pastor A, Kundu B, Chaudhuri TK. Stable self-assembled nanostructured hen egg white lysozyme exhibits strong anti-proliferative activity against breast cancer cells. Colloids and Surfaces B: Biointerfaces. 2015;**130**:237-245

[87] Visan A, Cristescu R, Stefan N, Miroiu M, Nita C, Socol M, Florica C, Rasoga O, Zgura I, Sima LE, Chiritoiu M, Chifiriuc MC, Holban AM, Mihailescu IN, Socol G. Antimicrobial polycaprolactone/polyethylene glycol embedded lysozyme coatings of Ti implants for osteoblast functional properties in tissue engineering. Applied Surface Science. 2017;**417**:234-243

[88] Grumezescu V, Holban AM, Sima LE, Chiritoiu MB, Chiritoiu GN, Grumezescu AM, Ivan L, Safciuc F, Antohe F, Florica C, Luculescu CR, Chifiriuc MC, Socol G. Laser deposition of poly(3-hydroxybutyric acid-co-3-hydroxyvaleric acid) – Lysozyme microspheres based coatings with anti-microbial properties. International Journal of Pharmaceutics. 2017;**521**(1-2):184-195

[89] Mazzei L, Cianci M, Musiani F, GáborLente MP, Ciurli S. Inactivation of urease by catechol: Kinetics and structure. Journal of Inorganic Biochemistry. 2017;**166**:182-189

[90] György E, Sima F, Mihailescu IN, Smausz T, Hopp B, Predoi D, Sima LE, Petrescu SM. Biomolecular urease thin films grown by laser techniques for blood diagnostic applications. Materials Science and Engineering C. 2010;**30**:537-541

[91] Stanisçuaski F, Carlini CR. Plant ureases and related peptides: understanding their entomotoxic properties. Toxins. 2012;**4**:55-67

[92] Raza Shah M, Soomro ZH. In: Sharma R, editor. Urease Inhibition, Enzyme Inhibition and Bioapplications. InTech. Available from: http://www.intechopen.com/books/enzyme-inhibition-and bioapplications/urease-inhibition; 2012 ISBN: 978-953-51-0585-5

[93] Doğaç YI, Teke M. Synthesis and characterisation of biocompatible polymer-conjugated magnetic beads for enhancement stability of urease. Applied Biochemistry and Biotechnology. 2016;**179**(1):94-110

[94] Gyorgy E, Mihailescu IN. Enzyme-based biosensors for trace detection, chapter 9. In: Vaseashta A, editor. Life Cycle Analysis of Nanoparticles: Risk, Assessment, and Sustainability. DEStech Publications, eBook; 2015. p. 231-248

[95] Saxena A, Bhattachary A, Kumar S, Epstein IR, Sahney R. Biopolymer matrix for nanoencapsulation of urease—A model protein and its application in urea detection. Journal of Colloid and Interface Science. 2017;**490**:452-461

[96] José R. Siqueira Jr, Denise Molinnus, Stefan Beging, Michael J Schoning Incorporating a hybrid urease-carbon nanotubes sensitive nanofilm on capacitive field-effect sensors for urea detection. ACS Analytical Chemistry. 2014;**86**:5370-5375

Chapter 4

Structural Modifications of KrF Excimer Laser-Ablated Zirconium Correlated to the Surface and Mechanical Properties

Nisar Ali, Umm-i-Kalsoom, Shazia Bashir and Narjis Begum

Additional information is available at the end of the chapter

http://dx.doi.org/10.5772/intechopen.70952

Abstract

The structural modifications of KrF excimer laser-ablated zirconium (Zr) have been investigated in correlation with its surface and mechanical properties after ablation in deionized water and ethanol. KrF excimer laser of pulse duration of 20 ns, wavelength of 248 nm, and repetition rate of 20 Hz has been utilized for this purpose. Irradiation of Zr was carried out for varying number of laser pulses ranging from 500 to 2000 for laser fluence value of 3.6 J/cm^2. The structural and chemical analyses were performed by using X-ray diffraction (XRD), Raman spectroscopy, and energy dispersive X-ray spectroscopy (EDS) techniques. Scanning electron microscope (SEM) and Vickers hardness tester were utilized for the analysis of surface morphology and hardness of laser-irradiated Zr targets. Presence of surrounding liquids played substantial role in structural, chemical, and mechanical modifications of Zr targets after irradiation. Pressure gradients and convective bubble motion owing to the confinement effects of the surrounding liquids, several thermal and chemical phenomena produced by heating through laser at the solid-liquid interface results in the generation of various hydrides and oxides of Zr, which are responsible for the development of various surface features and increase in hardness of irradiated Zr.

Keywords: zirconium, laser ablation, ambient environment, crystallinity, surface morphology, hardness

1. Introduction

Laser-induced micro and nanostructuring of materials is a versatile approach to modify the material properties for enhanced thermal properties [1], optical absorption [2], field emission properties [3], biocompatibility [4], hydrophobicity [5], and tribological performance [6]. The interaction of laser radiations with solid materials, accompanied by a variety of instabilities, may result in the formation of various surface features, for example, laser-induced periodic surface structures (LIPSSs), cracks, nanodots, nanopores or nanocavities, and grains [7, 8].

Nanofabrication based on laser ablation strongly depends on parameters of laser beam, such as number of pulses, pulse duration, wavelength, fluence, and material's nature as well as the environmental conditions and their nanochemical effects [8]. With increasing number of pulses, the incubation effects become dominant, causing the reduction in ablation threshold of the irradiated material. For metallic targets, incubation effects are related to the accumulation of plastic deformation produced by laser-induced thermal stresses [9]. Materials ablation in liquid-confined environment has become an important technique for the production of debris-free microstructures on material's surface. The presence of a dense medium adjoining to the molten layer of the target causes various instabilities. Various kinds of structures can be grown due to the viscous flow of the vapors of the surrounding liquid and its chemical reaction with the metals. Shock wave generation, confinement and shielding effects, different cooling and quenching rates, and different chemical reactivity of the environment significantly affects the ablation efficiency of the target material [1, 8] without momentous effect on the stoichiometry of the target, and results in the development of cavitation bubbles and colloids owing to the formation of metal alcohols and oxides. Several physical processes such as formation of plasma, condensation, agglomeration, nucleation and coalescence are accountable for the formation and growth of nanostructures in the presence of reactive environment [10]. Particles and nanostructures produced by this process show liquid's dense vapors interaction with the molten layer produced on the surface of target. Such interactions are attributable to hydrodynamic instabilities under such circumstances. The molten layer acts as a source for the formation of nanoparticles and nanostructure. When beam focus passes through liquids, several nonlinear effects [11] as well as higher harmonic generation, filamentation and self-phase modulation take place. Considerable literature is already reported on liquid-based pulsed laser ablation of metals. Barminaa et al. reported the improvement in thermionic properties of tungsten surface after ablation in liquid environment [12]. Ablation of materials in the presence of surrounding liquid includes heat transfer through convection/conduction and motion of bubbles, which is responsible for the removal of particles redeposited over the surface and reduces the oxidation of debris [13]; as a result, the enhanced coupling of energy to the target is directed to hygienic and clean target surface [14].

The purpose of structural, surface, and mechanical modification of Zr by irradiation with nanosecond laser is to develop and alter its widespread functional, physical, chemical, and mechanical characteristics such as wear, hardness, and corrosion resistance etc. The development of several surface structures can be well associated with the increase of chemical reactivity of Zr. To achieve the enhanced chemical and mechanical properties, Zr targets were irradiated with KrF excimer laser for varying number of pulses ranging from 500 to 2000, in the presence of deionized water and ethanol. Various diagnostic techniques such as X-ray diffractometry (XRD), energy dispersive X-ray spectroscopy (EDS), Raman spectroscopy, scanning electron microscope (SEM), and Vickers hardness techniques were utilized to relate the surface morphology and hardness with the variations in chemical composition and crystallinity after irradiation.

2. Experimental setup

KrF excimer laser (EX 200/125-157 GAM Laser, USA) of pulse duration of 20 ns, pulse energy of 70 mJ, repetition rate of 20 Hz, and wavelength of 248 nm was employed for irradiation of Zr targets. The unfocused beam was rectangular with size 11 × 7 mm. The beam was focused by using a circular lens of 50 cm focal length and the focused beam spot appeared elliptical with major and minor axis of 2.5 × 1 mm, respectively.

Zr was bought from Alfa Aesar with purity 99.9%. EDS of the untreated and laser-irradiated samples was performed to check the elemental composition of targets before and after irradiation. **Table 1** shows the wt.% of elements present in untreated and treated samples. Zr targets rectangular in shape (dimensions: 8 × 8 × 2 mm) were grinded and polished to get mirror-like surface in order to decrease the surface roughness and to enhance the quality of the surface. The polished samples were cleaned ultrasonically for 30 minutes with acetone. These polished samples were then placed in quartz cuvette (Cuvette dimensions: 10 × 10 × 45 mm). **Figure 1** shows the schematic diagram of the experimental setup. Threshold fluence for Zr was calculated by using following relation [15].

$$F_{th} = \rho L_v \sqrt{a\, t_{e.}} \tag{1}$$

where "L_v" is the specific heat of vaporization per unit mass, "ρ" is the density of sample, "a" is the thermal diffusivity (a = K/ρC, where C and K are the specific heat and thermal conductivity respectively), and t_e is the pulse duration of laser and their values are, ρ = 6.52 g/cm^3, L_v = 6375 J/g, a = 0.1246 cm^2/s, t_e = 20 ns by substituting the values in above equation, we get the value of threshold fluence, which comes out to be 2.08 J/cm^2. Zr targets were exposed for several number of overlapping laser pulses ranging from 500 to 2000, for laser fluence of 3.6 J/cm^2, which is higher than the threshold fluence. The ablation threshold fluence is the minimum fluence to generate a surface damage that can be seen under the SEM. For nanostructuring of

Elements	Untreated (wt. %)	Deionized water (wt.%)	Ethanol (wt.%)
Zr	87.52	70.78	74.81
B	2.78	2.75	4.08
C	2.82	2.59	3.36
O	4.91	21.52	15.55
Re	1.97	2.36	2.20

Table 1. The comparative EDX analysis of the unirradiated and central-ablated area of zirconium target after excimer laser irradiation at a fluence of 3.6 J/cm^2, wavelength of 248 nm, pulse duration of 20 ns, and repetition rate of 20 Hz for 1000 laser pulses under ambient environments of deionized water and ethanol.

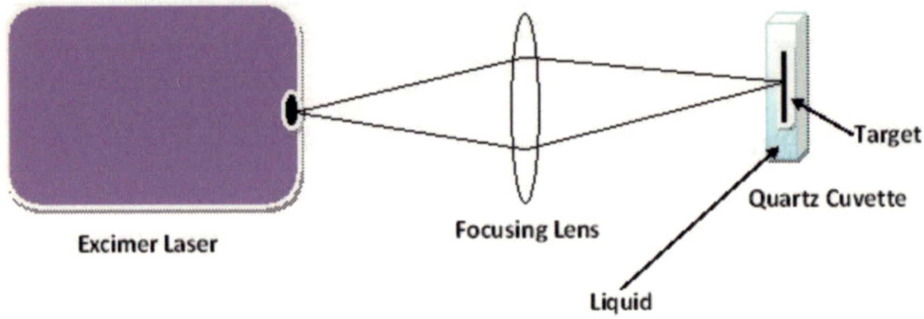

Figure 1. Schematic diagram of experimental setup.

materials, fluence should be higher than the threshold fluence and in order to avoid intense heating/melting and burning of the samples, fluence should not be very high.

Absorption coefficient for deionized water has value of 5.05×10^{-3}/cm [16], whereas for ethanol its value is 4×10^{-2}/cm [17]. It shows that ethanol will absorb more energy than deionized water; so, less energy will reach the target surface in case of ethanol. This shows that the threshold fluence value will be more for ethanol than deionized water. Experiment was conducted in two steps: (a) nanosecond laser ablation of Zr in ambient environment of deionized water, (b) nanosecond laser ablation of Zr in ambient environment of ethanol. Before exposure to laser, every time fresh liquid was filled in the cuvette with 4 mm thickness over the sample.

To evaluate the crystal structure and phase analysis of the irradiated Zr targets X-ray diffractometer (X'Pert PRO (MPD)) was employed. For chemical analysis, energy dispersive X-ray spectroscopy (EDS-S3700 N) was used. Raman spectroscopic analysis was performed to determine the structural modification of laser-ablated Zr. Scanning electron microscope (SEM-JEOL JSM-6480 LV) was employed to study the surface morphology of ablated targets. Hardness was evaluated by employing the Vickers microhardness tester (HV-1000A).

3. Results and discussion

3.1. Effect on surface microstructure

Figure 2 shows the XRD pattern of unirradiated Zr target. Unirradiated sample shows the presence of Zr (100), (002), (101), (102), (110), (103), (004), and (104) plane reflections.

Figure 3a represents the XRD diffractograms of laser-irradiated Zr targets, in an ambient environment of deionized water, for varying number of laser pulses ranging from 500 to 2000. **Figure 3b** shows the variation in crystallite size and residual stresses with increase in number of laser pulses. New phases of Zr oxides (zirconia): ZrO_2 (−111) (111), (212), (321), (220), Zr_3O (201), (200), (217) and Zr hydrides: ZrH_2 (112), ZrH (123) are also detected together with the original phases of Zr.

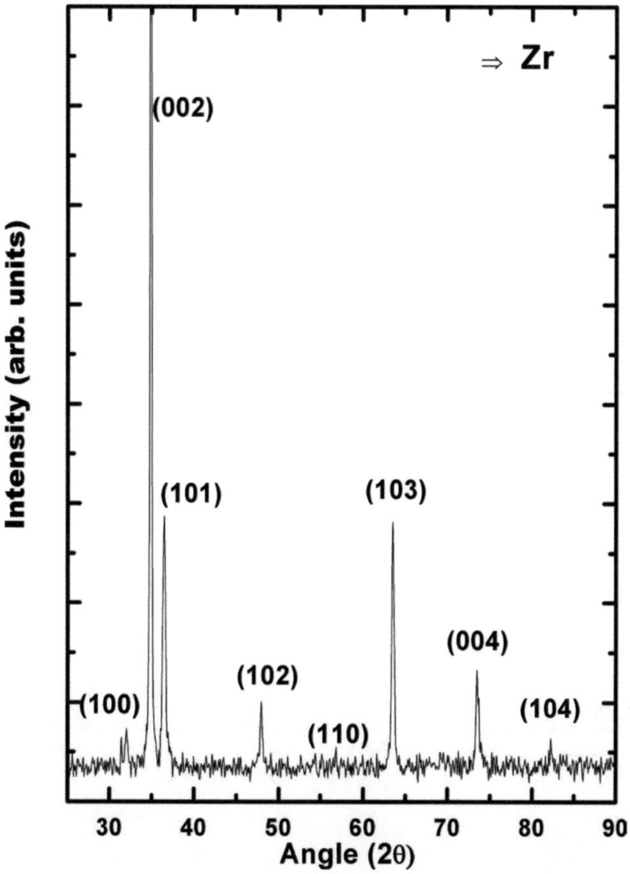

Figure 2. XRD patterns of unirradiated zirconium.

For ZrO_2 (212) plane reflection, the crystallite size is calculated by using Scherrer's formula [18].

$$\text{Crystallite Size (D)} = \frac{0.9\,\lambda}{\text{FWHM}\cos\theta} \tag{2}$$

where D is crystallite size, λ is the wavelength of X-rays (1.5406 Å), FWHM is full width at half maximum, and θ is the angle of diffraction.

Following relation is used to evaluate the residual strains [18].

$$\text{Strain}\ (\varepsilon) = \frac{d - do}{do} \tag{3}$$

where d is the observed, d_0 is the standard plane spacing, and ε is the induced strain.

By using the relation given below, the induced stresses are calculated

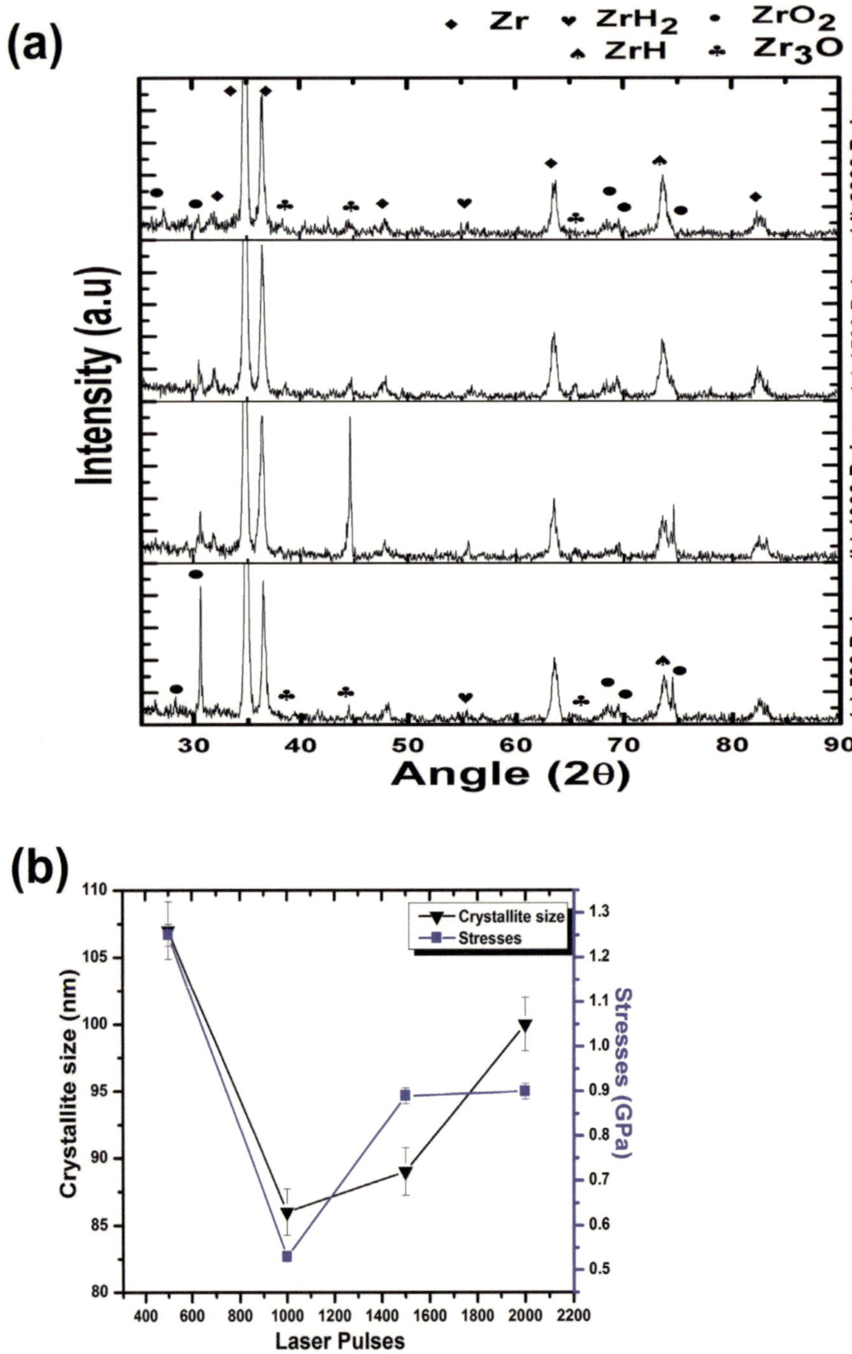

Figure 3. (a) XRD patterns of excimer laser-irradiated zirconium samples, (b) variation in the crystallite size and residual stresses of laser-irradiated zirconium for various number of overlapping laser pulses in an ambient environment of deionized water.

$$\text{Stress}\ (\sigma) = \varepsilon E \tag{4}$$

where ε is the strain, E is the young's modulus for ZrO_2 and its value is 186.21 GPa [19].

Intensity of peak and crystallite size of ZrO_2 (212) reflection plane first decreases and then increases with increase in number of laser pulses, **Figure 3a, b**. Initial reduction in the intensity of peak is due to the recrystallization of the irradiated targets during resolidification after melting.

After irradiation with laser, the larger crystallites breakup into smaller ones (**Figure 3b**) that also results in decrease in peak intensity [20]. The decrease in size of the crystallites is also credited to the interstitial diffusion of the gas atoms, which causes the peak broadening owing to the variation of d-spacing [21]. The ablated Zr atoms chemically react with atoms of the surrounding gas (oxygen/hydrogen) and redeposit on the target surface after the formation of oxides and hydrides. Formation of theses oxides and hydrides is attributable for the decrease in crystallite size [22]. The possible defect production and growth of residual stresses might also be considered to be other reasons for decrease in the crystallite size [22]. An increase in stacking faults and structural disorder owing to the diffusion of hydrogen/oxygen widens the XRD peaks [23]. Increases in density of point defects affect the crystallinity and grain boundary mobility, which in turns causes a linear increase in the FWHM of XRD peak. The increase in intensity of the peak and crystallinity of ZrO_2 (212) reflection plane is due to the increased oxygen diffusion into the surface across the boundaries of grains and enhancement in the X-ray's diffraction from the surface of the target [24]. The diffusion of oxygen in Zr after irradiation is exhibited in EDX analysis depicted in **Table 1**. The microstructure and Bragg's conditions of diffraction of Zr surface gets modified after the interaction with nanosecond laser. These variations in the microstructure and Bragg's conditions of diffraction are attributable to the variation in d-spacing and intensity of diffraction lines. The variation in laser-induced residual stresses for a variety of number of laser pulses is exhibited in **Figure 3b**.

Laser-induced thermal shocks and lattice defects generated by oxygen ions incorporation into the lattice may cause residual stress variation. Compressive stresses are generally due to ion implantation, whereas laser-induced thermal shocks are responsible for tensile stresses [25]. For 500 number of laser pulses, high tensile stresses are present, which relax on increase in number of laser pulses to 1000. The defects and stresses produced by initial pulses are annealed and relaxed after successive pulses. When number of pulses are increased up to a value of 2000, thermally induced shocks results into increasing tensile stresses [24].

XRD patterns displayed in **Figure 4a** confirms the presence of oxides (zirconia) and hydrides of zirconium after irradiation in an ambient environment of ethanol for various number of overlapping laser pulses of (a) 500, (b) 1000, (c) 1500, and (d) 2000. New phases of zirconium oxides (zirconia): ZrO_2 (111), (212), (321), (114), Zr_3O (201), Zr_3O (217), and zirconium hydrides: ZrH (123) and ZrH_2 (112) (appeared after 1000 laser pulses) are observed along with the original phases of zirconium.

Decrease in intensity (**Figure 4a**) and crystallinity (**Figure 4b**) of ZrO_2 (212) plane reflection is observed with increasing number of laser pulses up to 2000. Decrease in peak intensity and crystallite size is attributable to interstitial diffusion of oxygen atoms/ions. Interstitial diffusion of oxygen atoms/ions induces microstrain defects and these induced defects causes an increase in FWHM and reduction in the peak intensity and crystallite size.

Figure 4b also represents variation in induced stresses of zirconium after irradiation with various number of laser pulses. For 500 number of laser pulses, tensile stresses are observed. They relax with increasing number of laser pulses up to 2000. This relaxation is due to reduction in crystallite size. Diffusion of oxygen/hydrogen atoms into the surface causes the decrease in crystallite size and hence causes the highly tensile stresses to be relaxed.

Comparison of XRD data of ablated zirconium in deionized water (**Figure 3**) and ethanol (**Figure 4**) shows the presence of oxides (zirconia) and hydrides of zirconium for both media.

Energy dispersive X-ray spectroscopy (EDS) was carried out for chemical analysis of unirradiated and irradiated zirconium targets. Unirradiated sample shows following contents: zirconium ~ 87.52 wt.%, balanced to 100% by B (~2.78 wt.%), C (~2.82 wt.%), O (~4.91), Re (~1.97 wt.%). **Table 1** shows the comparative EDX analysis of the unirradiated and central-ablated area of Zr target after excimer laser irradiation under ambient environments of deionized water and ethanol. The increase in the content of oxygen from 4.91 wt.% to a value of 21.52 wt.% is observed in case of deionized water, whereas less diffusion of oxygen (15.55 wt.%) but increase in content of carbon from 2.82 to 3.36 wt.% is observed in case of ethanol. Multipulse irradiation induced heating of Zr causes efficient diffusion of atomic oxygen/carbon into the target surface; therefore, oxides and carbides are formed [26].

Figure 5a, b shows the Raman spectra of Zr samples irradiated for various number of laser pulses at a fluence of 3.6 J/cm^2 in ambient environments of deionized water and ethanol. Raman modes arise after laser ablation of Zr due to the formation of oxides and hydrides of Zr on the metallic surface.

For deionized water (**Figure 5a**), Raman peaks identified at 320, 337 cm^{-1} represents the Zr—Zr bond. Peak identified at 535 cm^{-1} corresponds to O—O bond [27], while peaks appearing at 363, 372, 420, 480, and 640 cm^{-1} are characteristic bands of monoclinic ZrO_2 [28] and the peak identified at 269 cm^{-1} can be assigned to the presence of tetragonal ZrO_2 [29]. The inset of **Figure 5a** represents the identified peak of zirconium hydride (ZrH) at 1597 cm^{-1}.

Figure 5b represents Raman spectra for ethanol environment. Peaks identified at 351, 422, and 538 cm^{-1} correspond to monoclinic ZrO_2 [28]. Raman peaks identified at 269 and 312 cm^{-1} can be assigned to the presence of tetragonal ZrO_2 [29]. Peak identified at 476 corresponds to O—O bond [27]. The inset of **Figure 5b** represents identified peak of C=O at 1760 cm^{-1}.

The Zr samples after irradiation in deionized water and ethanol consist of a mixture of monoclinic and tetragonal ZrO_2 phases. The presence of hydride is confirmed from Raman spectra of samples irradiated in deionized water and the presence of carbon diffusion is confirmed only in case of ethanol. These Raman results are in good agreement with the EDS and XRD results.

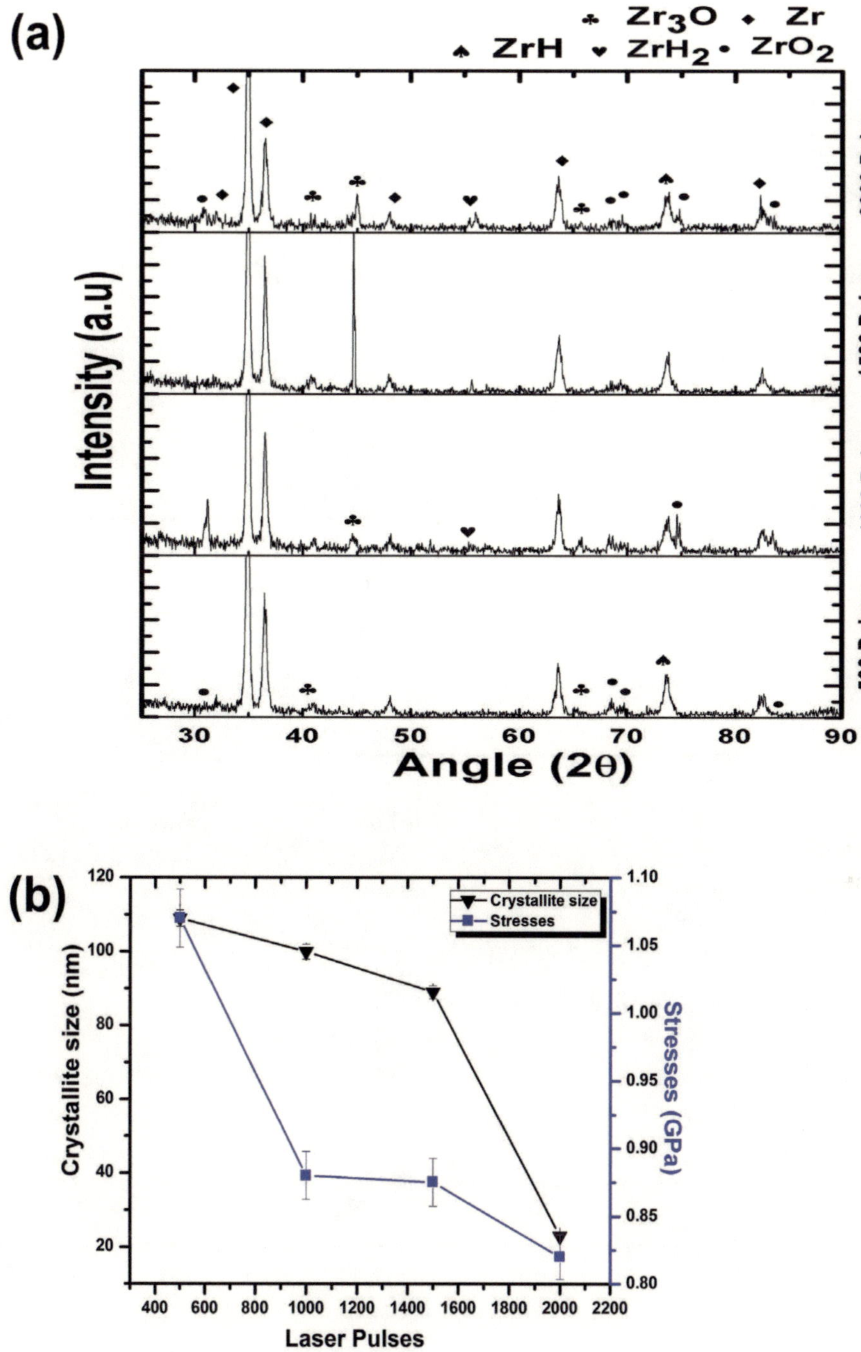

Figure 4. (a) XRD patterns of excimer laser-irradiated zirconium samples, (b) variation in the crystallite size and stresses of laser-irradiated zirconium for various number of overlapping laser pulses in an ambient environment of ethanol.

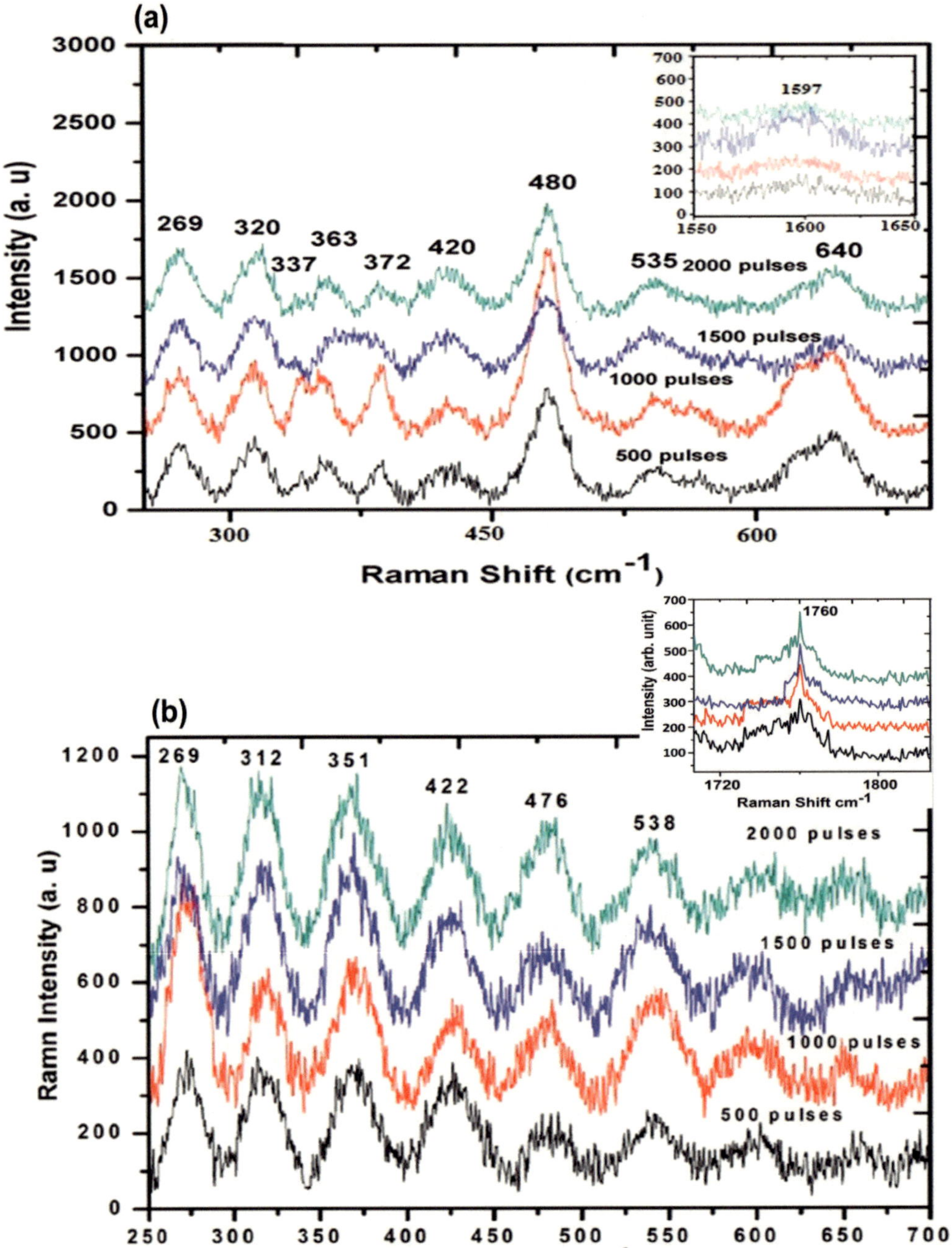

Figure 5. Raman spectroscopy analysis of excimer laser-irradiated zirconium in an ambient environment of (a) deionized water, (b) ethanol.

3.2. Effect on surface morphology

Figure 6a–e represents the surface morphology of Zr sample after irradiation with various number of overlapping laser pulses at fluence of 3.6 J/cm^2 in an ambient environment of deionized water.

Figure 6a exhibits the surface morphology of Zr sample after irradiation with 500 laser pulses. No significant redeposition of the material around the edges of the craters is observed because the ablated species are taken away by the surrounding liquid. Rapid temperature rise during laser-induced plasma expansion/formation generates bubbles in the liquid. The bubble motion is responsible for the removal of ablated particles and hence no redeposition on the target surface is observed [30]. The liquid has played considerable role in ablation of Zr and it effectively cools the target preventing the excessive heat accumulation after laser irradiation.

Figure 6b–e reveals enlarged SEM images of the central-ablated area of Zr in an ambient environment of deionized water for various number of pulses, that is, (b) 500, (c) 1000, (d) 1500, and (e) 2000. **Figure 6b** reveals the uniform distribution of large number of droplets over the porous surface. Small-sized cavities and cracks are also present with some signs of melting. Further increase in laser pulses up to 1000, few large-sized droplets along with cavities are revealed (**Figure 6c**). No cracks are observed for 1000 laser pulses. For 1500 number of laser pulses, density of pores and cracks increases significantly (**Figure 6d**). For 2000 number of pulses, highly porous surface is observed, whereas the droplets and cavities are completely vanished (**Figure 6e**).

Pulsed laser irradiation of Zr in liquid-confined environment creates high temperature and high pressure plasma, and consequently, high density of excited electrons. These electrons transport their energy to phonons during electron–phonon relaxation. The energy is redistributed during lattice vibrations and as a result, heat is conducted in the Zr target. This heat might melt or vaporize the sample. It is worth noting that the surface morphology of the irradiated Zr consists of large number of pores and cavities. An explosive melt expulsion in the confined environment could be a prevailing process for the laser ablation of Zr in deionized water [31]. These results are well correlated with the change in the crystallinity of the irradiated samples. For 500 number of laser pulses, highly tensile stresses are present (**Figure 3b**), which are responsible for the presence of cavities and pores over the surface. For 1000 pulses, refilling of cavities by melted and shock-liquefied material causes vanishing of pores and cavities representing the relaxation in tensile stresses [32]. For 2000 laser pulses, highly porous surface shows the presence of enhanced tensile residual stresses after irradiation (**Figure 3b**).

Figure 7 represents the surface morphology of overall and central-ablated area of Zr samples after irradiation with various overlapping numbers of laser pulses at 3.6 J/cm^2 in an ambient environment of ethanol. **Figure 7a** is an overall view of ablated area exposed to 500 laser pulses. Appearance of the overall ablated area is less distinct or faint as compared to

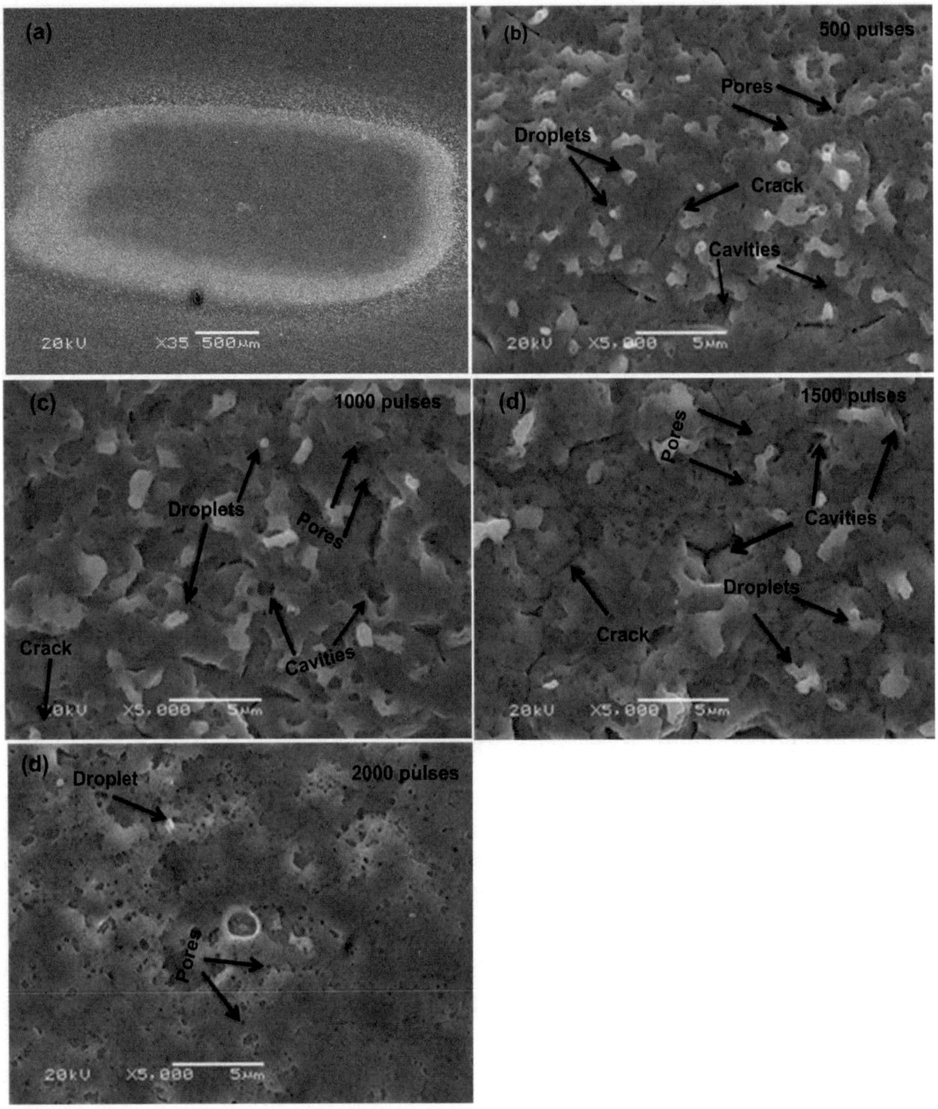

Figure 6. SEM images revealing the surface morphology of (a) overall view of the ablated crater after excimer laser irradiation with 500 laser pulses at a fluence of 3.6 J/cm^2, (b–e) magnified SEM images of central-ablated area revealing the variation in surface morphology of zirconium after irradiation under ambient environment of deionized water for various number of overlapping laser pulses, that is, (b) 500, (c) 1000, (d) 1500, and (e) 2000.

the overall image of ablation spot in case of deionized water. From literature, the observed values of absorbance and absorption coefficient are 0.00086 and 5.05 × 10^{-3}/cm for deionized water [16], whereas 0.0055 and 4 × 10^{-2}/cm for ethanol [17]. This difference in the appearance of ablation spot is attributable to more absorption of energy (0.0055) by ethanol, then deionized water.

Figure 7b–e reveals enlarged view of SEM images of the central-ablated area of Zr in an ambient environment of ethanol for various number of pulses, that is, (b) 500, (c) 1000, (d) 1500, and (e) 2000. **Figure 7b** represents granular morphology with distinct grain boundaries with average grain size of 2.20 μm. With increase in the number of laser pulses up to 1000 increase in size of grains up to 3.06 μm is observed with distinct and wider boundaries (**Figure 7c**). Further increase in number of pulses up to 1500 shows globules instead of distinct grains. The presence of cracks and large number of pores is also evident (**Figure 7d**). For 2000 number of laser pulses, material segregates into distinct grains with average grain size of 3.26 μm and much wider grain boundaries (**Figure 7e**).

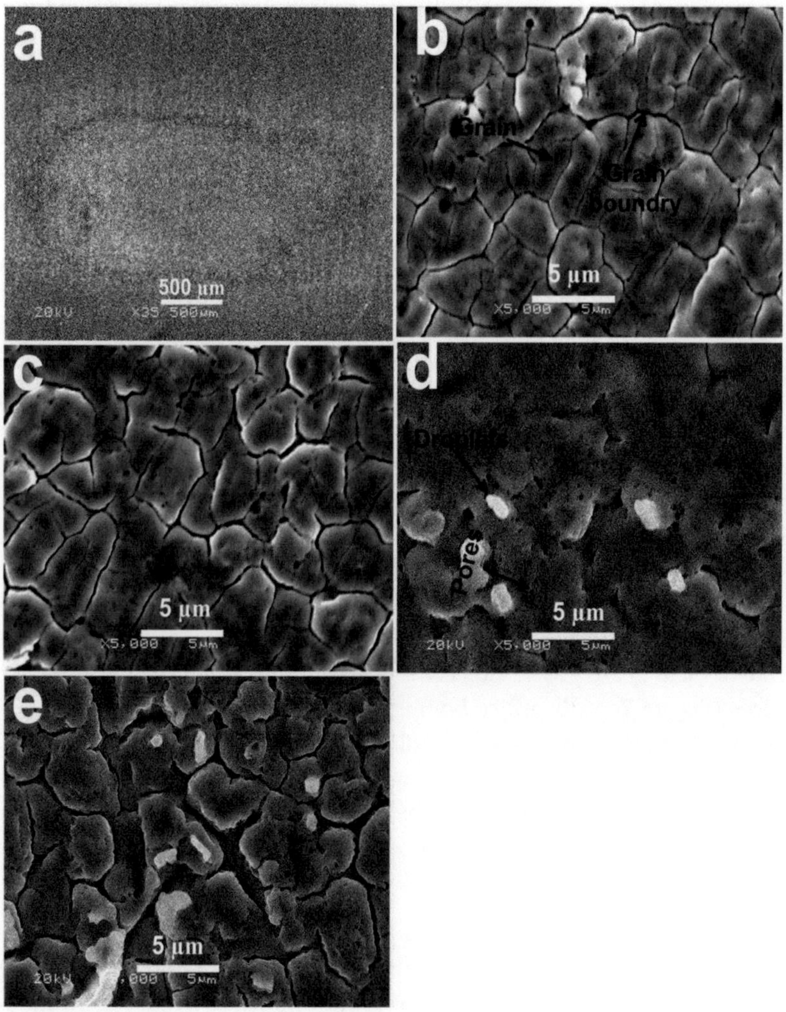

Figure 7. Magnified SEM images of excimer laser-irradiated zirconium at a fluence of 3.6 J/cm^2 (a) overall view of sample irradiated with 500 overlapping pulses (b–e) Magnified SEM images of central-ablated area revealing the variation in surface morphology of zirconium after irradiation under ambient environment of ethanol for various number of overlapping laser pulses, that is, (b) 500, (c) 1000, (d) 1500, and (e) 2000.

Laser-induced heating and cooling, the temperature gradient, and laser-induced residual stresses are responsible for growth of such grains [26]. When the laser pulse is over, the molten material generated by fast heating is supercooled and acts as an effective heat sink that quickly removes any released latent heat and causes the crystal growth. The presence of pores and cracks across the grain boundaries confirms the presence of tensile residual stresses after ablation in ethanol environment.

The presence of various crystal planes is property of the material. Variations in surface morphology can be related to the motion of the crystal planes after laser irradiation due to the production of defects owing to diffusion, which causes the formation of various new crystal planes (ZrO_2, Zr_3O, ZrH, ZrH_2 etc). During irradiation, diffusion of oxygen/carbon/hydrogen caused the increase in density of point defects, stacking faults, and structural disorder, which results in higher angle shifting of planes, which causes the production of compressive stresses. Laser-induced thermal shock causes the peak shift to lower angular position, which confirms the production of tensile residual stresses [25]. Variations in stresses are responsible for the production of various type of surface features, for example, cavities, pores, cracks, droplets, bumps, hillocks, protrusion, etc. after irradiation [21].

In case of ethanol, tensile stresses are present for all laser pulses (from 500 to 2000), but their behavior is not in agreement with the surface morphology especially for 1500 number of laser pulses. We have calculated stress variations for only one plane, that is, ZrO_2 (212). But other phases also play role for surface modification after irradiation, for example, with the increase of number of pulses from 500 to 1000 formation of new phases (Zr_3O, ZrH_2, ZrO_2) is observed due to the enhanced diffusion of oxygen/hydrogen across the grain boundaries, which results in increase in grain size (**Figure 7b**) along with wider grain boundaries. For further increase of number of pulses from 1000 to 1500, some peaks get demolished (ZrO_2, Zr_3O), whereas growth of some other phases (ZrH, ZrH_2, Zr_3O) is observed on the target surface. This shows that energy transferred to the target surface after 1500 number of laser pulses causes the laser annealing on target surface. After laser annealing during recovery process, rapid cooling due to the presence of surrounding liquid, some phases do not get sufficient energy to grow, whereas some of them get to grow. This is the reason for different structures over the surface for 1500 pulses (**Figure 7c**, subsurface boiling, globules, pores, and cracks). For 2000 number of pulses after annealing, due to recrystallization phenomena reduction in peak intensity and increase in FWHM, is observed for almost all phases along with the growth of demolished phases due to the diffusion of oxygen/hydrogen across the boundaries is observed. This variation is responsible for further variation in surface morphology of irradiated zirconium (**Figure 7e**).

Laser-induced rapid heating, cooling, and a large temperature gradient are considered to be responsible for grain growth. Grain growth occurs as a result of the recrystallization process, which arises during irradiation in order to minimize the absorbed strain and surface energy. The inter planar atomic motion reorients the lattice at the expense of this energy, which leads to the growth of large and smooth grains [33]. Cavities are formed by the liberation of adsorbed gases beneath the surface during laser irradiation, due to volume boiling [34].

Melting of the target surface after laser irradiation activates reaction between the molten surface and surrounding media. After laser irradiation, recrystallization takes place and oxygen, carbon//hydrogen disseminates into the surface (oxygen insertion is confirmed by EDS analysis and hydrogen diffusion is confirmed by XRD analysis) particularly at the grain boundaries, attack and weaken them. The attacked and weakened grain boundary in turn enhances the effect of tensile stress and causes the widening of the grain boundaries [35]. Successive heating by overlapping laser pulses enhances diffusion of oxygen/hydrogen into the molten surface and causing the formation of oxides and hydrides of Zr (confirmed from XRD analysis) [26].

Comparison of two media, deionized water (**Figure 6**) and ethanol (**Figure 7**), shows significant dissimilarities. For laser ablation in deionized water porous surface features are observed with large number of pores and cavities over the surface. For ethanol distinct, well defined and small-sized grains are developed. This difference in surface morphology is attributable to the difference in absorption offered by the liquids and difference in chemical reactivity of Zr with both liquids. During irradiation of Zr in liquid environment, plasma plume from the solid target will be generated at the solid-liquid interface and remains confined. A shock wave is created in the plasma plume by the laser-induced plasma due to the confinement of liquid. This shock wave will induce an extra pressure in the laser-induced plasma called plasma-induced pressure [36]. The plasma-induced pressure will lead to an additional temperature increase of the laser-induced plasma. Therefore, the shock wave generated by the expansion of laser-induced plasma, under confinement of liquid pushes the laser-induced plasma into a thermodynamic state of the higher temperature, higher pressure and higher density, and the plume species react with surrounding liquid molecules in this extreme state to form various compounds like ZrO_2 (tetragonal and monoclinic), Zr_3O, ZrH, and ZrH_2 [37].

Intermolecular forces determine the physical and chemical characteristics of a substance, such as boiling and melting points, viscosity, solubility, volatility, and surface tension. Ethanol (C_2H_6O) is more volatile than water (H_2O), and it boils at a much lower temperature (78.24°C) than water (100°C). The strength of the intermolecular forces in ethanol is much less than water. In water, intermolecular hydrogen bonding takes place due to the presence of highly polar O—H bond in the H_2O molecule. Ethanol can also participate in hydrogen bonding, but not as successfully as water because it has a non-polar region. Heat of vaporization of water is 44 kJ/mol and its vapor pressure is 3.17 kPa at 25°C, whereas for ethanol, the value of heat of vaporization is 38.56 kJ/mol and its vapor pressure is 5.95 kPa at 25°C. The molar mass of water is 18 g/mol and for ethanol, its value is 46.07 g/mol, which is more than twice the mass of water that means, when 1 g of water evaporates, it absorbs more than two times as much heat as 1 g of ethanol evaporating. So, ethanol boils at a lower temperature than water, and its equilibrium vapor pressure is higher at all temperatures. That is why ethanol is more volatile than water and is responsible for the formation of various compounds of Zr. Physical properties of both liquids are mentioned in **Table 2**.

XRD and Raman results reveal the formation of zirconium oxide (ZrO_2, Zr_3O), zirconium hydride (ZrH, ZrH2) in case of ablation in deionized water and ethanol. Diffusion of carbon

Physical properties of liquids		
Properties	**Deionized water**	**Ethanol**
Melting point	0°C	−114.14°C
Boiling point	100°C	78.24°C
Viscosity	0.890 mPa·s	1.074 mPa·s
Density	0.9970474 g/cm³	0.7893 g/cm³
Refractive index	1.3330	1.3611
Heat of vaporization	44 kJ/mol	38.56 kJ/mol
molar mass	18 g/mol	46.07 g/mol
Absorption coefficient	5.05×10^{-3}/cm	4×10^{-2}/cm
Vapor pressure	3.17 kPa (at 25°C)	5.95 kPa (at 25°C)
Strength of the intermolecular forces	High due to presence of highly polar O–H bond	Low due to presence of nonpolar region
Volatility	Low	High

Table 2. Physical properties of liquids (deionized water and ethanol).

along with oxygen (C=O) is only confirmed through the Raman analysis of samples ablated in ethanol environment (EDS supports the results of XRD and Raman, it confirmed the diffusion of oxygen in case of deionized water and diffusion of oxygen/carbon in case of ethanol). Pulsed laser ablation in liquid is an extremely nonequilibrium process. When the nanosecond laser pulses interact with Zr at the liquid-solid interface, the absorption is granted by electrons in the metal. In terms of time scaling, absorption takes place after 10^{-14} s. The hot carriers can transfer their energy to the target lattice leading eventually to a solid-liquid phase transition. This energy transfer leads to an equilibrium on a timescale, $\tau E = 10^{-12} - 10^{-11}$ s and the mass removal will occur from equilibrium phase transition [38]. The ejected species will have large kinetic energy after irradiation and a plasma plume is formed at the interface. The plasma plume expands adiabatically while the liquid confines it. The confinement effect of liquids result in much higher temperature and pressure inside the plasma plume [37] and the plume species react with the surrounding liquid molecules in this extreme state. Moreover, the cooling effect of the liquids cause quenching of the plasma plume and the plume species solidifies to preserve materials in the final yield. In our case zirconium oxide ZrO_2 (monoclinic and tetragonal phases), Zr_3O, and zirconium hydride (ZrH, ZrH_2) in case of ablation in both deionized water and ethanol, whereas in case of ethanol C=O are synthesized and redeposited over the surface after fast quenching of the plasma plume at room temperature. Furthermore, the laser energy absorption in case of deionized water is less than ethanol. High absorption in ethanol for 248 nm laser than water implies small rise in surface temperature during irradiation, which supports grain growth. In case of water due to less energy absorption for 248 nm laser, average rise in surface temperature will be high, which causes the melting and resolidification of Zr targets. This results in the formation of cavities, pores, cracks, and droplets over the surface after irradiation. The strength of the intermolecular forces in ethanol is much less than water, that is why ethanol is more volatile than water and is responsible for the diffusion of carbon along with hydrogen and oxygen. The observed diffusion of carbon may also be responsible for different structure in case of ablation in ethanol.

Moreover, the cooling effect of the liquids causes quenching of plasma plume and the plume species will be redeposited over the target surface. This is the reason for variation in concentration of various elements after ablation as compared to unirradiated Zr (EDS analysis, **Table 1**). The formation of ZrO_2 (monoclinic and tetragonal phases), Zr_3O, and zirconium hydride (ZrH, ZrH_2) in case of ablation in both liquids and formation of C=O in case of ethanol plays considerable role in the formation of various microstructures. During exposure in liquid environments, the sample surface gets cooled in the presence of liquids, which results in less pronounced thermal effects and the surface morphologies become much cleaner and smoother.

3.3. Effect on surface mechanical property

Untreated Zr sample shows hardness value of 850 MPa. **Figure 8** shows the variation in hardness of nanosecond laser-irradiated Zr targets as a function of increasing number of laser pulses. **Figure 8a** shows the variation in hardness after irradiation with nanosecond laser in ambient environment of deionized water. For 500 laser pulses, hardness value of 964 MPa is observed, which increases (up to 1125 MPa) with increase in number of laser pulses up to 1000. Further increase in number of laser pulses up to 2000 shows continuous decrease in hardness (up to 995 MPa).

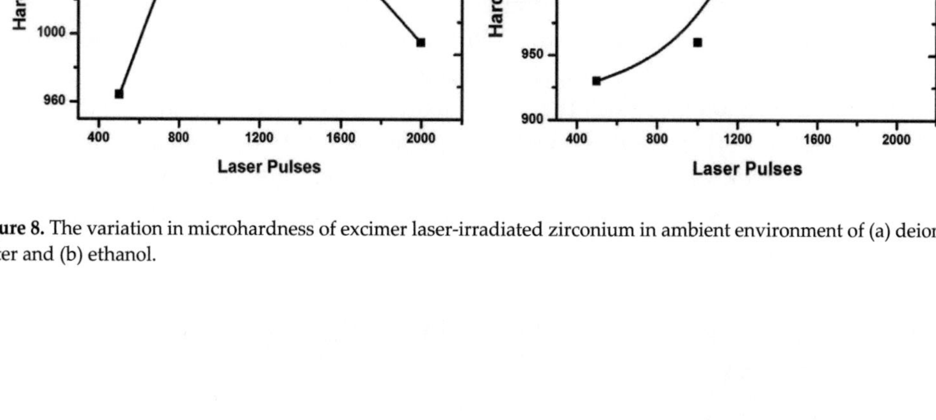

Figure 8. The variation in microhardness of excimer laser-irradiated zirconium in ambient environment of (a) deionized water and (b) ethanol.

Variations in the value of hardness depend on a number of factors including the lattice defects, density of oxide contents, phase composition, size and distribution of grains and crystal structure [39]. Increase in hardness is attributable to interstitial diffusion of oxygen into the lattice, which results in decrease in crystallite size and tensile residual stresses (**Figure 3b**) [40]. Smaller crystallite size is efficient in obstructing the dislocation movement and is responsible for high strength and hardness. Decrease in value of hardness, of irradiated targets can be explained on the basis of increasing crystallite size and enhanced tensile residual stresses. Oxygen diffusion across the grain boundaries results in larger crystallites (**Figure 3b**) that are more prone to coarsening, leading to lower crystallite density, and hence lower hardness [40, 41].

Figure 8b shows the variation in surface hardness after irradiation in ambient environment of ethanol. For 500 number of laser pulses, hardness of 960 MPa is observed, which increases with the increase in number of laser pulses up to 2000 (up to 1130 MPa).

4. Conclusion

The structural modifications of KrF excimer laser-ablated Zr has been investigated in correlation with its surface and mechanical properties after ablation in ambient environments of deionized water and ethanol. Both ambient environments have played significant role in structural, chemical, and mechanical modifications of the irradiated zirconium surface.

XRD results show diffusion of oxygen and hydrogen into the surface, which cause the formation of Zr_3O, monoclinic, and tetragonal phases of zirconium oxide (ZrO_2) and hydrides of zirconium (ZrH, ZrH_2) on the metallic surface. EDS analysis confirmed the increase in content of oxygen for both liquids. Raman analysis supports the EDS and XRD results. It confirms the formation of zirconia and hydrides of zirconium after ablation. The presence of liquids helps in removing the ablation debris and reducing the thermal damages, and offered cleaner surface during laser ablation of Zr. As a result, the surface morphologies of the craters produced are much smoother. For ablation in deionized water, porous surface morphology is observed for whole range of laser pulses, whereas fine grains are formed in case of ethanol environment. Formation of ZrO_2 (monoclinic and tetragonal phases), Zr_3O, and zirconium hydride (ZrH, ZrH_2) in case of ablation in both liquids and formation of C=O in case of ethanol and different absorption offered by both liquids plays considerable role in the formation of various microstructures. Different interacting mechanisms such as recoil pressures, production of bubbles, and laser-induced surface melting of Zr also plays considerable role in the formation of various microstructures. First, increase and then decrease in hardness is observed in case of deionized water, whereas continuous increase in hardness is observed for ethanol. Variations in hardness depend on a number of factors including the lattice defects, density of oxide contents, phase composition, size and distribution of grain, and crystal structure.

Author details

Nisar Ali[1,3,4]*, Umm-i-Kalsoom[2,3,4], Shazia Bashir[3] and Narjis Begum[5]

*Address all correspondence to: chnisarali@gmail.com

1 Department of Basic Sciences and Humanities, University of Engineering and Technology Lahore, Faisalabad, Pakistan

2 Department of Basic Sciences and Humanities, University of Engineering and Technology Lahore, Lahore, Pakistan

3 Centre for Advanced Studies in Physics, GC University, Lahore, Pakistan

4 Department of Physics, GC University, Lahore, Pakistan

5 Department of Physics, COMSATS Institute of Information Technology, Islamabad, Pakistan

References

[1] Barmina EV, Serkov AA, Stratakis E, Fotakis C, Stolyarov VN, Stolyarov IN, Shafeev GA. Nano-textured W shows improvement of thermionic emission properties. Applied Physics A. 2012;**106**:1

[2] Yang Y, Yang J, Liang C, Wang H. Ultra-broadband enhanced absorption of metal surfaces structured by femtosecond laser pulses. Optical Express. 2008;**16**:11259-11265

[3] Vorobyev AY, Makin VS, Guo C. Brighter light sources from black metal: Significant increase in emission efficiency of incandescent light sources. Physics Review Letters. 2009;**102**:234301

[4] Yang Y, Yang J, Liang C, Wang H, Zhu X, Zhang N. Surface microstructuring of Ti plates by femtosecond lasers in liquid ambiences: A new approach to improving biocompatibility. Optics Express. 2009;**17**:21124

[5] Bekesi J, Kaakkunen J, Michaeli W, Klaiber F, Schoengart M, Ihlemann J, Simon P. Fast fabrication of super-hydrophobic surfaces on polypropylene by replication of short-pulse laser structured molds. Applied Physics A. 2010;**99**:691

[6] Bonse J, Koter R, Hartelt M, Spaltmann D, Pentziea S, Höhm S, Rosenfeld A, Krüger J. Tribological performance of femtosecond laser-induced periodic surface structures on titanium and a high toughness bearing steel. Applied Surface Science. 2015;**336**:21

[7] Jelani M, Bashir S, Akram M, Yousaf D, Afzal N, Ahmad S. Mechanical behaviour of excimer laser irradiated polycrystalline zirconium. Physica Scripta. 2014;**89**:025703-025710

[8] Bashir S, Rafique MS, Husinsky W. Liquid assisted ablation of zirconium for the growth of LIPSS at varying pulse durations and pulse energies by femtosecond laser irradiation. Nuclear Instruments and Methods in Physics Research B. 2015;**349**:230

[9] Umm-i-Kalsoom S, Bashir N, Ali M, Rafique S, Husinsky W, Nathala CSR, Makarov SV, Begum N. Effect of fluence and ambient environment on the surface and structural modification of femtosecond laser irradiated Ti. Chinese Physics B. 2016;**25**(1):0181011-0181017

[10] Bashir S, Rafique MS, Nathala CS, Husinsky W. Surface and structural modifications of titanium induced by various pulse energies of a femtosecond laser in liquid and dry environment. Applied Physics A. 2014;**114**:243

[11] Major A, Aitchison JS, Smith PWE, Druon F, Georges P, Viana B, Aka GP. Applied Physics B: Lasers and Optics. 2005;**80**:199-201

[12] Barminaa EV, Stratakis E, Barberoglou M, Stolyarove VN, Stolyarove IN, Fotakisb C, Shafeev GA. Laser-assisted nanostructuring of tungsten in liquid environment. Applied Surface Science. 2012;**258**(15):5898-5902

[13] Ren J, Kelly M, Heeselink L. Laser ablation of silicon in water with nanosecond and femtosecond pulses. Optics Letters;**2005**(30):1740-1742

[14] Bashir S, Vaheed H, Mahmood K. Nanosecond pulsed laser ablation of brass in a dry and liquid-confined environment. Applied Physics A. 2013;**110**(2):389-395

[15] Cabalin LM, Lasema JJ. Experimental determination of laser induced break-down thresholds of metals under nanosecond Q-switched laser operation. Spectrochimica Acta B. 1998;**53**:723-730

[16] Park HK, Kim D, Grigoropoulos CP, Tam AC. Pressure generation and measurement in the rapid vaporization of water on a pulsed-laser-heated surface. Journal of Applied Physics. 1996;**80**(7):4072-4081

[17] Duley WW. UV Lasers: Effects and Applications in Materials Science. New York, United States of America: Cambridge University Press; 2005

[18] Ashraf M, Akhtar SMJ, Khan AF, Ali Z, Qayyum A. Effect of annealing on structural and optoelectronic properties of nanostructured ZnSe thin films. Journal of Alloys and Compounds. 2011;**509**(5):2414-2419

[19] Shackelford JF, Alexander W. CRC Materials Science and Engineering Handbook. 3rd ed. California, USA: CRC Press; 2000

[20] Rafique MS, Khaleeq-ur-Rehman M, Firdos T, Aslam K, Shahbaz Anwar M, Imran M, Latif H. XRD and SEM analysis of a laser-irradiated cadmium. Laser Physics. 2007; **17**(9):1138-1145

[21] Umm-i-Kalsoom, Bashir S, Ali N, Akram M, Mahmood K, Ahmad R. Effect of ambient environment on excimer laser induced micro and nano-structuring of stainless steel. Applied Surface Science. 2012;**261**:101-109

[22] Ikhlaq U, Ahmad R, Saleem S, Shah MS, Kalsoom U, Khan N, Khalid N. Argon gas concentration effects on nanostructured molybdenum nitride layer growth using 100 Hz pulsed dc glow discharge. The European Physical Journal Applied Physics. 2012;**59**: 20801

[23] Noyan IC, Cohen JB. Residual Stress-Measurement by Diffraction and Interpretation. New York: Springer-Verlag; 1987

[24] Khan IA, Hassan M, Ahmad R, Qayyum A, Murtaza G, Zakaullah M, Rawat RS. Nitridation of zirconium using energetic ions from plasma focus device. Thin Solid Films. 2008;**516**(23):8255-8263

[25] Gurarie VN, Otsuka PH, Jamieson DN, Prawe S. Crack-arresting compression layers produced by ion implantation. Nuclear Instruments and Methods in Physics Research B. 2006;**242**(1-2):421-423

[26] Mahmood K, Farid N, Ghauri IM, Afzal N, Idrees Y, Mubarik FE. Effects of laser irradiation on the mechanical response of polycrystalline titanium. Physica Scripta. 2010;**82**:045606-045613

[27] Kim BK, Hamaguchi HO. Mode assignments of the Raman spectrum of monoclinic zirconia by isotopic exchange technique. Physica Status Solidi (b). 1997;**203**:557-563

[28] Tan D, Lin G, Liu Y, Teng Y, Zhuang Y, Zhu B, Zhao Q, Qiu J. Synthesis of nanocrystalline cubic zirconia using femtosecond laser ablation. Journal of Nanoparticle Research. 2011;**13**:1183-1190

[29] Li M, Feng Z, Xiong G, Ying P, Xin Q, Li C. Phase transformation in the surface region of zirconia detected by UV Raman spectroscopy. The Journal of Physical Chemistry B. 2001;**105**:8107-8111

[30] Bashir S, Vaheed H, Mahmood Khaliq. Nanosecond pulsed laser ablation of brass in a dry and liquid-confined environment. Applied Physics A. 2012

[31] Karimzadeh R, Zamir Anvari J, Mansour N. Nanosecond pulsed laser ablation of silicon in liquids. Applied Physics A: Materials Science and Processing. 2009;**94**:949-955

[32] Douglas BC, Graham KH. Pulsed Laser Deposition of Thin Films. Singapore, New York, chichester, Brisbane, Toronto: Wiley-interscience Publication; 1994

[33] Passchier CW, Trouw RAJ. Microtectonics. Berlin: Springer, Verlag; 1998

[34] Craciun V, Craciun D. Subsurface boiling during pulsed laser ablation of Ge. Physical Review B. 1998;**58**(11):6787-6790

[35] Borgstedt HU, editor. Liquid Metal Systems: Material Behavior and Physical Chemistry in Liquid. Springer Science and Business Media; 2012. p. 419

[36] Yang GW. Laser ablation in liquids: Applications in the synthesis of nanocrystals. Progress in Materials Science. 2007;**52**:648-698

[37] Amendola V, Meneghetti M. Laser ablation synthesis in solution and size manipulation of noble metal nanoparticles. Physical Chemistry Chemical Physics. 2009;**11**:3805-3821

[38] Lam J. Pulsed Laser Ablation in Liquid: Towards the Comprehension of the Growth Processes, in Physique. 2006, Universit'e Claude Bernard Lyon 1: Ecole Doctorale de Physique et d'Astrophysique de Lyon

[39] Warcholinski B, Gilewicz A. Journal of Achievements in Materials and Manufacturing Engineering. 2009;**37**:498

[40] Umm-i-Kalsoom S, Ahmad R, Ali N, Khan IA, Saleem S, Ikhlaq U, Khan N. Effect of power and nitrogen content on the deposition of CrN films by using pulsed DC magnetron sputtering plasma. Plasma Science and Technology. 2013;**15**(7):666-672

[41] Corengia P, Ybarra G, Moina C, Cabo A, Broitman E. Microstructural and topographical studies of DC-pulsed plasma nitrided AISI 4140 low-alloy steel. Surface and Coatings Technology. 2005;**200**:2391-2397

Section 2

Properties and Characterization

Chapter 5

Sequentially Timed All-Optical Mapping Photography for Real-Time Monitoring of Laser Ablation: Breakdown and Filamentation in Picosecond and Femtosecond Regimes

Keiichi Nakagawa, Takakazu Suzuki and
Fumihiko Kannari

Additional information is available at the end of the chapter

http://dx.doi.org/10.5772/intechopen.71524

Abstract

To investigate ultrafast phenomena, a novel, ultrafast imaging technique was developed. Sequentially timed all-optical mapping photography (STAMP) performs single-shot image acquisition without the need for repetitive measurements and without sacrificing high-temporal resolution and image quality. The principle of this imaging method is based on the all-optical approach, and therefore it overcomes the temporal resolution in conventional high-speed cameras. Also, STAMP's single-shot movie-shooting capability allows us to obtain sequential images of non-repetitive ultrafast dynamic phenomena. Here, we present the motion pictures of early stage dynamics during femtosecond laser ablation captured by two types of STAMP setup. Breakdown was induced by intense femtosecond laser pulse and monitored with a frame interval of 15.3 ps and a total of six frames. The movie clearly shows the plasma generation and expansion on glass surface. Also, filamentation was generated inside a glass and observed with a frame interval of 230 fs and total of 25 frames. These phenomena have previously only been observed by pump-probe imaging. STAMP is a powerful tool to understand precise processes of complex dynamics in ultrashort laser ablation.

Keywords: laser ablation, ultrafast imaging, STAMP, laser breakdown, filamentation

1. Introduction

The dynamics of laser ablation is important for laser machining [1] and laser surgery [2]. In particular, research on a femtosecond laser ablation has been performed to utilize its advantages:

high-energy efficiency, low thermal effects, precise structuring due to the multiphoton absorption process, etc. Recent studies investigate more complex mechanism in laser ablation induced by multiple burst pulses [3], optical vortex [4], and so on. For understanding these dynamics, a method for direct observation of ultrafast phenomena has been required.

In this chapter, we introduce an ultrafast photographic technique, called "sequentially timed all-optical mapping photography (STAMP)" [5] for sub-nanosecond single-shot imaging and then present motion pictures of ultrafast dynamic events in ultrashort laser ablation.

2. High-speed imaging techniques for monitoring fast process of ultrashort laser ablation

High-speed imaging is powerful tool for discovering and studying dynamic phenomena [6, 7]. Many techniques and instruments have been developed to satisfy a great demand for observation of fast events. We need to choose an appropriate method according to the phenomena what we observe. **Figure 1** shows the timescale of physical events in ultrashort laser ablation of transparent materials [8]. The dynamics of femtosecond laser ablation is characterized differently on various time scales. First, laser ablation occurs by the material's absorption of photons at t = 1 to ~100 fs. At t = ~100 fs to ~10 ps, the energy of the electrons is transferred to the lattice via carrier-phonon scattering. At t = ~10 ps to ~10 ns, a shockwave and plume are formed at and propagate from the focal volume. Also, on the same time scale, the thermal

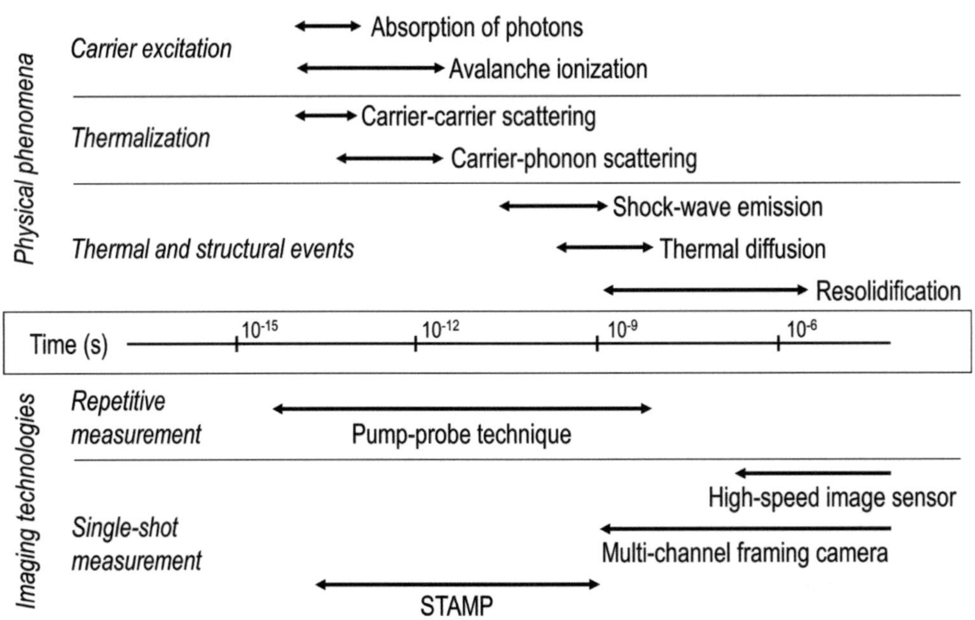

Figure 1. Timescale of physical events in ultrashort laser ablation and high-speed imaging technologies. The upper part of this figure is illustrated based on [8].

energy diffuses from the focal volume. In the bottom half of **Figure 1**, the characteristic of imaging technologies is summarized. Here, we review pump-probe technique and conventional high-speed cameras and introduce STAMP.

2.1. Conventional techniques for capturing ablation dynamics

Time-resolved imaging based on the pump-probe technique is a popular method for studying ultrafast dynamic events in laser ablation. Its principle is the construction of a time-resolved motion picture from repetitive measurements with different time delays between the trigger pulse (pump) and measurement pulse (probe). Although the pump-probe imaging allows us to capture ultrafast phenomena with the high-temporal resolution without the need for a fast detector, it requires that the event under observation to be relatively simple and easily reproducible. It falls short in studying non-repetitive or difficult-to-reproduce events such as those found in laser ablation of biological materials and carbon fiber reinforced plastic and multi-pulse laser ablation [3].

In many fields of basic science or applied technology, the versatile video camera based on charged-coupled device (CCD) or complementary metal-oxide-semiconductor (CMOS) technology has been widely used for study of dynamic phenomena. For achieving high temporal resolution in imaging, specialized instruments (for example, the *in-situ* storage image sensor [9], rotating mirror camera [10], and multi-channel framing camera [11]) have been developed. The time response of these instruments is ~1 ns, limited due to technical limitations in mechanical and electrical components of the camera such as mechanical scanning, data readout, and shutters.

A streak camera [12] is also used for monitoring ultrafast events with a sub-nanosecond resolution. The streak camera which operates by electronically sweeping the spatial profile of the target onto an image sensor can run faster than the 2D high-speed cameras. Although the streak camera can provide a femtosecond temporal resolution, its operation is limited to 1D imaging.

2.2. Sequentially timed all-optical mapping photography for sub-nanosecond single-shot imaging

STAMP was developed in order to overcome the limitations in conventional high-speed imaging technologies and to observe non-repetitive ultrafast phenomena. It performs continuous, single-shot, burst image acquisition without the need for repetitive measurements, yet with equally high-temporal resolution and image quality.

In ultrafast events, the information we need is compressed in the time domain. When we acquire a movie in the time domain, we have to use a device or material that has a very fast response speed to resolve and take the compressed information. All of these direct approaches must face to limitations in the response speed of the device or material. In order to avoid these limitations, STAMP performs "mapping" of the target's time-varying spatial profile onto a burst stream of photographs with spatiotemporal dispersion (**Figure 2**). The camera's all-optical operation eliminates the speed bottleneck that exists in conventional burst cameras and hence enables ultrafast, multi-dimensional motion picture photography.

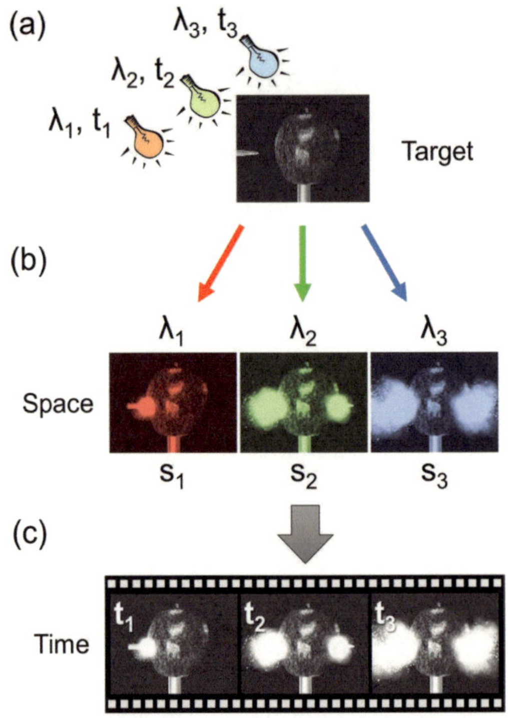

Figure 2. Principle of STAMP. (a) The target is illuminated by light with wavelength of λ_1, λ_2, and λ_3 at time t_1, t_2, and t_3. (b) The light modulated by the target is spectrally separated and detected on different positions s_1, s_2, and s_3. (c) Based on both the correspondence relation between time t and wavelength λ and the correspondence between wavelength λ and space s, the images are reconstructed as a movie.

Basic configuration of STAMP is schematically shown in **Figure 3**. It consists of an ultrashort pulse source, a temporal mapping device, a spatial mapping device, an image sensor, and a computer. The key components for realizing STAMP are a temporal mapping device and a spatial mapping device. The operation of STAMP is as follows. First, an ultrashort laser pulse generated by the laser source is temporally and spectrally shaped into a series of discrete daughter pulses by the temporal mapping device. Here, the daughter pulses correspond to different spectral bands of the original pulse. The daughter pulses are incident onto the target as successive "flashes" for stroboscopic image acquisition (which can be conducted in either the reflection or transmission mode, depending on the reflectivity profile of the target). The image-encoded daughter pulses are "passively" and "optically" separated by the spatial mapping device and incident onto different areas of the image sensor. Even though the events are very fast, we can use highly sensitive image sensors that permit microscopic imaging with low-intensity illumination. The data recorded on the image sensor are digitally processed on the computer to reconstruct motion pictures with the frame interval and exposure time calibrated from the settings of the pulse stretcher and pulse shaper in the temporal mapping device. It can be tuned, depending on the time scale of the dynamical event of interest [13]. The total number of frames can also be changed, depending on the bandwidth of the original laser pulse and the settings of the temporal and spatial mapping

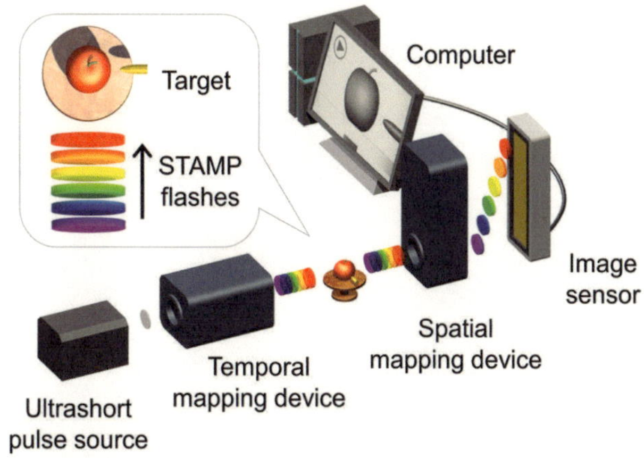

Figure 3. Basic configuration of STAMP.

devices. The speed barriers of mechanical and electrical operation can be eliminated completely, since only light emitted from the pulse source travels to the image sensor through the passive optical system.

As described above, the main feature of STAMP is its capability of sub-nanosecond multi-dimensional image acquisition with high image quality and without repetitive measurements. Additionally, in the STAMP operation, various imaging techniques such as interferometry, polarization imaging, and shadowgraph imaging can be employed for visualizing phenomena. On the other hand, there are some limitations in STAMP. We have to consider the potential negative effects of samples on STAMP's operation. STAMP has difficulty in being applied to photograph luminous objects. Also, we assume that the sample does not have any strong dependence on wavelength in STAMP's spectral range. In order to capture target phenomena, we need to understand these features and design its optical setup properly.

3. Real-time observation of early stage plasma evolution during femtosecond laser ablation

Here, we present movies of early stage plasma evolution during femtosecond laser ablation captured by STAMP. We focused on the picosecond timescale, where the physical dynamics in laser ablation of a transparent material transits from carrier excitation and thermalization to structural events (see **Figure 1**).

3.1. Experimental setup of ultrashort laser ablation and STAMP with spectral shaper

As shown in **Figure 4**, the surface of a glass plate was ablated by a focused femtosecond laser pulse and its resultant dynamics was observed with STAMP in transmission mode with a time resolution of a picosecond order. The ablation conditions and the detailed setup of imaging system are described below.

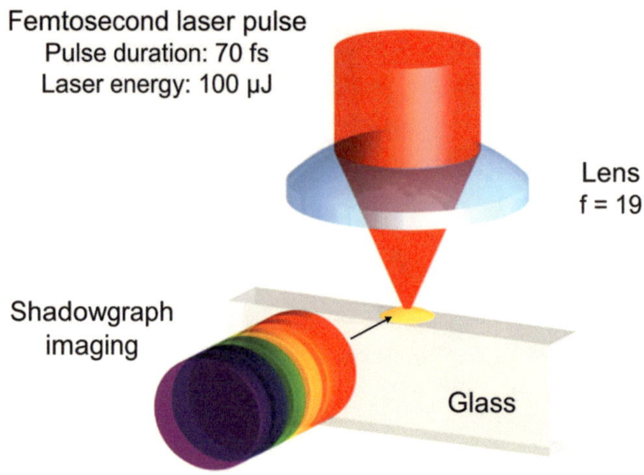

Figure 4. Experimental setup for performing and visualizing a breakdown on a glass surface in femtosecond laser ablation.

For an ultrafast laser source, we used Ti:Sapphire laser with a chirped pulse amplifier (CPA) system. The center wavelength, pulse duration, and repetition rate were 810 nm, 70 fs, and 1 kHz, respectively. The repetition rate was decreased with optical chopper as the rate of 36 Hz, and then a single pulse was generated by a mechanical shutter. This pulse was split into an excitation pulse of laser ablation and a STAMP pulse for capturing sequential images. The time difference between these two pulses was adjusted with an optical delay line. The excitation pulse with the energy of 100 μJ was focused by a lens (f = 19 mm) on the side surface of the ultrathin glass (t = 50 μm) placed in air. The power density of the laser pulse was approximately 3×10^{15} W/cm^2 at the focal point.

We developed a STAMP system that enables ultrafast motion picture photography with six frames. The developed system was composed of a femtosecond laser source which is also used for ablating materials, temporal mapping device consisting of a pulse stretcher and pulse shaper, spatial mapping device which performs spectral imaging, cooled CCD camera, and computer (**Figure 5**). In the pulse stretcher, the picosecond chirped pulse was produced from a femtosecond pulse through a grating pair for monitoring dynamic events in picosecond regime. The pulse shaper, based on the 4*f* optical configuration [14], was used as an amplitude modulator to tailor the intensity of each daughter pulse. For visualizing laser ablation dynamics, shadowgraph technique was employed. The STAMP's daughter pulses coming from the temporal mapping device illuminated the ablation area. The phenomena were imaged with an objective (10×, NA = 0.25). Here, the effects of chromatic aberration and wavelength dependence can be negligible because the difference of wavelength in each daughter pulse is around 3 nm. And then, the daughter pulses are separated by a spectral shaper [15]. The spectral shaper we developed is one of the snapshot multispectral imaging methods which perform image separation with passive optical elements [16]. In the spectral shaper, a periscope array is placed at the Fourier plane in the 4*f* configuration of the spectral shaper (**Figure 6**). The daughter pulses are diffracted by the first diffraction optics and then are focused onto the

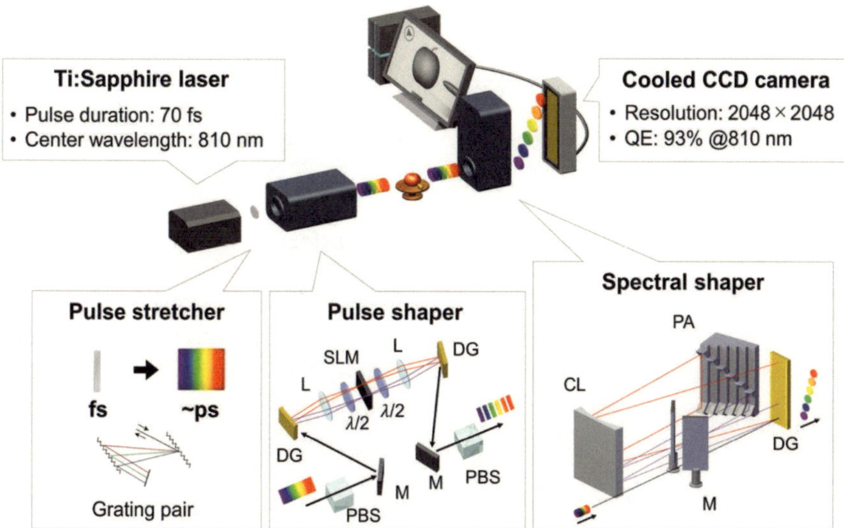

Figure 5. The schematic of STAMP utilizing the spectral shaper. CL: cylindrical lens, DG: diffraction grating, L: lens, M: mirror, PA: periscope array, PBS: polarizing beam splitter, SLM: spatial light modulator, $\lambda/2$: half wave plate.

periscope array. Each different spectral component reflects back, but comes out at a different height. Then, the daughter pulses are recombined at the second diffraction grating, but exit at different heights, keeping the spatial profile in each spectral component intact. Here, we used the periscope array having six periscopes. These daughter pulses were detected on the different position of the high-sensitive image sensor (2048 × 2048 pixels). Finally, detected images were arranged in order corresponding to the arrival time at the target.

The time difference of arrival (corresponding to the frame interval) and pulse duration (corresponding to the exposure time) of each daughter pulse was measured by the technique of cross correlation based on sum frequency generation. The optical setup is shown in **Figure 7**.

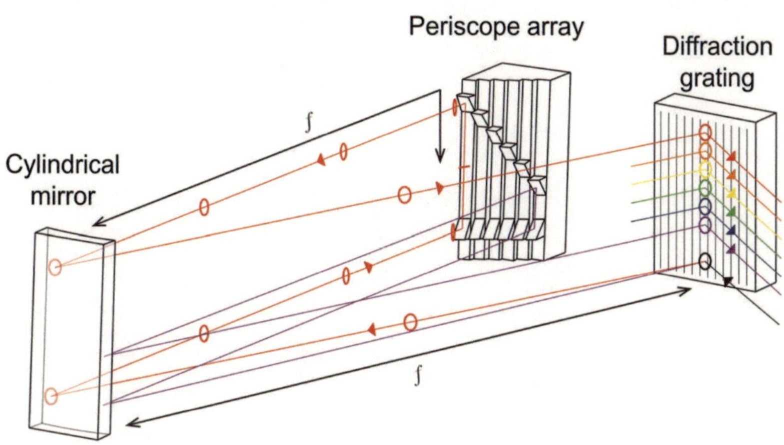

Figure 6. Configuration of the spectral shaper.

Here, each daughter pulse was produced by the pulse shaper and characterized one by one. **Figure 8** shows the temporal profile. The averaged frame interval and exposure time in this STAMP setup were 15.3 and 13.8 ps, respectively.

Figure 9 shows the microscopic images of a micrometer. The field of view and pixel resolution were 148 × 122 μm² and 680 × 560 pixels, respectively.

3.2. Plasma generation and expansion around glass surface

Figure 10 shows STAMP's motion picture of the generation and expansion of plume. Frames 1–3 indicate the generation of a flat plume, whereas frames 4–6 indicate its propagation in the upper direction. In frames 2–6, the process of carrier-phonon scattering is indicated by the

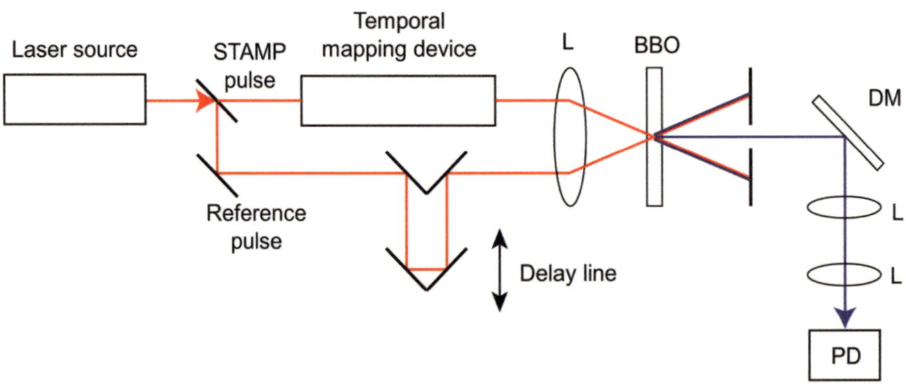

Figure 7. Schematic of the optical setup for characterizing the STAMP pulses. DM: dichroic mirror, L: lens, PD: photodiode.

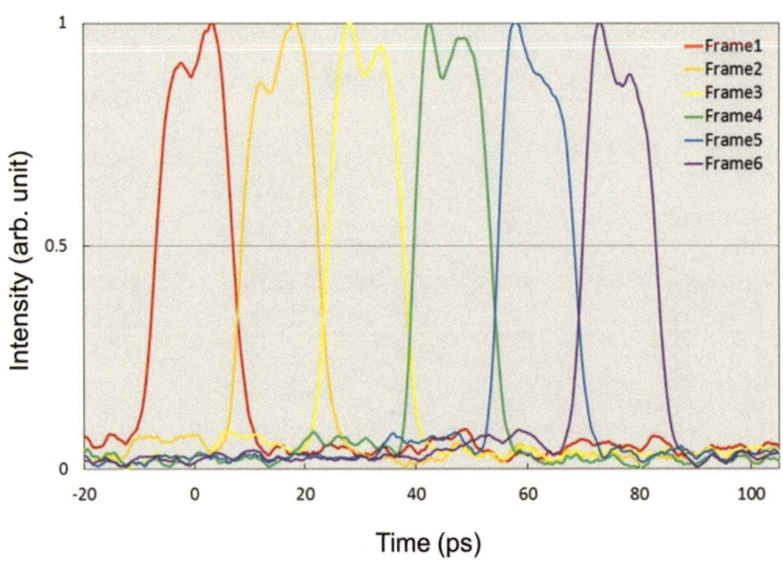

Figure 8. Temporal profile of STAMP pulses. An average frame interval and an exposure time are 15.3 and 13.8 ps, respectively [5].

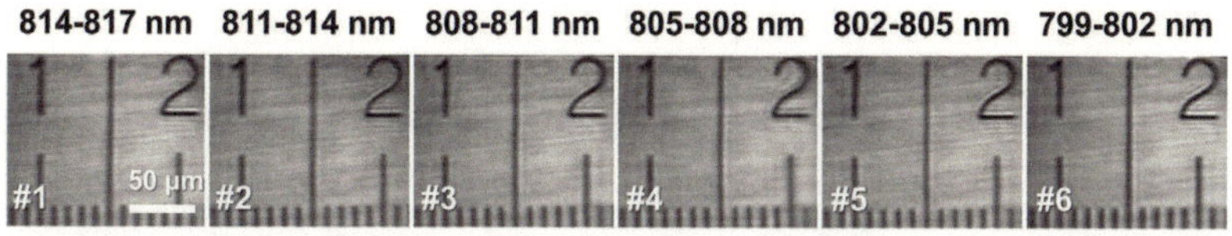

Figure 9. Image of test target. The images were captured and processed with the resolution of 680 × 560 pixels. The scale bar shows 50 μm [5].

dark region (white arrow in **Figure 10**) that corresponds to the high density of electrons in the plasma state caused by the material's absorption of light. This observation agrees well with previous reports [17].

The detailed analysis shows a slight asymmetry in the plume's wavefront profile (**Figure 11**). This is due to the fact that the ablation pulse's angle of incidence was not exactly perpendicular to the air-glass interface. The error bars in **Figure 11** are due to the ambiguity in image contrast. From the movie, the propagation speed of the plume's wavefront is found to be ~100 km/s (90°), which agrees with previously reported values [18, 19].

Figures 12 and **13** are the motion pictures of the dynamic events in femtosecond laser ablation. Although the experiment parameters such as laser intensity, beam profile, and material were not changed, the difference in evolution of a plasma plume was observed. A fluctuation

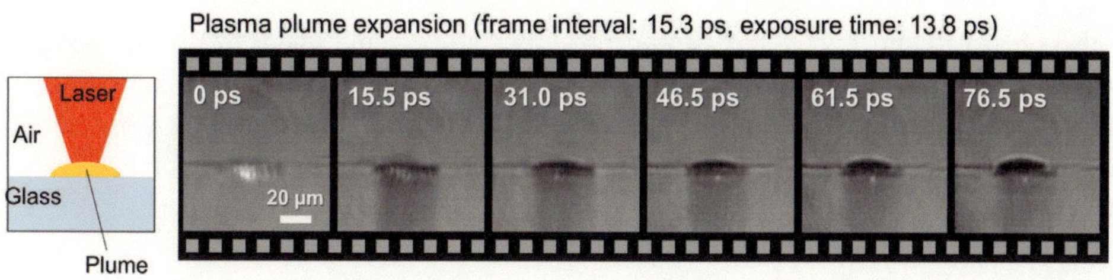

Figure 10. Motion picture of plasma generation and expansion in femtosecond laser ablation [5].

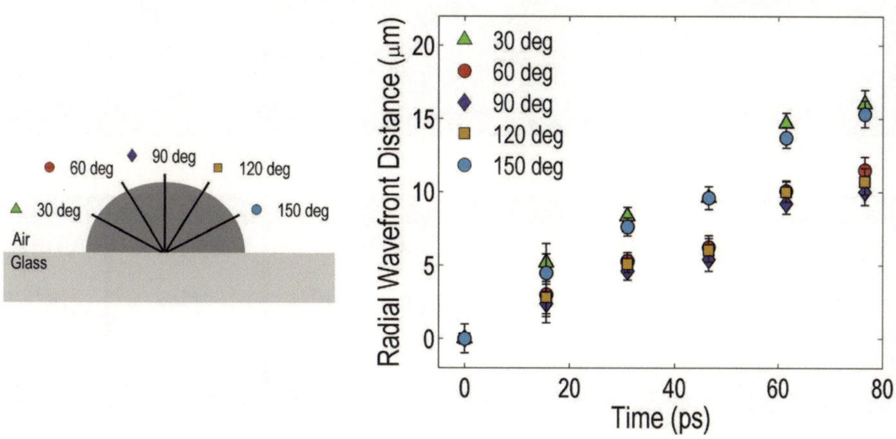

Figure 11. Analysis of evolution of the plume wavefront [5].

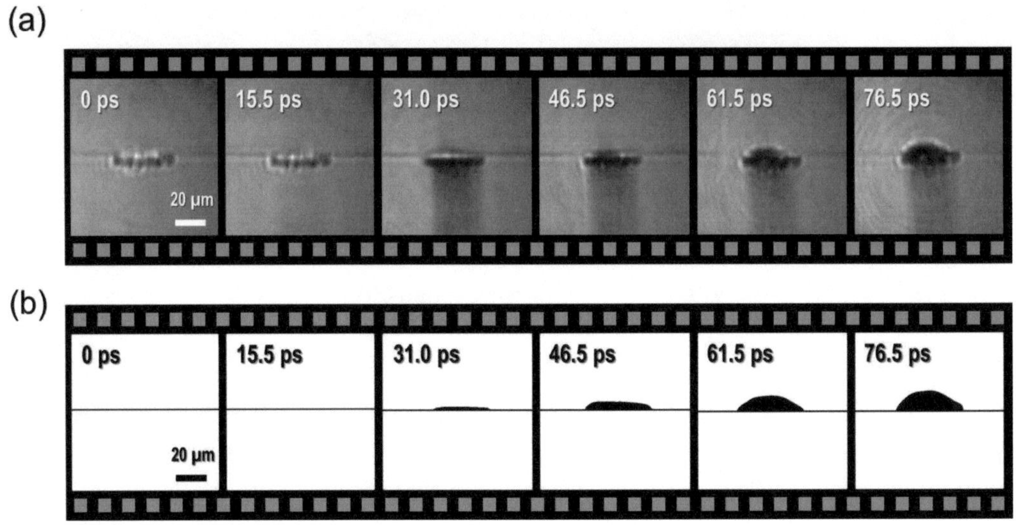

Figure 12. Motion picture of laser ablation (a) and its binarized image (b).

of the laser pulse and non-uniformity of the surface roughness of the glass often decrease the reproducible of laser ablation. STAMP is an effective tool to study a series of transitions in such difficult-to-reproduce phenomena.

We also observed shockwave emission caused as a subsequent event of plasma generation. The shock front is clearly visualized in each frame. The shock speed is quite slow comparing to the frame interval of this experimental setup and hence the difference between each image is not large (**Figure 14**).

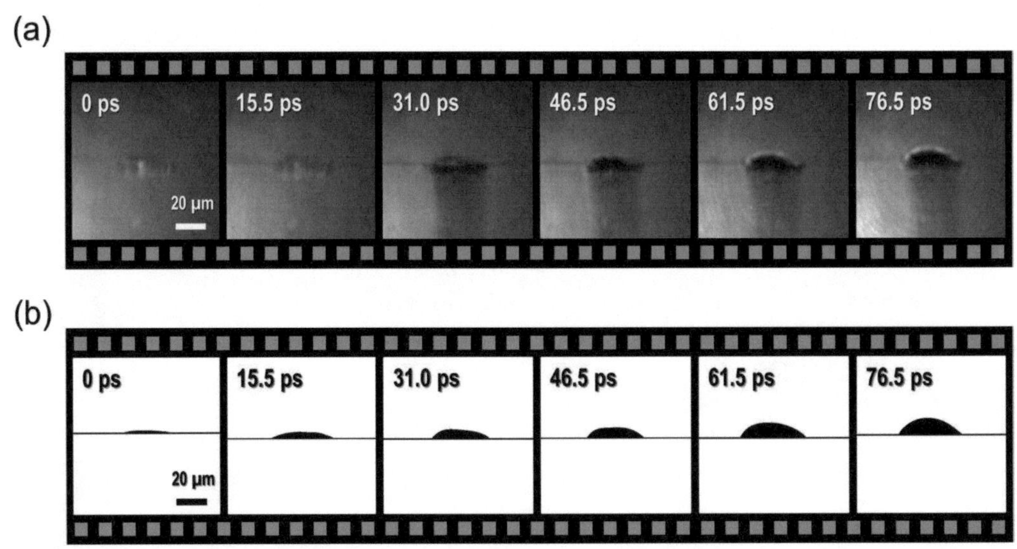

Figure 13. Motion picture (a) and binarized image (b) of laser ablation induced under the same conditions as those in the experiment of **Figure 12**.

http://dx.doi.org/10.5772/intechopen.71524

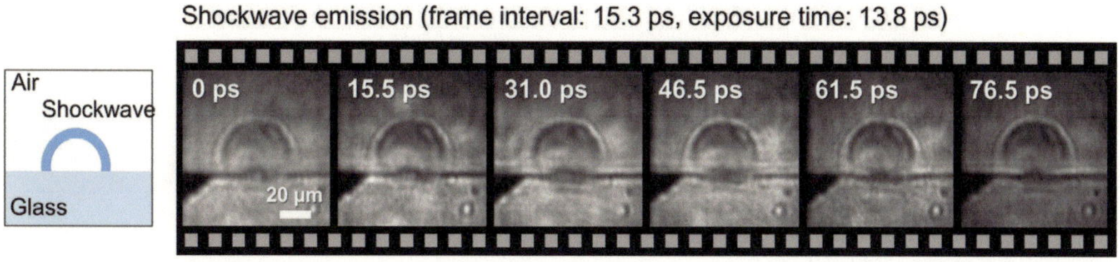

Figure 14. Motion picture of shockwave emission after breakdown.

4. Real-time monitoring of plasma filament generated by femtosecond laser inside a glass

In STAMP, the temporal resolution can be increased to the femtosecond timescale by tuning the pulse stretcher. For observing a detailed process in filamentation induced by intense femtosecond laser, the number of frame was also increased. Here, we present movies of plasma filament generated inside a glass acquired by an improved STAMP system, called SF-STAMP (STAMP utilizing spectral filtering) [20–22].

4.1. Experimental setup of filamentation and STAMP with spectral filtering

Figure 15 shows the schematic of experimental setup. A single pulse was produced from the Ti: Sapphire laser system as the same way described in Section 3.1. The center wavelength, pulse duration, and energy of excitation pulse were 800 nm, 50 fs, and 30 μJ, respectively. The femtosecond pulse was split into a STAMP pulse and an excitation pulse, and the excitation pulse was focused inside the glass with a lens (f = 8 mm).

For capturing the filamentation with 25 frames and higher temporal resolution, we improved our STAMP system by utilizing spectral broadening with an Ar-gas filled hollow core fiber and spectral filtering with a combination of a diffractive optical element (DOE) and band-pass filter (BPF). As shown in **Figure 16**, the STAMP pulse was focused into the Ar-gas filled hollow

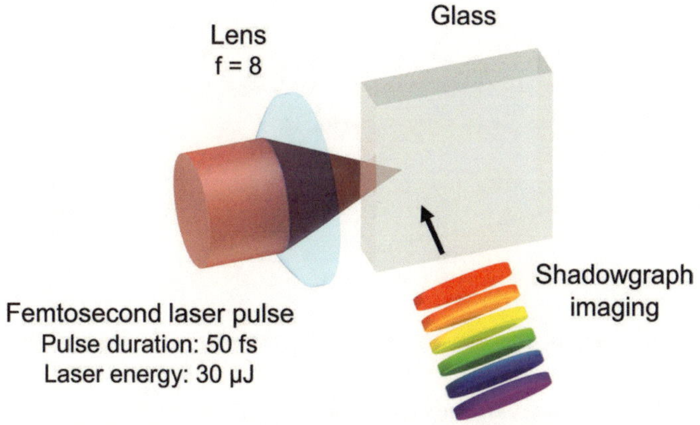

Figure 15. Experimental setup for monitoring filamentation generated by femtosecond laser.

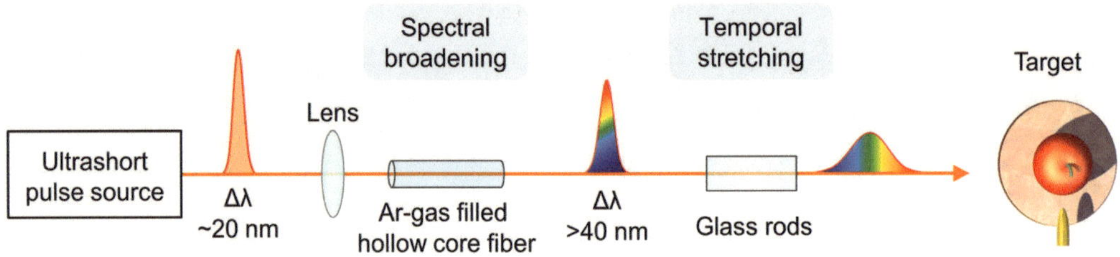

Figure 16. Temporal mapping device with an Ar-gas filled hollow core fiber for spectral broadening and with glass rods for recording with high temporal resolution.

core fiber (the length and core diameter are 400 mm and 126 μm, respectively) to produce a broadband pulse in the process of self-phase modulation. We used spectral components from 785 to 825 nm (bandwidth of 40 nm) in the broadband pulse which had nearly the uniform intensity in this spectral range. For monitoring the filamentation with higher temporal resolution, the pulse was stretched with glass rods. With the dispersion parameter and length of the glass roads, the time window and frame interval are calculated to be 5.6 ps and 230 fs, respectively. STAMP utilizing spectral filtering (SF-STAMP) has another type of the spatial mapping device which is based on the scheme of "spatially and temporally resolved intensity and phase evaluation device: full information from a single hologram (STRIPED FISH)" [23, 24]. The optical setup of the spatial mapping device in SF-STAMP was composed of a DOE, a BPF (a center wavelength of 830 nm and a bandwidth of 2.2 nm), and two lenses ($f_1 = f_2 = 50$ mm) arranged in the 4*f* configuration as shown in **Figure 17**. Since the bandwidth of the STAMP pulse was sufficiently broadened, the number of frames in this STAMP setup was determined to be 25 frames by the diffraction property of the DOE which produces 25 array beams from a single beam. While the target profile was transferred on an image sensor by the two lenses, the image-encoded chirped pulse was split into 25 pulses by the DOE place at the Fourier plane. And then these diffracted pulses were resolved by the tilted BPF depending on their

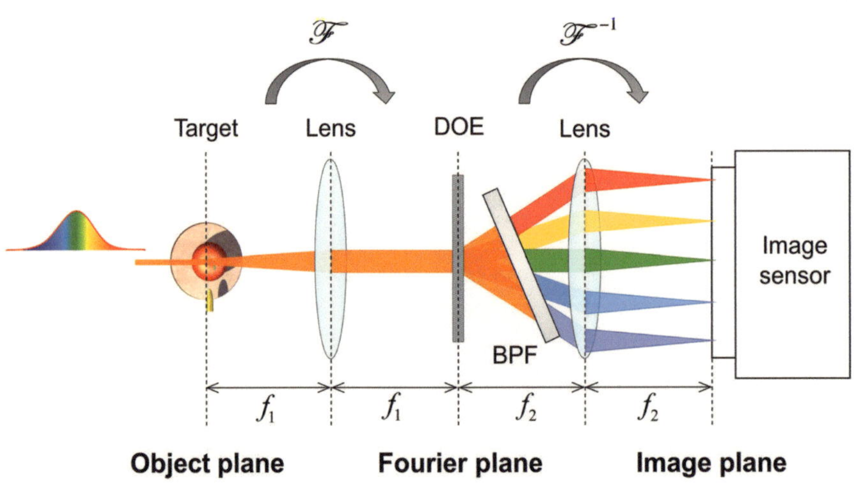

Figure 17. Spectral filtering. BPF: band-pass filter, DOE: diffractive optical element.

http://dx.doi.org/10.5772/intechopen.71524

incident angles without losing the image information. As a result, the time-varying target profiles were mapped on the different position of the image sensor. Although the much energy of an illumination pulse is needed at the target in SF-STAMP comparing to STAMP utilizing the spectral shaper, the spatial mapping device of SF-STAMP becomes compact and easy to use.

The dynamic phenomena were visualized with shadowgraph technique. **Figure 18** shows the image(s) of a resolution test chart detected by the SF-STAMP with a cooled CCD camera (4872 × 3248 pixels). The image of the left end is frame 1 of a motion picture, and the image of the right end is frame 25 projected by the spectral component which arrived at the target 5.6 ps after the arrival of the first spectral component. These frames were clipped as movie frames with the pixel resolution of 740 ×480 pixels.

4.2. Femto-to-picosecond dynamics of plasma filaments inside a glass

Figure 19 is the 25-frame movie showing the change of the refractive index inside the glass induced by the focusing of femtosecond laser. In frame 5, a dielectric breakdown appears at the center of the image. In frames 7–9, the change of intensity gradually increases. From frame 11, this dark region grows to the shape of a filament. In frame 25, the dark spot that corresponds to the high electron density in plasma state and also filamentation around the spot are clearly observed. This phenomenon occurring in the femtosecond-to-picosecond timescale has previously only been observed by pump-probe imaging in which a sample is replaced in every measurement. This demonstration highlights the utility of STAMP's single-shot movie-shooting capability.

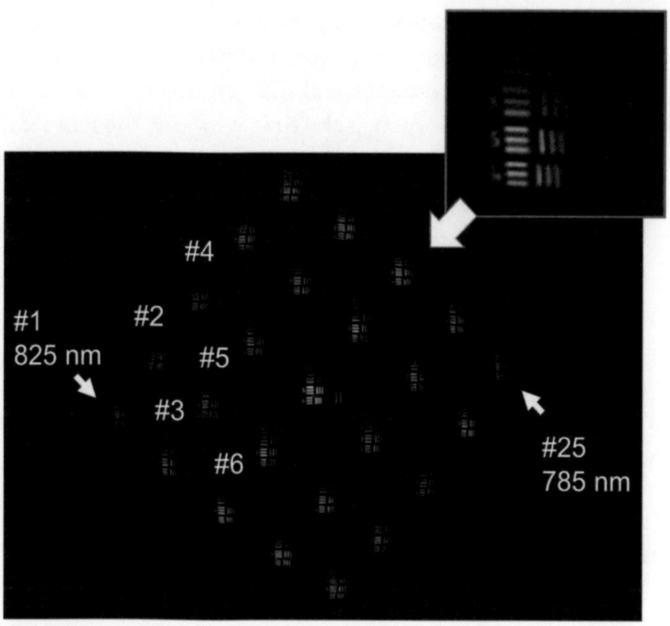

Figure 18. Image of a resolution chart.

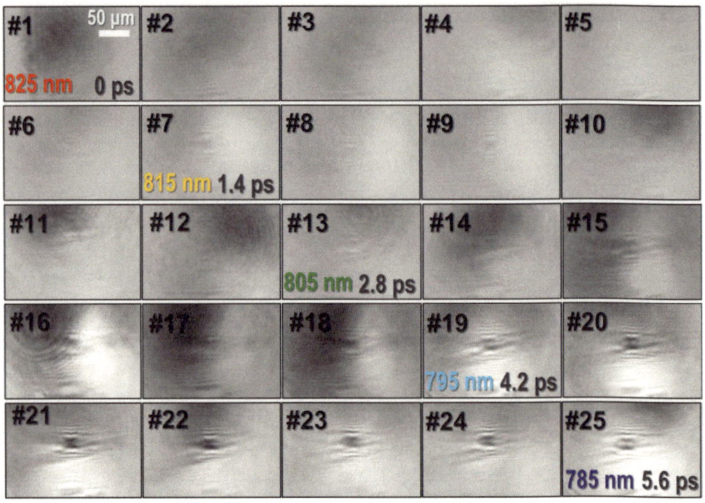

Figure 19. Motion picture of plasma filaments [20].

5. Conclusion

In this chapter, we introduced the motion picture camera called STAMP and presented motion pictures of femtosecond laser ablation. STAMP performs single-shot image acquisition without the need for repetitive measurements and without sacrificing high temporal resolution and image quality. The principle of this imaging method is based on the all-optical mapping of time-varying spatial profiles of target. We introduced two examples of the STAMP configurations: STAMP utilizing the spectral shaper and utilizing the spectral filtering and spectral broadening. The recording parameter such as the time window, the temporal resolution, and the number of frames can be tuned by changing the configuration of STAMP components, especially a temporal mapping device and a spatial mapping device.

By using STAMP, we obtained sequential images of sub-nanosecond dynamic events under the process of femtosecond laser ablation. Laser breakdown induced by intense femtosecond laser pulse was monitored with a frame interval of 15.3 ps and a total of six frames. The movie clearly shows the plasma generation and expansion on glass surface. Plasma filamentation generated inside a glass was also visualized with a frame interval of 230 fs and total of 25 frames. As demonstrated in this chapter, STAMP is a useful tool to understand precise processes of complex dynamics in ultrashort laser ablation.

Acknowledgements

The present work was supported in part by the Translational Systems Biology and Medicine Initiative from the Ministry of Education, Culture, Sports, Science and Technology (MEXT),

Japan, the Photon Frontier Network Program of MEXT, a Grant-in-Aid from the Japan Society for the Promotion of Science, and a Grant-in-Aid for the Program of Leading Graduate Schools of Keio University for "Science for Development of Super Mature Society" from MEXT.

Author details

Keiichi Nakagawa[1]*, Takakazu Suzuki[2] and Fumihiko Kannari[2]

*Address all correspondence to: kei@bmpe.t.u-tokyo.ac.jp

1 University of Tokyo, Japan

2 Keio University, Japan

References

[1] Momma C, Chichkov BN, Nolte S, Alvensleben F, Tunnermann A, Welling H, Wellegehausen B. Short-pulse laser ablation of solid targets. Optics Communication. 1996;**129**(1-2):134-142. DOI: 10.1016/0030-4018(96)00250-7

[2] Vogel A, Venugopalan V. Mechanisms of pulsed laser ablation of biological tissues. Chemical Reviews. 2003;**103**(2):577-644. DOI: 10.1021/cr010379n

[3] Ilday FÖ, Kerse C, Kalaycioglu H, Elahi P, Yavas S, Kesim D, Akçaalan Ö, Çetin B, Öktem B, Asik M, Hoogland H, Holzwarth R. Ablation-cooled material removal with ultrafast bursts of pulses. Nature. 2016;**537**:84-88. DOI: 10.1038/nature18619

[4] Toyoda K, Takahashi F, Takizawa S, Tokizane Y, Miyamoto K, Morita R, Omatsu T. Transfer of light helicity to nanostructures. Physical Review Letters. 2013;**110**(14):143603. DOI: 10.1103/PhysRevLett.110.143603

[5] Nakagawa K, Iwasaki A, Oishi Y, Horisaki R, Tsukamoto A, Nakamura A, Hirosawa K, Liao H, Ushida T, Goda K, Kannari F, Sakuma I. Sequentially timed all-optical mapping photography (STAMP). Nature Photonics. 2014;**8**:695-700. DOI: 10.1038/nphoton.2014.163

[6] Jussim E, Kayafas G, Edgerton H. Stopping Time: The Photographs of Harold Edgerton. New York, NY: Harry N. Abrams; 1987. p. 167

[7] Ray SF. High Speed Photography and Photonics. Bellingham, Washington, USA: SPIE Publications; 2002. p. 424

[8] Gattass RR, Mazur E. Femtosecond laser micromachining in transparent materials. Nature Photonics. 2008;**2**:219-225. DOI: 10.1038/nphoton.2008.47

[9] Tochigi Y, Hanzawa K, Kato Y, Kuroda R, Mutoh H, Hirose R, Tominaga H, Takubo K, Kondo Y, Sugawa S. A global-shutter CMOS image sensor with readout speed of 1-Tpixel/s

burst and 780-Mpixel/s continuous. IEEE Journal of Solid-State Circuits. 2013;**48**:329-338. DOI: 10.1109/JSSC.2012.2219685

[10] Gelderblom EC, Vos HJ, Mastik F, Faez T, Luan Y, Kokhuis TJ, van der Steen AF, Lohse D, de Jong N, Versluis M. Brandaris 128 ultra-high-speed imaging facility: 10 years of operation, updates, and enhanced features. The Review of Scientific Instruments 2012;**83**:103706. DOI: 10.1063/1.4758783

[11] Versluis M. High-speed imaging in fluids. Experiments in Fluids. 2013;**54**:1458. DOI: 10.1007/s00348-013-1458-x

[12] Feng J, Shin H, Nasiatka R, Wan W, Young A, Huang G, Byrd J, Padmore H. An X-ray streak camera with high spatio-temporal resolution. Applied Physics Letters. 2007;**91**:134102. DOI: 10.1063/1.2793191

[13] Tamamitsu M, Nakagawa K, Horisaki R, Iwasaki A, Oishi Y, Tsukamoto A, Kannari F, Sakuma I,Goda K. Design for sequentially timed all-optical mapping photography with optimum temporal performance. Optics Letters. 2015;**40**(4):633-636. DOI: 10.1364/OL.40.000633

[14] Weiner AM, Leaird DE, Patel JS, Wullert JR. Programmable shaping of femtosecond optical pulses by use of 128-element liquid crystal phase modulator. IEEE Journal of Quantum Electronics. 1992;**28**:908-920. DOI: 10.1109/3.135209

[15] Hashimoto K, Mizuno H, Nakagawa K, Horisaki R, Iwasaki A, Kannari F, Sakuma I, Goda K. High-speed multispectral videography with a periscope array in a spectral shaper. Optics Letters. 2014;**39**:6942-6945. DOI: 10.1364/OL.39.006942

[16] Hagen NA, Kudenov MW. Review of snapshot spectral imaging technologies. Optical Engineering. 2013;**52**:090901. DOI: 10.1117/1.OE.52.9.090901

[17] Mao X, Mao SS, Russo RE. Imaging femtosecond laser-induced electronic excitation. Applied Physics Letters 2003;**82**:697-699. DOI: http://dx.doi.org/10.1063/1.1541947

[18] Hu W, Shin YC, King G. Early-stage plasma dynamics with air ionization during ultrashort laser ablation of metal. Physics of Plasmas 2011;**18**:093302. DOI: http://dx.doi.org/10.1063/1.3633067

[19] Zhao X, Shin YC. Coulomb explosion and early plasma generation during femtosecond laser ablation of silicon at high laser fluence. Journal of Physics D: Applied Physics. 2013;**46**:335501. DOI: 10.1088/0022-3727/46/33/335501

[20] Suzuki T, Hida R, Ueda R, Isa F, Nakagawa K, Kannari F. Single-shot ultrafast 2D-burst imaging by STAMP utilizing spectral filtering (SF-STAMP). In: Santa Fe . New Mexico, USA: Santa Fe Community Convention; 2016. UTh4A.18

[21] Suzuki T, Isa F, Fujii L, Hirosawa K, Nakagawa K, Goda K, Sakuma I, Kannari F. Sequentially timed all-optical mapping photography (STAMP) utilizing spectral filtering. Optics Express. 2015;**23**:30512-30522. DOI: 10.1364/OE.23.030512

[22] Suzuki T, Hida R, Yamaguchi Y, Nakagawa K, Saiki T, Kannari F. Single-shot 25-frame burst imaging of ultrafast phase transition of $Ge_2Sb_2Te_5$ with a sub-picosecond resolution. Applied Physics Express. 2017;**10**:092502. DOI: 10.7567/APEX.10.092502

[23] Gabolde P, Trebino R. Single-frame measurement of the complete spatiotemporal intensity and phase of ultrashort laser pulses using wavelength-multiplexed digital holography. Journal of the Optical Society of America B. 2008;**25**(6):A25-A33. DOI: 10.1364/JOSAB.25.000A25

[24] Guang Z, Rhodes M, Davis M, Trebino R. Complete characterization of a spatiotemporally complex pulse by an improved single-frame pulse-measurement technique. Journal of the Optical Society of America B. 2014;**31**(11):2736-2743. DOI: 10.1364/JOSAB.31.002736

Chapter 6

Optical Properties of Complex Oxide Thin Films Obtained by Pulsed Laser Deposition

Valentin Ion, Andreea Andrei, Maria Dinescu and
Nicu Doinel Scarisoreanu

Additional information is available at the end of the chapter

http://dx.doi.org/10.5772/intechopen.70803

Abstract

The market for thin films of complex oxides obtained by different deposition techniques is increasing exponentially in last decades due to large variety of possible application such as high-efficient solar cell, optoelectronic devices, etc. pulsed laser deposition (PLD) is a versatile growth technique and recently became more attractive for industrial applications due to the possibility to obtain crystalline thin films on a large area. Laser processing techniques were successfully used to obtain thin films with good optical properties starting from simple oxides, such as Sm_2O_3, ZrO_2, etc., to more complex lead-free materials: $Sr_xBa_{1-x}Nb_2O_6$ (SBN) and $Na_{1/2}Bi_{1/2}TiO_3$–x%$BaTiO_3$, or superconductive oxide $YBa_2Cu_3O_{7-\delta}$. When oxide thin films are designated for electronic and optoelectronic devices or for solar cells, the optical properties and the thickness must be well known. For this purpose, the spectroscopic ellipsometry technique was developed. Ellipsometry is a powerful technique to determine the optical properties of thin films especially when the thicknesses of thin films are in a nanometer range.

Keywords: spectroscopic ellipsometry, pulsed laser deposition, thin films, complex oxides, magnetic alloy, high temperature superconductor

1. Introduction

A thin film is a layer of material, which can be created by various processes, one of the most common being the condensation of particles (atoms, molecules, ions, aggregates) on a substrate. The thickness of the layer can vary between few angstroms (monolayer) and several micrometers. The market for thin films obtained by different deposition techniques had an exponential increase in the last decades due to the large variety of possible applications, such as hard coatings, semiconductor devices, optical coatings, energy generation and storage, memory devices, sensors, actuators, etc.

Thin film deposition techniques are divided in two major categories: chemical vapor deposition (CVD) and physical vapor deposition (PVD) [1, 2]. Pulsed laser deposition (PLD), developed by Maiman, Smith, and Turner, in the 1960s, is a PVD technique based on laser-material interaction [3–5]. It is a versatile technique and recently became more attractive for industrial applications due to the possibility to obtain crystalline thin films on large areas. Laser processing techniques were successfully used to obtain thin films from all kind of materials, starting from alloys (NdFeB), simple oxides (Sm_2O_3, ZrO_2, etc.), to more complex materials: high temperature superconductors ($YBa_2Cu_3O_{7-\delta}$), ferroelectric $Sr_xBa_{1-x}Nb_2O_6$ with tungsten bronze structure or solid solution of $Na_{1/2}Bi_{1/2}TiO_3$–x%$BaTiO_3$, etc.

The basic setup for pulsed laser deposition is relatively simple, but the physical phenomena of laser-matter interaction and film growth are very complex [5]. Because thin films have a large surface area to bulk volume ratio, their physical and chemical properties can be different from the bulk materials [6]. The substrate properties and the deposition parameters during the grown process can drastically influence the properties of thin films. By carefully choosing these parameters, thin films with enhanced physical properties can be obtained.

The properties of thin films can be investigated by various analysis techniques. X-ray diffraction (XRD) and transmission electron microscopy (TEM) are used for structural investigation, scanning electron microscopy (SEM) and atomic force microscopy (AFM) for morphological analysis, secondary ion mass spectroscopy (SIMS) for compositional analysis, etc. Electrical and optical properties are also very important when thin films are designated to be used in electronic, opto-electronic, or optical applications. For optical investigation, ellipsometry is a very precise and fast method, especially when the thickness of the thin films ranges from subnanometers to hundreds of nanometers.

In Section 2, a short overview of the principles and analysis procedures of ellipsometry are presented. In Section 3, a review of the optical investigation of thin films obtained by PLD is presented. Finally, conclusions are given in Section 4.

2. Ellipsometry

The history of ellipsometry is difficult to trace precisely. It can be said that it starts with the "wave theory of light" of Robert Hooke followed by the contribution of many scientists: Christian Huygens, Augustin-Jean Fresnel, Michael Faraday, James Clerk Maxwell, etc. In 1902, Paul Drude published the "Polarization" chapter in the book "The Theory of Optics," which still serves as a modern introduction to the study of polarized light and ellipsometry [7]. Ellipsometry developed slowly for long time after the Drude ellipsometer, using a single wavelength in the 1960s. After 1970, due to the necessity of faster measurements for thin films used in computers or the semiconductor industry, the ellipsometry techniques developed rapidly [8, 9]. Nowadays, there are many types of ellipsometers, such as null, photometric, spectroscopic, etc.

Ellipsometry is a tool that gives information about the optical properties and microstructural parameters of materials, such as layer thicknesses, porosity, and crystalline state. Ellipsometry is a nondestructive, fast, sensitive, and accurate technique used to determine the optical constants of a large range of materials: metals, glasses, complex oxide structures, polymers, organic materials, etc. It can be used for ex-situ or in-situ analysis, both for inorganic and organic materials. However, a major disadvantage of this technique is that it is an indirect analysis, and an accurate optical model is required for data analysis.

2.1. Principle and analysis procedure of ellipsometry

In the past decades, many books and scientific papers regarding the principle, data analysis procedure, and applications of ellipsometry were published [8, 10–13]. In this section, we present a brief overview of the ellipsometry principles and data acquisition and analysis procedure.

In ellipsometry, the change of the polarization state of linearly polarized light is measured upon reflection at the surface of a sample. The polarization states after reflection depend on properties of the investigated surface and, in case of multilayer thin film system, on the properties of each layer. Two experimental parameters, ψ and Δ, called ellipsometric angles, are usually measured (**Figure 1**).

Here, ψ is the relative amplitude and Δ is the phase difference for p- and s-polarized lights, and the quantities ψ and Δ are described by the fundamental equation of ellipsometry (Eq. (1)),

$$\rho = \frac{\tilde{R}_p}{\tilde{R}_s} = \tan(\Psi)e^{i\Delta} = \left(\frac{E_{rp}/E_{ip}}{E_{rs}/E_{is}}\right), \tag{1}$$

where R_p and R_s are the Fresnel reflection coefficients; E_{rp}, E_{is}, E_{rp}, and E_{rs} represent the p- and s-components of the incident and reflected light waves. The p- and s-polarization terms represent the parallel (p) and the perpendicular (s) components of electric field, related to the incidence plane.

In case of thin films, the incident light is reflected on the thin film surface and penetrates into the film and into the substrate. An obtained ellipsometric data include information about the

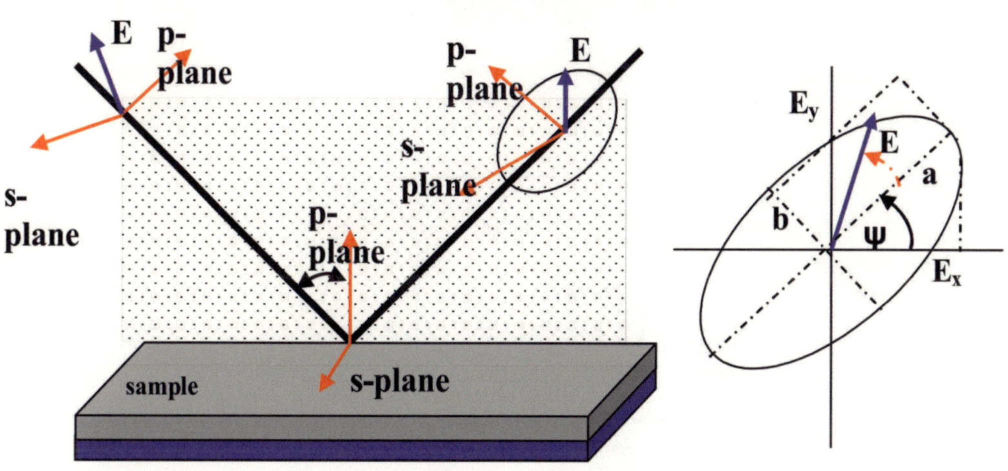

Figure 1. Scheme showing the basic principle of ellipsometry: linearly polarized light with p- and s-components at oblique incidence is reflected and it becomes elliptically polarized. Relative attenuation of s- and p-polarized lights etermines the tilt of the ellipse and the relative phase shift is related to ellipticity [12].

investigated materials. For a thin film deposited on a substrate, $\varrho = F(n_0, n_1, n_2, \theta, d)$ where n_0, n_1, and n_2 are the complex refractive indices of the ambient, the thin film, and the substrate, θ is the angle of incidence of the light beam, and d is the thickness of the thin layer (**Figure 2**).

The complex refractive index is composed by a real part "n" and by an imaginary part "k" The imaginary part is called the extinction coefficient and is directly related to absorption of the light in the material. The absorption coefficient is given by $\alpha = 4\pi k/\lambda$, where λ is the free-space wavelength of light. The complex dielectric function $\varepsilon = \varepsilon_1 - i\varepsilon_2$ and the complex index of refraction N are related to each other through the relation derived from Maxwell's equations [11, 12]: $\varepsilon = N^2$. The real and imaginary parts are: $\varepsilon_1 = n^2 - k^2$ and $\varepsilon_2 = 2nk$. Since ellipsometry measures changes in the polarization state, this makes it a highly accurate and reproducible technique. Ψ and Δ can be determined experimentally with high precision, but the extraction of information (optical constants and thickness) requires, for calculation, an accurate optical model.

In case of thin film system, the purpose of ellipsometry measurement is to determine the optical properties of the materials from the measured Ψ and Δ parameters. The refractive index, extinction coefficient, and thickness of each layer can be determined. **Figure 3** shows the process of ellipsometry analysis procedure.

The standard procedure of analysis includes the following steps.

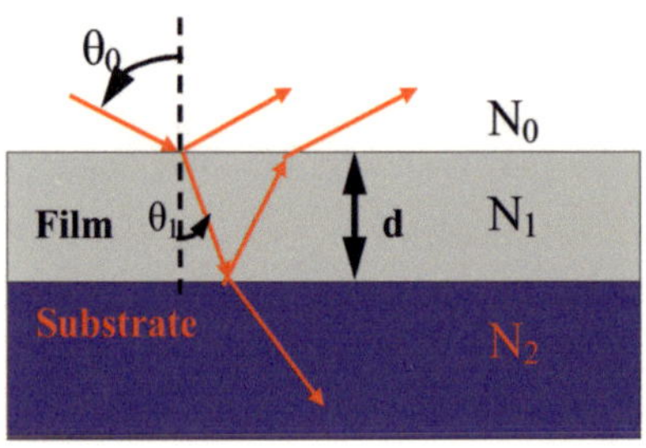

Figure 2. The optical model for thin film system.

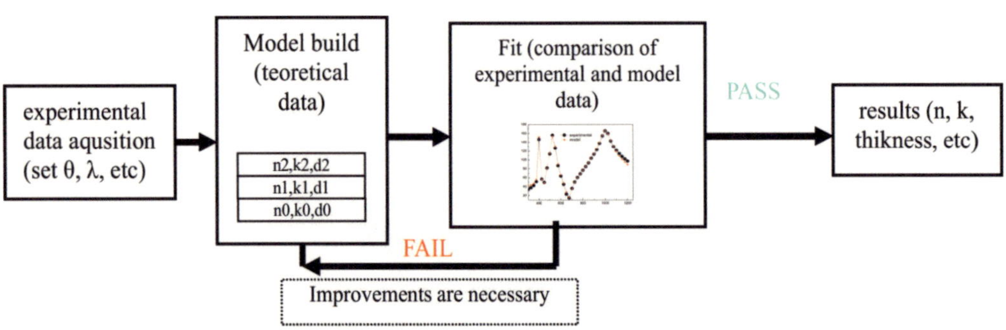

Figure 3. Process of ellipsometry data fitting and model analysis.

2.1.1. *The experimental measurement*

In this step, the routine of experimental data point acquisition is established. The set-up parameters include the wavelength, the number of experimental data points, and the angle of incidence. For spectroscopic ellipsometry, measurements are done in a large wavelength spectrum, ranging from deep ultraviolet (UV) to near infrared (IR) (i.e., 250–1700 nm).

2.1.2. *The optical model*

A suitable optical model is required for ellipsometric analysis. In case of thin films, the optical model is composed by a stack of several kinds of material layers with different thicknesses. The properties of material layers are described by a specific complex dielectric function. Usually, at this point, the dielectric function of the substrate is known. If not, a separate analysis is required for bulk substrate. In the optical model, some of the parameters are unknown, such as the thickness or the values of optical constants. For optical characterization of thin films, it is necessary to select a suitable dispersion model. Examples of dispersion models are Cauchy dispersion with Urbach tail, Sellimeier model, effective medium approximation (EMA), Drude model or Lorentz model, etc. In order to find what dispersion model should be selected for a thin film depends on the types of materials.

2.1.3. *Comparison of model and experimental data*

Using the optical model, the values of theoretical Ψ and Δ are generated and compared with the experimental data. If the graphs of both data match, the optical model can be considered correct. The following function is usually employed for the comparison (Eq. (2)):

$$\mathrm{MSE} = \frac{1}{2N - M}\sum_{i=1}^{N}\left[\left(\frac{\Psi_i^{\mathrm{mod}} - \Psi_i^{\mathrm{exp}}}{\sigma_{\psi,i}^{\mathrm{exp}}}\right)^2 + \left(\frac{\Delta_i^{\mathrm{mod}} - \Delta_i^{\mathrm{exp}}}{\sigma_{\Delta,i}^{\mathrm{exp}}}\right)^2\right], \tag{2}$$

where N is the number of (Ψ, Δ) pairs, M is the number of variable parameters in the model, and σ is the standard deviations on the experimental data points.

The MSE, as shown in the equation, represents a sum of the squares of differences between the measured and calculated data, with each difference weighted by the standard deviation of that measured data point. The unknown parameters in the optical model are varied in order to produce a good fit. The best fit will lead to minimum of the MSE value. In case of "bad" match of theoretical and experimental data, or high value of MSE, the optical model will be adjusted until the minimum value of MSE is reached. After obtaining the best fit and the minimum of MSE value, the values of the optical constants can be generated. The dispersion of optical constants needs to be physically reasonable regarding the analyzed thin film materials.

As mentioned before, for optical characterization of thin films, it is necessary to select a suitable dispersion model, and this model depends on the analyzed material. The values of the refractive index, the extinction coefficients, and the thickness of the thin film are considered as unknown parameters. Also, the optical model will usually include a distinct top layer, which accounts for the roughness of the thin film.

In many cases, the analyzed materials are UV-absorbing, such as semiconductors, metal oxides, or ferroelectric or multiferroic materials. For these materials, a common routine for ellipsometric analysis can be given. When the film is transparent over a portion of measured spectral range, it is possible to determine the thickness of the film with high accuracy using a Cauchy or Sellmeier dispersion model. In the next step, the thickness and Cauchy parameters values will be kept fixed, and the experimental data will be fitted point by point across the entire measured spectral range. In this way, a set of optical constants is obtained. The final values of optical constants (or dielectric function) will be obtained by replacing the Cauchy model with a sum of oscillators. Fitting again the experimental data, the parameters of the oscillators will be determined. For example, in case of Lorentz or Gauss oscillators, the parameters are as follows: the oscillation amplitude, the broadening, and the position of oscillation. After the oscillator parameters are known, the thickness will be fitted again for final adjustment. Ellipsometry is quite sensitive to the surface structure and is necessary to incorporate a rough top layer in the optical model in the data analysis.

Once the unknown parameters of thin film are determined, the value of thickness and dielectric function of the rough top layer will be calculated. It is rather difficult to estimate the complex refractive index of the roughness layer. The rough top layer is composed by a mix of thin film materials and air or voids.

If we apply the effective medium approximation (EMA), the complex refractive indices of surface roughness layer can be calculated [14]. In case of thin films, a percent of 50:50 (voids: material) in Bruggeman effective medium approximation (BEMA) is a good approximation for the composition of the rough top layer. The Bruggeman effective medium approximation (BEMA) is expressed by the following equation (Eq. (3)):

$$f_A \frac{\tilde{\varepsilon}_A - \tilde{\varepsilon}}{\tilde{\varepsilon}_A + 2\tilde{\varepsilon}} + f_B \frac{\tilde{\varepsilon}_B - \tilde{\varepsilon}}{\tilde{\varepsilon}_B + 2\tilde{\varepsilon}} = 0, \tag{3}$$

where f_A and f_B are the volume ratios of media A and B, and ε_A and ε_B are the complex dielectric functions of media A and B.

In the next section, we will present some examples of ellipsometric analysis on thin films obtained by pulsed laser techniques, starting from simple oxide materials, magnetic, to more complex ferroelectric thin films, evidencing in this way the capability of this technique.

3. Results and discussion

3.1. Experimental set-up

3.1.1. Pulsed laser deposition (PLD)

The experimental set-up for PLD experiment is based on two main equipments: a laser system and a vacuum chamber. The used laser systems were a Nd:YAG laser and a ArF excimer laser. The targets and collector substrate are mounted in the chamber in a vertical position, at variable distance. The angle of incidence of the laser beam was set at 45°. The substrates were fixed on a heating system. During deposition, the targets were rotated and the laser beam was translated on surface of the targets. The film deposition was performed in a dynamic ambient background gas (O_2, N_2, etc.) and the flow rates of gases were precisely controlled through a MKS mass flow controller. All substrates used in the experiments were commercial substrates and before the experiments were cleaned in an ultrasonic bath using acetone or methanol as cleaning agents, then dried under pressurized nitrogen gas.

3.1.2. *Spectroscopic ellipsometry*

Optical measurements were performed with a Woollam variable angle spectroscopic ellipsometer (VASE) system, equipped with a high pressure Xe discharge lamp incorporated in an HS-190 monochromator. The ellispometry measurements and analysis were done using the VASE32 software. Usually, the experimental data of Ψ and Δ were acquired in the 250–1700 nm range of wavelengths, with a step of 2 nm. The angle of incidence of the light beam was set in function of the analyzed materials, from 60 to 75° with a step of 5°.

3.1.3. *Samarium oxide*

Samarium oxide (Sm_2O_3) is a lanthanide oxide with a monoclinic structure. Sm_2O_3 is used as a neutron absorber in control rods for nuclear power reactors and in optical and infrared absorbing glass to absorb infrared radiation or as a gate dielectric in metal-oxide-semiconductor (MOS) devices [15, 16].

The target used in PLD experiments was prepared by pressing the Sm_2O_3 powder of 99.9% purity. For ellipsometry measurements, the substrate used was silicon. A laser beam from a pulsed Nd:YAG laser system (266 nm wavelength) was focused on the Sm_2O_3 target, with a laser spot of 0.8 mm^2 with 10 Hz repetition rate. The substrates were kept at room temperature during the depositions. The deposition was performed under the oxygen pressure range between 0.02 and 0.6 mbar.

Spectroscopic ellipsometry measurements were performed in the UV-visible and near-IR region of the spectrum (250–1350 nm), a step of 10 nm, at 60 and 65° angles of incidence. The optical model consists of five layers: the silicon substrate, the Si-SiO_2 interface, the native SiO_2 layer, the Sm_2O_3 layer, and a rough top layer. The rough top layer is set to be half air and half Sm_2O_3. The refractive indices for the substrate and the native oxide are from Herzinger et al. [17]. For optical constants of Sm_2O_3, we used the Cauchy dispersion model for transparent zone of wavelength, with Urbach tail for low absorbing region (below 400 nm). The Cauchy-Urbach model is given by the following equations (Eqs. (4) and (5)):

$$n(\lambda) = A_n + \frac{B_n}{\lambda^2} + \frac{C_n}{\lambda^4} \tag{4}$$

and

$$k = A_k \cdot e^{B(E-E_b)}, \tag{5}$$

where A_n, B_n, C_n, A_k, and B are constants and E is energy in eV ($E = 1240/\lambda$; $E_b = 1240/\lambda_b$).

Fitting the experimental data with Cauchy-Urbach dispersion model and assuming that the top layer consists of 50% air (voids) and 50% Sm_2O_3, we have obtained the parameters listed in **Table 1**:

Figure 4 presents the SE spectra of a 59.8 ± 0.3 nm thick Sm_2O_3 on silicon substrate with a roughness of 6.5 ± 0.3 nm, the optical constants n = 1.94 and k = 6 × 10^{-3}, at λ = 550 nm, and the MSE = 4.55.

The surface topography was analyzed by AFM. The roughness of the thin film was found to be in nanometer range, excluding the droplets present on the film surface [18].

The obtained Sm_2O_3 thin films are highly transparent in the 400–1350 nm wavelength range. Below 400 nm, the film becomes optically absorbing, with a maximum extinction coefficient k ~ 3 × 10^{-2} (at 250 nm). The Sm_2O_3 is a high refractive index material, with a value of n = 2.09 at a wavelength of 589.3 nm and has a monoclinic crystalline phase. This value is with 0.1 larger than our value. The amorphous materials have a smaller refractive index than the crystalline material [11, 12], and our smaller value indicates an amorphous phase of Sm_2O_3 in the thin film, which is consistent with the XRD measurements [18].

As a short conclusion of this case: using a simple model such as Cauchy dispersion with Urbach tail, it is possible to obtain valuable information regarding the dispersion of optical constants, thickness, and the crystalline phase of the film. However, the Cauchy dispersion model is limited at the transparent range of wavelength, and the Urbach tail is limited at low absorption zone. To sum up: a correct value of optical band gap E_g cannot be given using this model.

3.1.4. Zirconia

Zirconia (ZrO_2) (zirconium dioxide) is a dielectric material with a dielectric constant of about 20 (in the low frequency domain). Because of its low leakage current level and high thermal

Sample	Thickness (nm)	Roughness (nm)	An	Bn	Ak	Bk	MSE
Sm_2O_3/Si	59.8 ± 0.22	6.5 ± 0.36	1.932 ± 0.002	0.016 ± 0.003	0.002 ± 0.001	1.343 ± 0.28	4.555

Table 1. The Cauchy-Urbach fit parameters for Sm_2O_3 thin film.

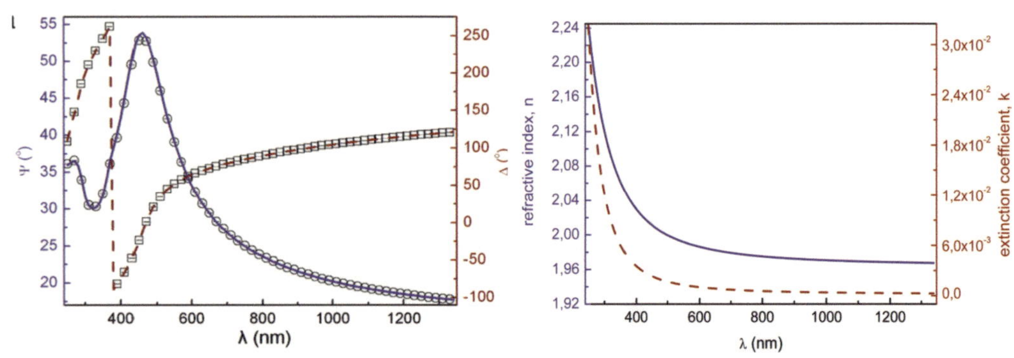

Figure 4. SE spectra on a Sm_2O_3 film: (left) the open symbols represent experimental data, whereas the solid line (Ψ) and dashed line (Δ) are obtained from a fitting procedure using a five-layer model and (right) the refractive index (solid line) and the extinction coefficient (dashed line) for the Sm_2O_3 thin films [18].

stability, ZrO_2 is another candidate used for gate dielectrics in metal-oxide-semiconductor (MOS) [19]. Pure ZrO_2 exists as three polymorphs at different temperatures: monoclinic, tetragonal, and cubic.

A ceramic target of ZrO_2 was ablated with an ArF laser beam (λ = 193 nm) at a repetition rate of 40 Hz. The laser fluences were varied in the range of 2.0–3.4 J/cm^{-2}. The silicon (Si) substrate was kept at a distance of 4 cm from the target. The substrate temperature was between room temperature and 600°C, in the presence of reactive oxygen at a pressure of 0.1 mbar. The reactive oxygen was introduced in the deposition chamber by an additional RF discharge at a power of 100 W.

Ellipsometric measurements were carried out in the spectral range of 250–1700 nm, in steps of 2 nm, and incident angles of 60, 65, and 70°. The optical model comprises the silicon substrate, a layer of native oxide of silicon (3 nm), the ZrO_2 thin film, and a roughness layer, approximated by 50% voids and 50% ZrO_2. For the substrate, we used the dielectric function from literature [17].

The calculation of the complex refractive index and the thickness of the thin film and the rough layer required a two-step fitting procedure. In the first step, a simple Cauchy dispersion model was used in the 500–1700 nm range of wavelength, where the ZrO_2 is transparent [20]. The values of thickness and Cauchy parameters are listed in **Table 2** [21].

In order to fit the entire measured spectra and to calculate the extinction coefficients, the values of thickness were kept fixed and the Cauchy dispersion model was replaced by a single Gaussian oscillator. The imaginary part of the dielectric function for Gauss oscillator is given by the following equations (Eqs. (6) and (7)):

$$\varepsilon_2 = A^{e^{-\left(\frac{E-E_n}{\sigma}\right)^2}} - A^{e^{-\left(\frac{E+E_n}{\sigma}\right)^2}} \tag{6}$$

and

$$\sigma = \frac{Br_n}{2\sqrt{\text{In}(2)}}, \tag{7}$$

where A is the amplitude of the curve's peak, En refers to the centering energy or energy at the curve's peak, Br is the broadening, and σ is the standard deviation of the curve [12]. The fitted result parameters are listed in **Table 3**.

Values of MSE obtained in the second step of fitting increase slightly from 5.94 for Cauchy fit for the thin film grown at 300°C to 9.67 for Gauss fit, but the difference is still small enough to

Sample/temperature (°C)	Thickness (nm)	Roughness (nm)	An	Bn	MSE
ZrO_2/Si (RT)	64.177 ± 0.0391	0.904 ± 0.06	1.9882 ± 0.000591	0.014203 ± 0.000156	1.908
ZrO_2/Si (300°C)	190.812 ± 0.0537	1.403 ± 0.0481	2.1417 ± 0.000256	0.018515 ± 0.000153	5.944
ZrO_2/Si (600°C)	154.380 ± 0.0384	2.047 ± 0.0429	2.1535 ± 0.000211	0.01856 ± 0.000192	4.815

Table 2. The Cauchy fit parameters for ZrO_2 thin film.

be taken into consideration. This difference can arise from small discrepancy between the model and the experimental curve in the UV region caused by small defects on the surface, such as droplets or pores.

In **Figure 5**, we can observe that the higher values of refractive index were obtained for a substrate temperature of T = 600°C and also for the highest value of extinction coefficients. At room temperature, the smaller value of "n" indicates an amorphous phase, and this is in agreement with XRD results [21]. The best optical constants in terms of high "n" and lower "k" were obtained at 300°C. From **Table 2**, the values of thickness of ZrO_2 thin films grown by PLD in the presence of RF discharge were found to be 64 nm for room temperature, 190 nm for 300°C, and 154 nm for 600°C. At this point, we can conclude that the best temperature for growing transparent ZrO_2 thin films is 300°C. The XRD results indicate the presence of two crystalline phases, tetragonal and monoclinic, for ZrO_2 thin film in case of 600°C substrate temperature [21]. Coexistence of these two phases indicates a polycrystalline grown mode and explains the higher value of extinction coefficient (for 600°C).

As a short conclusion of this case, using a simple Gauss model, it is possible to determine the behavior of complex refractive index in function of PLD deposition parameters.

3.1.5. NdFeB

NdFeB is a permanent magnet based on rare earth components and has been widely investigated for its applications in micromagnetic, magneto-electronic, or magnetic recording media.

Sample/temperature (°C)	Amp	En (eV)	Br (eV)	MSE
ZrO_2/Si (RT)	19.849 ± 5.15	8.9932 ± 0.137	1.9394 ± 0.525	4.533
ZrO_2/Si (300°C)	20.196 ± 1.41	8.841 ± 0.0585	2.2913 ± 0.171	9.675
ZrO_2/Si (600°C)	16.331 ± 0.338	9.1236 ± 0.0346	2.9503 ± 0.0698	9.783

Table 3. The Gauss fit parameters for ZrO_2 thin film.

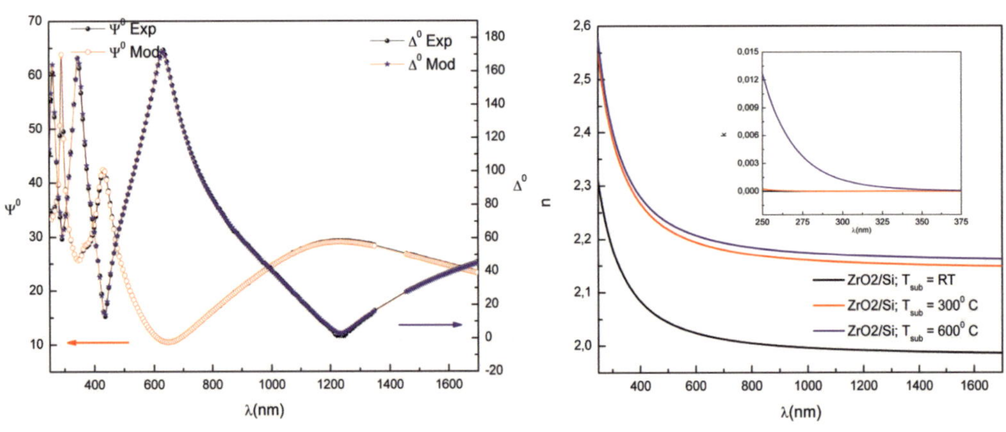

Figure 5. (Left) Experimental and Gauss-modeled curves for the parameters psi (Ψ) and delta (Δ) for an ZrO_2/Si(100) thin film deposited at 6000°C by RF-plasma-assisted PLD and (right) the dispersion of optical constants for ZrO_2 thin films grown at different substrate temperatures.

NdFeB thin films were deposited by various methods, e.g., magnetron sputtering, molecular beam epitaxy (MBE), and PLD [22–25].

In our experiments, the NdFeB thin films were obtained by PLD method [26] starting from an alloy target. A Nd:YAG laser (266 nm) was used to irradiate the targets with 20,000–40,000 pulses. The substrates were platinized silicon (Pt/Si). The samples were heated to 650°C, and the temperature was kept constant during the thin film growth.

Spectroscopic ellipsometry measurements were performed between 350 and 1700 nm spectral ranges, with a step of 2 nm at a fixed angle of incidence (75°). The initial optical model used to determine the complex refractive index consisted of a stack of five layers: the silicon substrate, the titanium layer (25 nm), the platinum layer (150 nm), the NdFeB layer, and the rough top layer. Because the platinum layer is thick enough, we considered this layer as a substrate and the optical model was reduced to three layers: Pt, NdFeB, and the rough layer. The values for the optical constants of NdFeB thin films were fitted using different procedures. The Cauchy-Urbach dispersion model used in case of Sm_2O_3 did not fit the NdFeB due to their limitation [11, 12]. The best fit was obtained with a Lorentz oscillator having the complex dielectric function (Eq. (8)) written as:

$$\varepsilon_{n\text{–Lorentz}} = \frac{A_n Br_n E_n}{E_n^2 - E^2 - iBr_n E}, \tag{8}$$

where A_n is an oscillation amplitude (dimensionless), Br is broadening (expressed in eV), and E_n is the position of oscillation (eV). The total dielectric function is given by (Eq. (9)):

$$\varepsilon = \varepsilon_1 + i\varepsilon_2 = e_{1\text{offset}} + \varepsilon_{n\text{–Lorentz}}, \tag{9}$$

where $\varepsilon_{1\ \text{offset}}$ is a purely real constant, equivalent to ε_{inf}. [12].

The Lorentz parameters and the thickness resulted from fitting experimental data with the proposed optical mode are presented in **Table 4**.

The thickness of a thin film deposited at 40,000 pulses, 266 nm laser wavelength, and 2 J/cm^2 laser fluency was found to be 93 nm with a roughness of 10 nm. For optical constants, a refractive index with a value of n = 2.43 was found for λ = 600 nm with a maximum of 2.46 at 360 nm. The extinction coefficient "k" is nonzero on the entire measured range of wavelength thin film that has high optical absorption [26].

The value of MSE is 31, and from **Figure 6**, it is easy to observe a mismatch between the experimental and the model data in UV-visible range. This mismatch indicates an improvement needed in the optical model. Even with this discrepancy, we have demonstrated that with a correct dispersion model (Lorentz in this case), it is possible to determine the complex optical constants for this magnetic thin film.

Thickness (nm)	Roughness (nm)	E_n (eV)	B_r (eV)	A_n	$e1_{offset}$	MSE
93.609 ± 0.307	10.353 ± 0.221	5.7013 ± 0.208	4.2273 ± 0.217	3.1234 ± 0.169	3.3783 ± 0.13	31.3

Table 4. Spectroscopic ellipsometry fitted data for an NdFeB thin film.

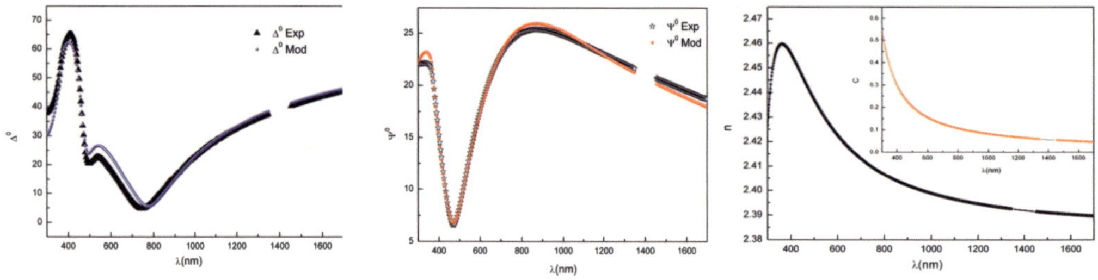

Figure 6. Experimental and model spectra for Δ (left) and Ψ (center) and refractive indices (n), and extinction coefficients (inset graph), for a thin film sample deposited at 266 nm laser wavelength [24].

3.1.6. Yttrium barium copper oxide ($YBa_2Cu_3O_{7-\delta}$)

$YBa_2Cu_3O_{7-\delta}$ (YBCO) is the famous high-temperature superconductive material, and it was extensively studied due to its superconducting properties. For appropriate oxygen content in the films, the maximum critical temperature T_c can reach a value of 92 K. The application of YBCO thin films varies from electronic industry (microwave filters), military to medical devices (magnetic resonance imaging, Josephson junctions), or in sensor industry as highly sensitive magnetic field sensors (*SQUIDs*) [27, 28] and IR radiation sensors (bolometer). The superconducting properties of $YBa_2Cu_3O_{7-\delta}$ thin films are highly sensitive to the oxygen content. When δ varies between $0 \le \delta \le 1$, two symmetry phases can be observed, the tetragonal phase and the orthorhombic phase. For $\delta = 1$, YBCO became an insulator, and for $\delta \approx 0.1$, this compound is a superconductor [29]. The epitaxial $YBa_2Cu_3O_{7-\delta}$ thin films have been obtained by PLD at high deposition temperature, usually around 780°C. For achieving the highest quality, the $YBa_2Cu_3O_{7-\delta}$ thin films need a postdeposition thermal treatment in oxygen atmosphere [30].

The deposition of YBCO thin films by pulsed laser deposition technique has been done by ablating a commercial YBCO ceramic target with an ArF excimer laser ($\lambda = 193$ nm) working at 5 Hz repetition rate. The thin films of YBCO was deposited on $SrTiO_3$ (001) single-crystal substrates [31]. Two sets of YBCO thin films have been deposited: first set was obtained by conventional PLD method and the second one by a hybrid technique where the PLD process is assisted by a radiofrequency-generated oxygen plasma plume (RF-PLD). For the first set of films, the following deposition parameters were used: laser fluency of 1.5 J/cm^2, oxygen pressure of 0.2 mbar, and substrate temperature of 780°C and target-substrate distance of 5 cm. The annealing treatments were done in the deposition chamber in two steps: 15 min at 650°C in 1 bar O_2 and 1 hour at 450°C in 1 bar of oxygen. The second set of samples was done by RF-PLD with or without postannealing treatments. The substrates were kept during deposition at 730°C, and the radiofrequency oxygen plasma beam working at 13.56 MHz and 200 W was used. The role of gas-discharge plasma (in oxygen) was to compensate the difference in deposition temperature.

Spectroscopic-ellipsometry measurements were performed between 300 and 1700 nm spectral range, with a step of 2 nm at 45, 60, and 75° angles of incidence. The optical model consists of three layers: the $SrTiO_3$ substrate, the YBCO layer, and a rough top layer considered to be

http://dx.doi.org/10.5772/intechopen.70803

composed by 50% air and 50% YBCO in the Bruggeman effective medium approximation [11, 12]. The bulk dielectric function for the STO substrate was taken from literature [32]. In case of YBCO material, the dielectric function was described in detail by Kumar et al. [33] and is written as a sum of Lorentz and Drude terms.

The experimental data of Ψ and Δ for YBCO thin films were fitted with a sum of Lorentz oscillators for UV-VIS wavelengths and a Drude oscillator for near IR. The thickness of samples was found in the range of 120–150 nm, whereas the thickness of the top layer was 20–40 nm with an MSE value higher than 20. The roughness was in the same range with AFM results [31]. Because the ellipsometry is very sensitive at surface roughness, the large value of MSE can be explained by high value of roughness. In **Figure 7**, the experimental data and the fit results using the Lorentz-Drude model for the thin film of YBCO obtained by PLD at 780°C substrate temperature are presented.

From **Figure 7**, it is easy to observe the difference in the optical behavior for YBCO thin film growth in different conditions. For sample growth at 780°C, the normal optical behavior was

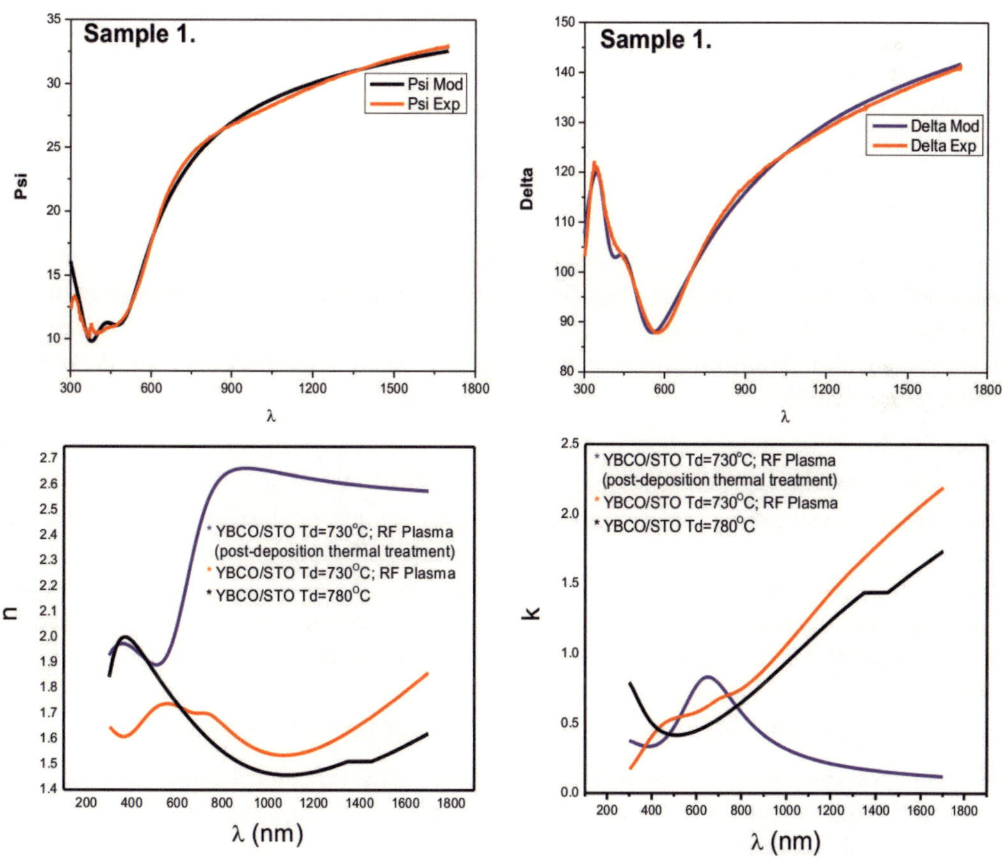

Figure 7. Comparison between the Lorentz and Drude fitting predictions and experimental measurements for YBCO thin film deposited by PLD and the refractive indices (left down) and extinction coefficient (right down) for YBCO thin films.

obtained [34]. In case of RF-PLD thin film growth at lower temperature (730°C) but in presence of oxygen RF discharge (without postannealing treatments), the value of refractive index and extinction coefficients are higher, and the optical behavior is different in the range of 250–600 nm. From XRD results, the YBCO films have an orthorhombic symmetry (a ~ 3.840 Å, b ~ 3.880 Å, and c ~ 11.688 Å) [31]. However, in case of RF-PLD deposition, the cell parameters were found to be a ~ 3.865 Å, b ~ 3.880 Å, and c ~ 11.70 Å, which means that the films were incompletely oxidized. The difference in the optical behavior can be explained by this difference in an oxidized state of YBCO layer.

For thin films deposited in the same way but with postannealing treatments in oxygen atmosphere, the dispersion of refractive index and extinction coefficients are completely different. This behavior is not characteristic for YBCO superconductor, and most probably, the amount of oxygen during deposition was too high.

As short conclusion of this study, the deposition parameters are very important especially when the properties of materials can be easily modified by the amount of certain chemical component, in our case the amount of oxygen in YBCO structure.

3.1.7. Strontium barium niobate

Strontium barium niobate (SBN) ($Sr_xBa_{1-x}Nb_2O_6$) is a ferroelectric material with tetragonal tungsten bronze structure without volatile chemical elements. SBN has pyroelectric activity as well as a large linear electro-optic (EO) coefficient(s) (r_{33} ~ 1300 p mV^{-1}) [34]. The high value of EO coefficient makes SBN an excellent candidate for electro-optic devices.

In the PLD experiment, the target of single-crystal SBN:60 (ce inseamna 60?) was ablated by a Surelite II Nd:YAG laser (λ = 265 nm). The repetition rate of the laser beam was set at 10 Hz. The laser fluency was varied in the 1.5–2.8 J/cm^2 range. The substrates, $MgO_{(001)}$ and $Si_{(100)}$, were placed at a distance of 4–5 cm from the target and were heated to the deposition temperature (600–700°C).

Spectroscopic ellipsometry measurements were performed in the visible and near-UV region of the spectrum, at energies between 1 and 5 eV, a step of 0.01 eV and at 65, 70, and 75° angles of incidence, with a step of 5°. The correlation between the energy and the wavelength scale of the spectrum is given by E = 1240/λ [12]. For the thin films of SBN deposited on Si substrates, the optical model consists of four layers: the silicon substrate, the native oxide (~3 nm), the SBN layer, and the rough top layer. In case of MgO substrate, the optical model was reduced to three layers: the MgO substrate, the SBN thin film, and the rough layer. In both cases, the dielectric function of the substrates is taken from literature and the rough top layer was approximated to consist of 50% air and 50% SBN in the Bruggeman approximation.

The film thickness and roughness were obtained by fitting the experimental data with a Cauchy dispersion model in the 1.2–2 eV range (620–1030 nm). In this range, the film is nonabsorbing (k ~ 0) and such dispersion is accepted. The thickness was found to be in the 150–250 nm range, and the MSE was found to be 3.8 for SBN/MgO and 2.5 for SBN/Si [35] (**Table 5**).

Samples	MSE	Thickness (nm)	An	Bn	Roughness (nm)
On MgO	3.741	165.935 ± 0.0418	2.0486 ± 0.000727	0.027885 ± 0.000728	2.215 ± 0.0428
On Si	2.559	227.304 ± 0.167	2.0418 ± 0.000943	0.024381 ± 0.000546	1.618 ± 0.121

Table 5. The fitting parameters obtained by Cauchy fit for SBN thin films [35].

In the next step, we replaced the Cauchy layer with a Tauc-Lorentz (T-L) oscillator. The T-L oscillator model was developed by Jellison and Modine and is used for amorphous materials [12], having the dielectric function (Eqs. (10) and (11)) written as:

$$\varepsilon = \varepsilon_1 + j\varepsilon_2 = (n + jk)^2 \tag{10}$$

with

$$\varepsilon_2 = \frac{AE_0C(E - E_g)^2}{(E^2 - E_0^2)^2 + C^2E^2}\frac{1}{E} \text{ for } E > E_g \text{ and } e_2 = 0 \text{ for } E \leq E_g \tag{11}$$

where A, E_0, and C are dispersion parameters in the following limits: $E_0 > E_g$ and $C < 2E_0$.

For Tauc Lorentz oscillator, the real part of the dielectric function ε_1 is a (the) complete analytical solution to the Kramers-Kroning integral [12] (**Figure 8** and **Table 6**).

For the SBN thin film deposited on MgO and Si substrates, the values of refractive index and extinction coefficients were found to be smaller than for the crystalline target. The optical band gap, Eg, that results from Tauc-Lorentz fit is Eg = 3.81 eV for SBN on MgO substrate and Eg = 3.74 for SBN on Si substrate. The Eg value for single crystal target is Eg = 3.85 eV. From the XRD analysis, it results that all the samples made from the single-crystal targets are amorphous [35]. In this way, the smaller value of optical constants and value of band gap (Eg) can be explained because amorphous materials have lower refractive index than the same crystalline ones [12].

3.1.8. $Na_{1/2}Bi_{1/2}TiO_3{-}x\%BaTiO_3$

The solid solution of $(Na_{1/2}Bi_{1/2})TiO_3$ (BNT) with $BaTiO_3$ (BT) (BNT-BT) is considered to be a good alternative lead-free material to the PZT perovskite materials. (1-x) NBT-x BT shows a morphotropic phase boundary (MPB) between the rhombohedral and the tetragonal phase, at x between 0.06 and 0.07 [36]. The electrical and optical properties of NBT-xBT can be modified due to the phase transition from rhombohedral (R) to tetragonal (T).

A ceramic target of BNT-5%BT was ablated by a Surelite II Nd:YAG pulsed laser with wavelength of 266 nm, pulse duration of 5 ns and frequency 10 Hz, and a fluency of 1.6 J/cm^2. The number of laser pulses was 36,000. The films were grown on $Pt/TiO_2/SiO_2/Si$ substrates, in reactive oxygen atmosphere at 0.1 mbar pressure. The distance between target and substrate was kept at 5 cm. During deposition, the substrates were heated at temperatures of 700 and 730°C.

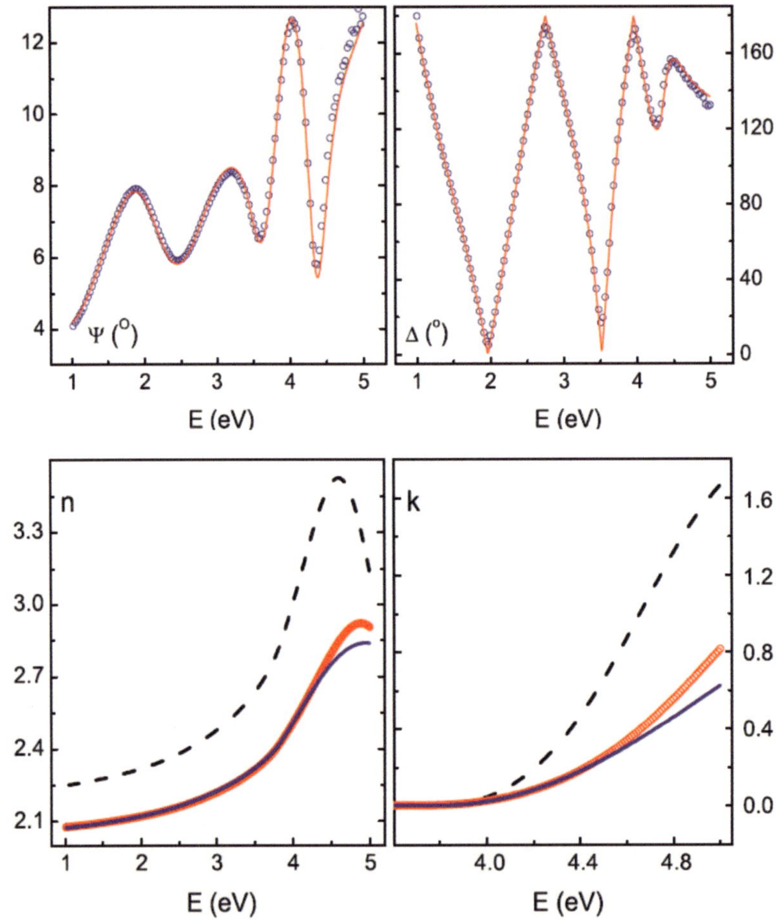

Figure 8. Ellipsometry spectrum for a-SBN on MgO. The circles represent the experimental data, whereas the solid lines result from the fitting procedure using the Tauc-Lorentz oscillator. (Down) Refractive index for crystalline target (dashed lines): a-SBN on MgO (red thick line) and a-SBN on Si (blue thin line) [35].

Samples	MSE	A	En	C	Eg	ε1
On MgO	8.094	193.1 ± 8.77	5.2978 ± 0.0393	3.2284 ± 0.162	3.8102 ± 0.0105	1.639 ± 0.0591
On Si	8.842	131.58 ± 8.34	5.2161 ± 0.0427	1.759 ± 0.163	3.7443 ± 0.0209	2.3611 ± 0.0893

Table 6. The fitting parameters obtained from Tauc-Lorentz fit [35].

For ellipsometric measurements, the angle of beam light incidence was set at 60–70° with a step of 5°. The experimental data of Ψ and Δ were acquired in the 250–1700 nm range of wavelength with a step of 2 nm. The optical model consists of three material layers: the platinum layer with a thickness of 150–200 nm, considered as a substrate; the BNT-5%BT thin film, and the roughness. As in the case of SBN, the fitting procedure consists of two steps: first, the thicknesses of thin films and the roughness were calculated from fitting the experimental data in the 600–1700 nm

http://dx.doi.org/10.5772/intechopen.70803

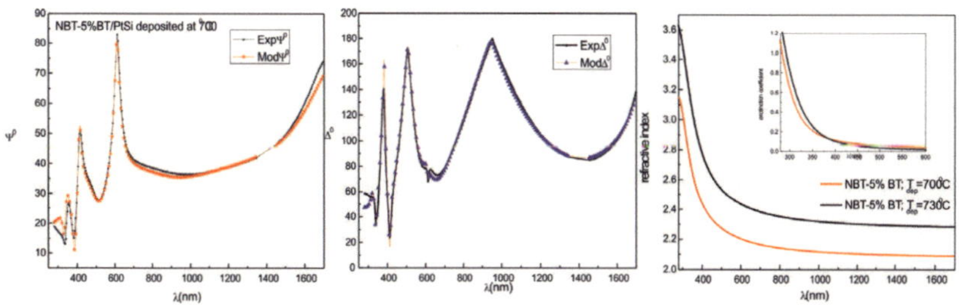

Figure 9. Ellipsometry spectrum for BNT-5%BT on PT Si (left and center). The black lines represent the experimental data, whereas the dots result from the fitting procedure using the Gauss and Cody-Lorentz oscillator; the calculated refractive index and extinction coefficient (inset) for BNT-5%BT (right) deposited at substrate temperature of 700 (dots) and 730°C (lines) [40].

range of wavelength, and second, the Cauchy layer was replaced by an oscillator. In case of BNT-5%BT, the best fit was obtained using a sum of Gauss and Cody-Lorentz oscillator [37], a well-known optical model for BNT and solid solution BNT−x%BT [38, 39]. The thickness of samples was found to be d = 131.9 nm for BNT-5BT growth at 700°C, with a roughness of 26.9 nm, and for a film growth at 730°C, the thickness was 241.2 nm with a roughness of 23.7 nm. The values of MSE was 78.8 in case of Tdep = 700°C and 91.34 for Tdep = 730°C. In **Figure 9**, the Cody-Lorentz fit result and the calculated optical constants are presented.

The results of XRD indicate a polycrystalline randomly oriented thin films of BNT-BT growth by PLD on Pt/Si substrate [40]. The polycrystalline nature of BNT-BT thin film with randomly oriented nanocrystals can explain the high value of MSE. From **Figure 9**, it is easy to observe the discrepancy in modeled and experimental data in UV range, and this discrepancy can be caused by existence of small defects in sample. In terms of refractive index, the highest value was obtained for higher deposition temperature. The extinction coefficients "k" increase fast in the wavelength ranging from 280 to 380 nm. In the 380–600 nm spectrum range, the "k" presents an Urbach tail and this can be explained by optical scattering on randomly oriented nanocrystals.

4. Conclusions

Thin films of amorphous and crystalline oxides (Sm_2O_3, ZrO_2, $Sr_xBa_{1-x}Nb_2O_6$ (SBN), $Na_{1/2}Bi_{1/2}TiO_3$−x%$BaTiO_3$ (NBT-x%BT), high-temperature superconductive oxide $YBa_2Cu_3O_{7-\delta}$ (YBCO), and magnetic alloy (NdFeB) have been grown by pulsed laser deposition. The optical properties of films were successfully determined by spectroscopic ellipsometry. High transparent thin film was obtained in case of ZrO_2. The films of SBN and NBT-x%BT show a high value of refractive index ($n > 2$) and low extinction coefficients ($k < 10^{-4}$) in visible and near-IR range of wavelength. In our optical models presented above, we have to choose every time a different oscillator to describe the behavior of complex refractive index, and this choice was conducted by applying two rules: the "rule" of MSE—the smallest values of MSE mean the better fit, and in order to obtain the smallest value, a sum of oscillators can be tried until the experimental data and the theoretical data are overlapped. Second, we have to look at the types and properties of an initial material. The Lorentz oscillator have a symmetric shape, and the high and low energy sides of the function decrease at the same rate and can be used for complex refractive index of conductive oxides or metal [41, 42]. The Lorentz oscillators have long asymptotic tail and cause unwanted absorption in transparent regions for some types of materials. The long tail of Lorentz oscillator was the reason for Lorentz fit in case of NdFeB alloy. In case of ferroelectric materials, the Lorentz does not fit very well because of the characteristics of the complex dielectric function at the near ultraviolet and ultraviolet (UV). For ferroelectric materials, the Tauc-Lorentz [43] or Cody-Lorentz oscillators are more flexible at higher energies in the ultraviolet range of the spectrum. The Tauc-Lorentz oscillator is used for amorphous ferroelectric materials, and this was the reason for choosing it in case of SBN thin film. For BNT-BT material, a Cody-Lorentz combined with a Gaussian oscillator will give the best results.

Acknowledgements

This work has been financed by the Romanian National Authority for Research and Innovation in the frame of Nucleus programs, contract no. 4 N/2016, and by a grant of the Romanian National Authority for Scientific research, UEFISCU, project number no. PED 99/2017 (PROFILE).

Author details

Valentin Ion*, Andreea Andrei, Maria Dinescu and Nicu Doinel Scarisoreanu

*Address all correspondence to: valentin.ion@inflpr.ro

National Institute for Laser, Plasma and Radiation Physics, Romania

References

[1] Maissel LI, Clang R. Handbook of Thin Film Technology. New York: McGraw-Hill; 1970

[2] Bunshah RF. Deposition technologies for Films and Coatings: Developments and Applications. Park Ridge, NJ: Noyes Publications; 1982

[3] Maiman TH. Optical and microwave-optical experiments in ruby. Physical Review Letters. 1960;**4**:564

[4] Smith HM, Turner AF. Vacuum deposited thin films using a ruby laser. Applied Optics. 1965;**4**:147-148. DOI: 10.1364/AO.4.000147

[5] Chrisey DB, Hubler GK. Pulsed Laser Deposition of Thin Films. Wiley-Interscience; 1 edition (June 7, 1994); ISBN-13: 978-0471592181

[6] Ghandhi SK. VLSI Fabrication Principles: Silicon and Gallium Arsenide. New York: John Wiley & Sons; 1983

[7] Losurdo M, Hingerl K. Ellipsometry at the Nanoscale. Berlin Heidelberg: Springer-Verlag; 2013. DOI: 10.1007/978-3-642-33956-1

[8] Azzam RMA, Bashara NM. Ellipsometry and Polarized Light. New York, NY, USA: North-Holland Publishing Co; 1977

[9] Adams JR, Bashara NM. Determination of the complex refractive index profiles in P+31 ion implanted silicon by ellipsometry. Surface Science. 1975;**49**(2):441-458

[10] Tompkins HG. A User's Guide to Ellipsometry. San Diego, CA, U.S.A: Academic Press Inc.; 1993. ISBN 978-0-126-93950-7

[11] Tompkins HG, Irene EA. Handbook of Ellipsometry. Norwich, NY, U.S.A.: William Andrew Publishing/Noyes; 2005. ISBN 978-0-815-51499-2

[12] Fujiwara H. Spectroscopic Ellipsometry: Principles and Applications. John Wiley & Sons, Ltd: Chichester, UK; 2007 978-0-470-01608-4

[13] Azzam RMA. Ellipsometry. In: Bass M, editor. Handbook of Optics. Geometrical and Physical Optics, Polarized Light, Components and Instruments. Vol. 1. 3rd ed. New York, NY, USA.: McGrow-Hill Companies, Inc; 2010. p. 16.1–16.25. ISBN 978-0-071-62925-6

[14] Aspnes DE, Theeten JB, Hottier F. Investigation of effective-medium models of microscopic surface roughness by spectroscopic ellipsometry. Physical Review B. 1979;**20**:3292-3302

[15] Dakhel AA. Dielectric and optical properties of samarium oxide thin films. Journal of Alloys and Compounds. 2004;**365**(1–2):233-239. DOI: 10.1016/S0925-8388(03)00615-7

[16] Yang H, Wang H, Luo HM, Feldmann DM, Dowden PC, DePaula RF, Jia QX. Structural and dielectric properties of epitaxial Sm_2O_3 thin films. Applied Physics Letters. 2008;**92**: 062905. DOI: 10.1063/1.2842416

[17] Herzinger CM, Johs B, McGahan WA, Woollam JA, Paulson W. Ellipsometric determination of optical constants for silicon and thermally grown silicon dioxide via a multi-sample, multi-wavelength, multi-angle investigation. Journal of Applied Physics. 1998; **83**:3323. DOI: 10.1063/1.367101

[18] Constantinescu C, Ion V, Galca AC, Dinescu M. Morphological, optical and electrical properties of samarium oxide thin films. Thin Solid Films. 2012;**520**(20):6393-6397. DOI: 10.1016/j.tsf.2012.06.049

[19] Wilk GD, Wallance RM, Anthony JM. High-κ gate dielectrics: Current status and materials properties considerations. Journal of Applied Physics. 2001;**89**:5243-5275. DOI: 10.1063/1.1361065

[20] Zhamg W, Gan J, Hu Z, Yu W, Li Q, Sun J, Xu N. Infrared and Raman spectroscopic studies of optically transparent zirconia (ZrO_2) films deposited by plasma-assisted reactive pulsed laser deposition. Applied Spectroscopy. 2011;**65**(5):522-527

[21] Cancea VN, Birjega R, Ion V, Filipescu M, Dinescu M. Analysis of zirconia thin films grown by Pulsed Laser Deposition. Physics AUC. 2012;**22**:50-62

[22] Yamashita S, Yamasaki J, Ikeda M, Iwabuchi N. Anisotropic Nd–Fe–B thin-film magnets for milli-size motor. Journal of Applied Physics. 1991;**70**:6627. DOI: 10.1063/1.349879

[23] Nakano M, Sahara M, Yanai T, Yamashita F, Fukunaga H. Nd–Fe–B thick film magnets with Nb additive prepared by vacuum arc deposition method. Journal of Applied Physics. 2011;**109**:07A755. DOI: 10.1063/1.3566061

[24] Constantinescu C, Scarisoreanu N, Moldovan A, Dinescu M, Petrescu L, Epureanu G. Thin films of NdFeB deposited by PLD technique. Applied Surface Science. 2007;**253** (19):8192-8196. DOI: 10.1016/j.apsusc.2007.02.165

[25] Liu W, Zhang Z, Liu J, Chen HL, Liu LY, Sun X, Sellmyer D. Exchange coupling and remanence enhancement in nanocomposite multilayer magnets. Advanced Materials. 2002;**14**(24):1832-1834. DOI: 10.1002/adma.200290012

[26] Constantinescu C, Ion V, Codescu M, Rotaru P, Dinescu M. Current Applied Physics. 2013;**13**:2019-2025. DOI: 10.1016/j.cap.2013.09.002

[27] Koelle D, Kleiner R, Ludwig F, Dantsker E, Clarke J. Reviews of Modern Physics. 1999;**71**:631

[28] Sekitani T, Miura N, Ikeda S, Matsuda YH, Shiohara Y. Upper critical field for optimally-doped $YBa_2Cu_3O_{7-\delta}$. Physica B: Condensed Matter. 2004;**346–347**:319. Bibcode:2004PhyB01.098. DOI: 10.1016/j.physb.2004

[29] Pedarnig JD, Göttlich H, Rössler R, Heckl WM, Bäuerle D. Patterning of $YBa_2Cu_3O_{7-\delta}$ films using a near-field optical configuration. Applied Physics A: Materials Science & Processing. 1998;**67**:403-405

[30] Kusumori T, Muto H. Influence of target composition on the quality of YBCO films fabricated by Nd:YAG pulsed laser deposition. Physica C. 2001;**351**:227-244

[31] Stanciu G, Scarisoreanu ND, Ion V, Moldovan A, Andronescu E, Dinescu M. Ba2Cu3O7-δ thin films deposited by pulsed laser deposition and radio-frequency assisted pulsed laser deposition. Journal of Optoelectronics and Advanced Materials. September–October 2012;**14**(9–10):852-857

[32] Palik ED. Handbook of Optical Constants of Solids. Vol. II. p. 1035-1048, Academic Press; 1 edition (April 4, 1991), ISBN-13: 978-0125444224

[33] Kumar AR, Zhang ZM, Boychev VA, Tanner DB, Vale LR, Rudman DA. Far-infrared transmittance and reflectance of $YBa_2Cu_3O_{7-s}$ films on Si substrates. Journal of Heat Transfer. 1999;**121**:844-851

[34] Hesselink L, Bashaw MC. Optical memories implemented with photorefractive media. Optical and Quantum Electronics. 1993;**25**(9):S611-S661. DOI: 10.1007/BF00444334

[35] Ion V, Galca AC, Scarisoreanu ND, Filipescu M, Dinescu M. Spectroscopic ellipsometry study of amorphous SrxBa1−xNb2O6 thin films obtained by pulsed laser deposition. Physica Status Solidi (c). 2008;**5**(5):1180-1183. DOI: 10.1002/pssc.200777818

[36] Takenaka T, Maruyama K, Sakata K. (Bi1/2Na1/2)TiO3-BaTiO3 system for lead-free piezoelectric ceramics. Japanese Journal of Applied Physics. 1991;**30**:2236

[37] Ferlauto AS, Ferreira GM, Pearce JM, Wronski CR, Collins RW, Deng X, Ganguly G. Analytical model for the optical functions of amorphous semiconductors from the near-infrared to ultraviolet: Applications in thin film photovoltaics. Journal of Applied Physics. 2002;**92**:2424. DOI: 10.1063/1.1497462

[38] Cernea M, Trupina L, Dragoi C, Galca A-C, Trinca L. Structural, optical, and electric properties of BNT–BT0.08 thin films processed by sol–gel technique. Journal of Materials Science. 2012;**47**(19):6966-6971. DOI: 10.1007/s10853-012-6646-1

[39] Dorywalski K, Lemée N, Andriyevsky B, Schmidt-Grund R, Grundmann M, Piasecki M, Bousquet M, Krzyzynski T. Optical properties of epitaxial Na0.5Bi0.5TiO3lead-free piezoelectric thin films: Ellipsometric and theoretical studies. Applied Surface Science. 2016:18. DOI: 10.1016/j.apsusc.2016.09.078

[40] Scarisoreanu N, Craciun F, Ion V, Birjega R, Dinescu M. Structural and electrical characterization of lead-free ferroelectric Na1/2Bi1/2TiO3–BaTiO3 thin films obtained by PLD and RF-PLD. Applied Surface Science. 2007;**254**(4):1292-1297. DOI: 10.1016/j.apsusc.2007.09.036

[41] Synowicki A. Spectroscopic ellipsometry characterization of indium tin oxide film microstructure and optical constants. Thin Solid Film. 1998;**313–314**:394-397. DOI: 10.1016/S0040-6090(97)00853-5

[42] Brevnov DA, Bungay C. Diameter-dependent optical constants of gold mesoparticles electrodeposited on aluminum films containing copper. The Journal of Physical Chemistry B. 2005;**109**(30):14529-14535. DOI: 10.1021/jp0511707

[43] Jellison GE, Modine FA. Parameterization of the optical functions of amorphous materials in the interband region. Applied Physics Letters. 1996;**69**:2371. DOI: 10.1063/1.118064

Chapter 7

Nanoparticles, Nanocrystals, and Nanocomposites Produced with Pulsed Laser Ablation and Their Applications

Bülend Ortaç, Elif Uzcengiz Şimşek and
Canan Kurşungöz

Additional information is available at the end of the chapter

http://dx.doi.org/10.5772/intechopen.70594

Abstract

In the last few decades, nanoparticles have become key components in a variety of applications in nanotechnology, nanoengineering, and nanoscience. Pulsed laser ablation in liquids (PLAL) method is frequently preferred for fast and pure nanoparticle generation. There exists a wide range of metal and semi-conductor nanoparticles that are successfully synthesized by PLAL method. In our research, nanoparticle synthesis of different materials and their applications are pursued. After nanoparticle synthesis, the application research proceeds and the scope of the research spans many subjects ranging from sensor realization to biological applications.

Keywords: nanoparticles, nanocrystals, nanocomposites, pulsed laser ablation method, nanoparticle applications, nanotoxicology, nanodevice

1. Introduction

In recent years, nanoparticles (NPs) have received great attention in various scientific areas due to their unique material properties and applications. The scope of nanoscience spans 100 nm or smaller scales, where the nanoparticles show different physical and chemical properties than their bulk counterparts as the scale goes down to nano. The unique properties of the nanomaterials led to a significant rise in the progress of nanoscience research. The nanoparticle generation methods become a crucial part in the research due to the application requirements. Nanoparticles could be synthesized in the laboratory environment by physical and chemical methods [1]. One of the most promising top-down approaches for nanomaterial generation is PLAL method due to its various advantages. In PLAL method, pure and uncontaminated nanoparticles could be produced by a relatively fast process and it is also a good candidate for mass production at a low cost.

1.1. Laser-material interaction and pulsed laser ablation in liquid method

Lasers offer a unique ability to directly deposit great amount of energy onto the specified regions of a material to fulfill the response in demand. When the laser energy is delivered onto the material surface, three main cases—reflection, transmission, and absorption—have a significant role in results of the interaction. In the first two cases, after laser beam hits the material, some amount of the incoming laser energy is reflected due to the real part of the index of refraction and the rest of the laser energy is transmitted through the material. The reflection from the material also depends on frequency (or wavelength) of the light source. For example, the reflection from metals varies from 0.4 to 0.95 in the spectral range from near UV to visible (UV–Vis) and 0.9–0.99 in the spectral range of infrared (IR) [2]. In the third case, some of the laser energy is absorbed by the material due to its absorption coefficient (α), and the laser intensity decreases exponentially as it travels inside the material. This ability can be explained by laser-material interaction, which depends on the internal properties of both lasers and materials. The effects of linear optical phenomena are taken into account in the approaches mentioned above. These cases for both material and laser are not necessarily the same for all conditions. For instance, some of the materials have strong nonlinear properties, which affect the refractive index variation for different wavelengths; thus, the whole process will be affected and altered such as self-focusing, defocusing, and soliton propagation [3]. Also, laser sources could act as an important role for the process. The laser source could be continuous wave (CW) or pulsed (ps, fs). While continuous wave (CW) and pulsed laser processing, the single photon absorption is considered; however, in the ps or fs laser processing, nonlinear phenomena could occur such as multi-photon absorption and optical breakdown, which cause a decrease in absorption depths [4]. These phenomena are very important for the nanomaterial generation, and they will be given in the PLAL. When the material absorbs the laser energy directly, the removal of the material is observed. This phenomenon is called ablation. Laser ablation is generally considered as a pulsed laser process; however, it is also applicable with CW lasers. The ablation of materials occurs when the laser energy becomes greater than the threshold fluence. Laser and material parameters determine the process characteristics. The main parameters that the laser depends on are wavelength, pulse duration, fluence, and pulse length [5]. The material properties such as absorption, defects inside the material, and general characteristics of the material are also important for the ablation. In the PLAL process, laser and material interact in the liquid surrounding and a series of physical phenomenon occur. The plasma plume is created when intense laser beam hits the solid target. The created plasma plume begins to cool and it becomes condensed. Finally, nanoparticles will be generated through this process. Although many synthesis and applications of nanoparticles have been demonstrated by using PLAL method in the past decade, the main physical mechanism behind PLAL method remains a puzzle. To understand the mechanism, Direct Simulation Monte Carlo (DSMC) calculations have been performed to simulate the ultra-short, laser-ablated plume dynamics and nanoparticle evolution under realistic experimental conditions [6].

1.2. Outline of book chapter

In this book chapter, a variety of nanomaterial generation through PLAL method will be given in terms of their characterization techniques, properties and applications. Nanoparticles, nanocrystals and nanocomposites synthesis could be successfully achieved since PLAL technique is applicable to a wide range of materials from metals to semi-conductors. Metal nanoparticles (Au, Ag, Ti, etc.) are ablated by intense laser beam and nanoparticle generation will be observed in organic solutions. Semi-conductor materials such as III nitride group (InN, GaN, and MoS2) could be generated in very small diameters and by-product free by PLAL method. Their physical properties such as light absorption, crystal structure, and photoluminescence (PL) could be used in many applications. Nanodevices such as photodetectors are fabricated successfully due to broadband light absorption of nanomaterials in different regions of spectrum from ultraviolet to infrared. High-sensitive strain sensor, solar cell, and memory devices have been realized with different nanomaterials synthesized by PLAL technique. Nanomaterial properties and their application areas will be resumed in this part. Moreover, nanoparticles produced by PLAL will be evaluated in terms of their toxicological effects on both the environment and ecosystem and the human health.

2. Nanoparticles, nanocrystals, and nanocomposites from pulsed laser ablation

2.1. Metal nanoparticle generation

The wide range of materials can be synthesized by PLAL method, and metals (Au, Ag, Ti, etc.) are the most commonly generated materials due to their unique properties unlike their bulk state. PLAL method allows a mass production due to easy and fast process. The chemical-process-free nanoparticle generation provides noncontaminated and pure nanoparticles which contribute to research areas such as biological and chemical sensing, medical applications and strain gauges [7–9].

2.1.1. Gold nanoparticles

The conduction property of chemically sensitized Au-NPs is frequently studied [10]. Chemically obtained Au-NP films tend to show exponential decay dependence due to their chemical ligands over the Au-NPs. This behavior can be explained by quantum-tunneling effect which depends on the width of the potential barrier. Due to this effect, it is indicated that the Au-NP films can be used for as highly sensitive strain gauges. In principle, the presence of chemical ligands may decrease the performance of the sensitivity. Au-NPs generated by PLAL method are used as films on PDMS substrate to demonstrate high gain sensitivity [9]. In **Figure 1a**, Au-NPs are represented as spheres, and the edges of the sensor are Pt contacts. The initial state of the Au-NP-deposited sensor is on the left-hand side where there is no strain and the resistance of the film is denoted as R. When the strain is applied to the sensor, the distance between Au-NPs begins to increase and the resistance of the film becomes ΔR as shown in the image at the right-hand side.

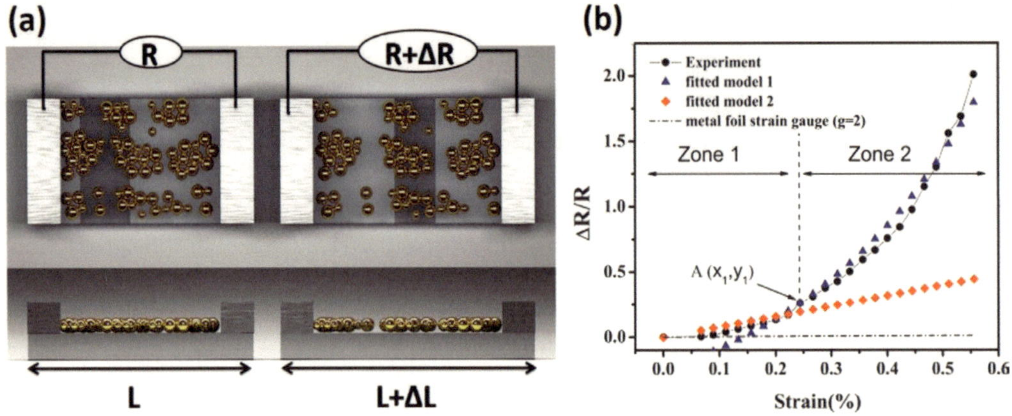

Figure 1. (a) Au-NP film-deposited strain sensors, (b) resistive response of the Au-NP film [9].

In **Figure 1b**, the resistance response of the Au-NP films when the strain applied is represented. The strain response of the Au-NP sensors is formulated as $\Delta R/R = \exp(g\varepsilon) - 1$ [11]. The G factor of Au-NP films is reported as ~300 for strains higher than 0.22%. The high sensitivity of the Au-NP generated by PLAL method used in the strain sensors is connected to the noncontaminated NP surfaces, the size of the NPs. The quantum-tunneling effect between nanoparticles enhances the gain sensitivity due to purely obtained Au-NP by PLAL method. In comparison to chemical NP generation method, PLAL method provides chemical-free and pure NPs. There would not be any contaminants over the surface of the Au-NPs. Due to this behavior, quantum-tunneling effect becomes stronger and more effective.

2.1.2. *Silver nanoparticles*

Nanoparticles exhibit quite different properties compared to their bulk counterparts. High specific surface areas and mobilities, which are the results of their small sizes (<100 nm), make them a potential risk for humans and environment [12]. Widespread production and use of nanoparticles in different applications in medical, biological, electronic and industrial fields might lead to unfavorable effects on both humans and a wide range of organisms [13]. Thus, nanoparticles should be evaluated in terms of ecological and toxicological aspects for a deeper understanding of the impacts on both humans and other organisms.

Silver nanoparticles (Ag-NPs) are one of the commonly used nanoparticles which are utilized in therapeutics, cosmetics, food additives, textiles and antimicrobial coatings on medical implants, catheters, wound dressings and so on [14]. Ag-NP production is mainly achieved by reducing the Ag salt but there are also other nanoparticle production methods for the synthesis of Ag-NP, such as electrolysis, photoreduction, pyrolysis, and sol-gel methods [15, 16]. The major drawback of these Ag-NPs produced by aforementioned methods is that the resulting Ag-NPs are coated either with a by-product due to the chemical reaction or with other molecules on purpose such as polyvinylpyrrolidone (PVP), antibodies, and surfactants. These coatings lead to different surface chemistries which masks the real Ag-NP toxicity. Since pure uncoated Ag-NPs are the commonly found types of Ag-NPs in nature and in industry as products or by-products, it is crucial to reveal their toxicity profile [17]. PLAL is the most suitable method to produce pure uncoated Ag-NP for the evaluation of Ag-NP toxicity. Our group studied different aspects of Ag-NP toxicity in two distinctive studies.

In both of the studies, Ag-NPs were produced in water by using a commercial nanosecond pulsed ND:YLF laser (Empower Q-Switched Laser, Spectra Physics) (527-nm wavelength, 100-ns pulse duration, 16-W average output power, 1-kHz pulse repetition rate and 16-mJ pulse energy). The Ag-NPs were crystalline in structure and the size distribution of these nanoparticles was determined as 5–50 nm although there were also larger nanoparticles up to micron sizes. The nanoparticles did not show an aggregation pattern as can be observed from both transmission electron microscopy (TEM) and scanning electron microscope (SEM) images. Energy-dispersive X-ray (EDX) analysis revealed that a pure Ag-NP solution was obtained by the nanosecond laser ablation method since the only peak was the silver peak apart from the carbon and copper peaks due to TEM grids. The characteristic UV–vis absorption peak at approximately 400 nm also confirms the presence of Ag-NP, which is consistent with the previous studies [18]. These characteristics are present in both the studies mentioned subsequently (**Figure 2**).

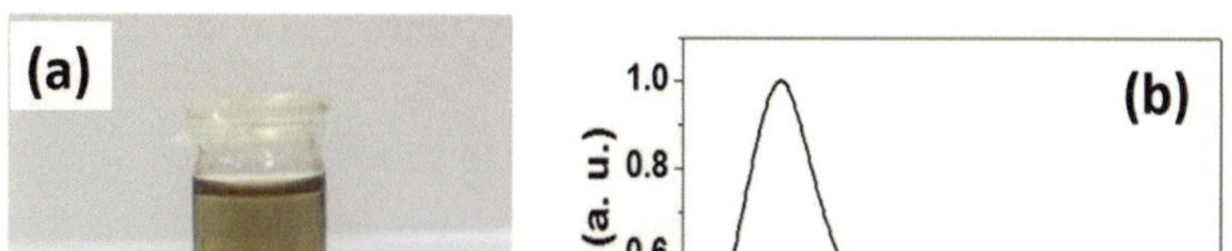

Figure 2. (a) Ag-NP solution produced by PLAL. No precipitation was observed at least for 4 months. (b) Absorption spectrum exhibited absorbance peak at approximately 400 corresponding to Ag-NP. (c) SEM image of Ag-NPs (2-mm scale bar). (d) TEM image of the Ag-NPs (20-nm scale bar); the inset represents EDX analysis of Ag-NP [21].

Firstly, the toxic effects of Ag-NPs on the environment were evaluated by using an aquatic macrophyte *Lemna minor* as the model organism. For this purpose, dose- and time-dependent toxicities were assessed. It was shown that *L. minor* amplifies rapidly in low concentration ranges; however, concentration-dependent toxicity is much less pronounced at higher concentrations. When Ag-NP concentrations increase slightly in lower concentrations, there is a remarkable increase in toxicity. This might be a significant result in ecological perspective since low amounts of Ag-NP, which might be released to the environment accidentally, might affect the ecosystem and aquatic macroflora in a significant manner [19].

Secondly, as indicated in the aforementioned application fields of Ag-NPs, humans frequently encounter these nanoparticles in daily life. NPs can reach the brain both by systemic circulation and through the olfactory bulb which leads to serious effects in central nervous system [20]. It was also demonstrated that Ag-NPs are detected in different brain regions after inhalation [21]. Our group aimed to determine the effects of Ag-NP on learning and memory, and thus hippocampus was chosen for this study. The distribution and the toxicity of the pure Ag-NPs produced by PLAL in the rat hippocampal slices were determined. Furthermore, it was revealed that the size of pure Ag-NPs was crucial for the cellular uptake mechanism of pure Ag-NPs. Phagocytosis was shown as the type of endocytotic pathway that governs the entry of larger Ag-NPs into the hippocampal neurons. This study demonstrated that PLAL is a substantially useful method for studying NP toxicity, since it provides pure NPs mimicking the ones frequently used in the industrial fields.

2.2. Semi-conductor nanoparticle generation

The generation of the semi-conductor NPs is intensely studied in the past decade because of the physical, chemical, electrical, and optical properties of the nanomaterials [22–24]. The optical absorption and photoluminescence properties of the SC-NPs are highly related with the size of the NPs due to quantum confinement effect and they can be widely used in the optics and optoelectronics applications, and multi-functional nanocomposites are developed thanks to the unique optical properties of the SC-NPs [25–27]. In this section, Si, GaN, InN, MoS2 NC synthesis and their unique properties will be discussed.

2.2.1. Germanium nanocrystals

The thin film technology is the most commonly used method for the development of NPs, SC-NPs and nanocomposites [28–31]. However, the downsides of this method include the control of the film thickness, homogeneity of the film, and it is also hard to fabricate high surface to materials. The unique properties of laser-material interaction lead to the wide variety of metal and SC material generation [32]. PLAL method offers pure and agent-free NP generation. The photoluminescent Ge-NCs synthesis was demonstrated by ns-PLAL method; then, the Ge-NCs were merged into the electrospun polymeric nanofiber matrix to generate functional nanofibrous composite web [33]. The generated functional nanocomposites are represented in **Figure 3**. Their optical properties UV–vis absorption spectra and PL spectra are shown in **Figure 4**.

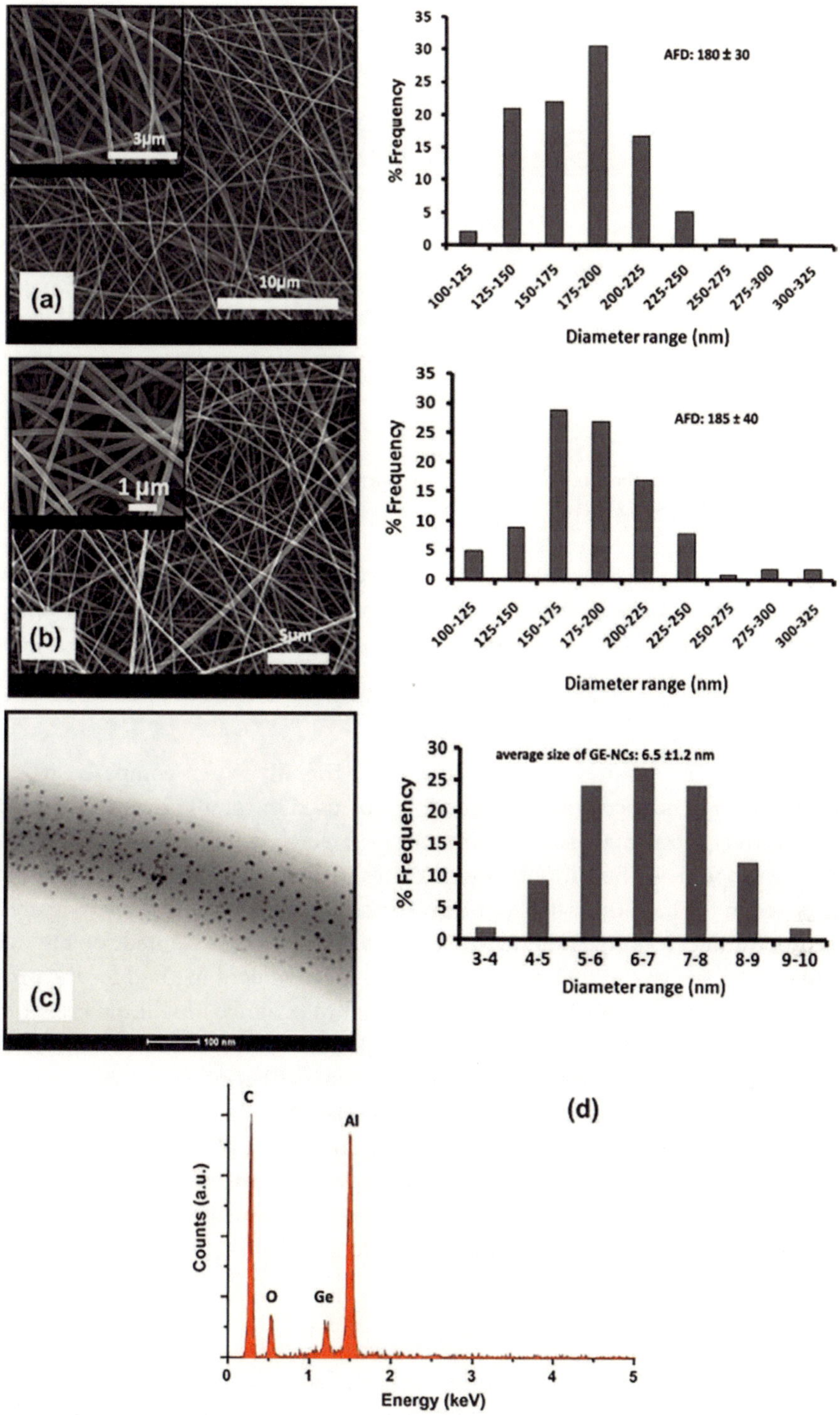

Figure 3. SEM images of (a) PVA nanofibers, (b) Ge-NCs/PVA nanofibers (the insets represent the high magnification SEM images), (c) STEM image of Ge-NCs/PVA nanofiber [33], (d) EDX data of Ge-NCs/PVA.

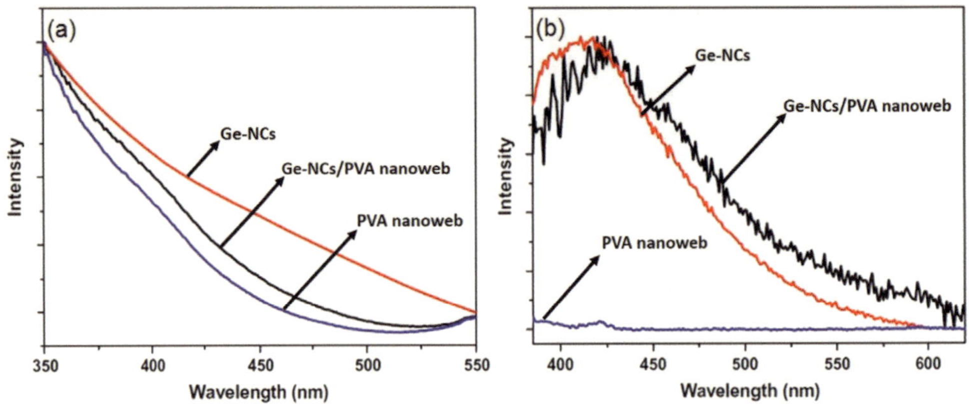

Figure 4. (a) Absorption spectra of Ge-NCs in acetone solution in UV–vis region, PVA nanoweb, and Ge-NCs/PVA nanoweb in the solid state, (b) photoluminescence (PL) spectra of Ge-NCs in acetone solution, PVA nanoweb and Ge-NCs/PVA nanoweb in the solid state [33].

The Ge-NCs/PVA nanocomposite represents similar photoluminescence spectra as Ge-NCs solution generated by PLAL method. Therefore, the functional polymeric nanoweb shows unique optical properties by integrated Ge-NCs generated by PLAL method in electrospun PVA nanofibres.

2.2.2. *Silicon nanocrystals*

Silicon is abundant in nature and environmentally friendly when compared to cadmium when considered as a nanocrystal quantum dot to be used in color-conversion light-emitting diodes. Moreover, ultra-small silicon nanoparticles were demonstrated to present strong luminescence properties. Thus, different techniques to produce silicon nanocrystals (Si-NCs) were suggested in the literature such as ion implantation, electrochemical etching and chemical vapor deposition. These techniques generally were shown as limited for the mass production. On the other hand, PLAL was proposed as a convenient method for the production of Si-NCs in the literature [34–36]. Our group aimed to produce blue luminescent colloidal Si NCs by using a two-stage process. Firstly, small particles (5–100 nm in diameter) were produced using femtosecond laser. Then, ultrasonic and filtering post-treatment processes were utilized as chemical-free post-treatment methods. The resulting Si-NCs were 1–5.5 nm in diameter, which indicated the production of ultra-small Si-NCs. Moreover, PL emission characteristic of untreated Si NCs was blue-shifted with a broadband PL emission after the post-treatment. Thus, PLAL might be a safe and alternative method to produce ultra-small Si-NCs to be used in biological applications owing to their chemical-free nature [37].

2.2.3. *GaN nanocrystals*

III-N-based semiconductors gained attention recently for their wide, direct and tunable band gap from IR to UV regions to be used in optoelectronic devices [38, 39]. Specifically, GaN-based materials were shown to be good candidates for devices in harsh environments due to their high strength, high power, high temperature, and high frequency resistance [40]. Different growth techniques were suggested for the production of nanocrystalline GaN in the literature [41–43]. However, these methods result in GaN nanostructures with chemical contaminants. PLAL, on the other hand, offers the production of chemical-free and clean nanomaterials which can be used in various applications, including biomedical fields [44, 45]. Our group showed the production of ultra-small GaN-NCs in ethanol with a one-step PLAL method by using a femtosecond laser. It was shown that the colloidal spherical-shaped GaN-NCs were successfully produced and their size distribution was shown as 4.2 ± 1.9 nm, which is smaller than the doubled exciton Bohr radius. X-ray diffraction (XRD) data confirmed the hexagonal wurtzite crystalline structure of GaN-NCs (**Figure 5**), which was further determined with selected area electron diffraction (SAED) pattern. Further X-ray photoelectron spectroscopy (XPS) characterization showed the GaN-NC production. Furthermore, PLAL method was repeated with a nanosecond laser, which resulted in the production of amorphous GaN-NPs with a 12.4 ± 7.0 nm in ethanol. The size distribution decreased to 6.4 ± 2.3 nm when nanoparticles were produced in PVP polymer, yet GaN-NPs were amorphous in structure. GaN-NCs produced by femtosecond PLAL exhibited a 1-eV shift in UV/Vis spectrum and a blue-shifted peak at 4.06 eV due to the quantum confinement effect when compared to bulk GaN. Thus, these significant optical absorption and PLAL properties of GaN-NCs produced by PLAL method imply that these nanocrystals might be powerful candidates for optical and biomedical devices such as biosensors and photodetectors [46].

2.2.4. InN nanocrystals

Besides GaN, with high mobility and high saturation velocity due to its low effective mass, InN materials represent potential to be used in high speed and high frequency electronic devices [47]. Indium nitride-based light-emitting diodes show narrow band-gap values (0.7–0.9 eV) at spectral wavelengths [48]. Moreover, InN-NCs might be used in biology and medicine due to their biocompatible nature and infrared emission properties [49]. There are a number of methods

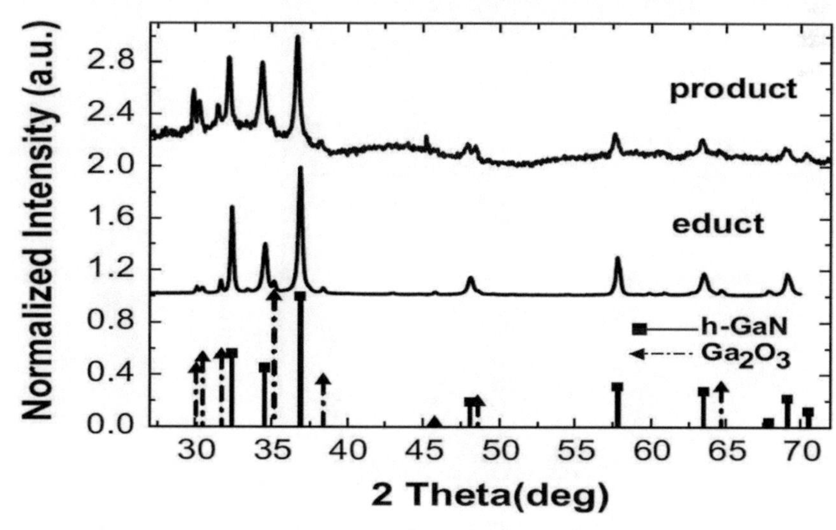

Figure 5. Result of the XRD analysis of GaN-NPs (product) produced by PLAL with a femtosecond laser in ethanol and the starting material GaN powder (educt). Square patterns display GaN wurtzite XRD data and triangles show few Ga_2O_3 peaks [46].

described in the literature for the production of InN-NCs. However, compared to other InN-NC synthesis methods, PLAL is a versatile method for the production of colloidal, highly pure and contaminant-free InN-NCs. Our group demonstrated two different InN-NC production studies by using PLAL, in which the starting material was changed.

In the first study, InN-NCs were produced in ethanol with nanosecond PLAL of HPCVD-grown InN thin film on GaN/sapphire template substrate with different pulse energies [50]. Spherical InN-NCs within size ranges of 5.9–25.3, 5.4–34.8, and 3.24–36 nm were successfully produced using laser pulse energy values of 8, 12, and 16 mJ. The average diameter of the produced InN-NCs was 10 nm (**Figure 6a–c**). TEM, Raman spectroscopy and XPS analysis confirmed the production of InN-NCs. The UV/Vis/NIR spectrum analysis indicated the strong absorption edge from NIR region to UV region, which makes InN-NCs suitable candidates for photonics, optoelectronics, and biological applications [49].

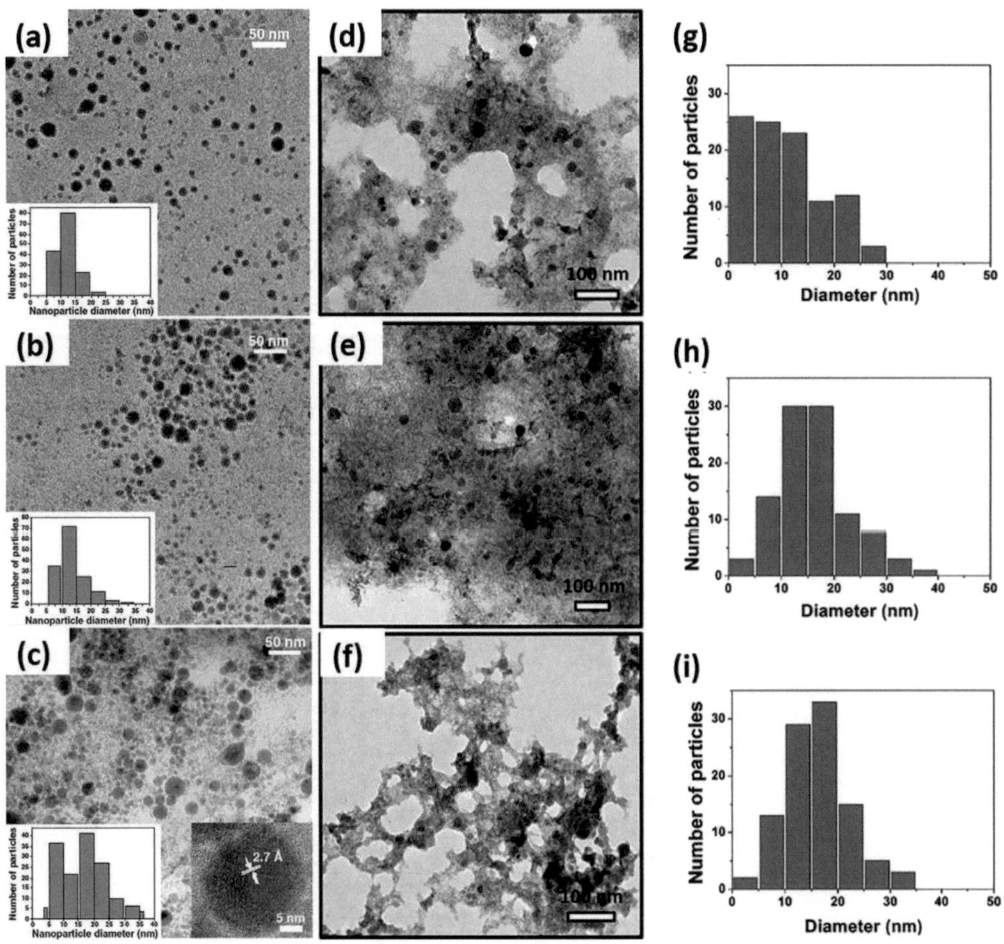

Figure 6. TEM images of InN-NCs produced with PLAL by using laser pulse energies of 8 (a), 12 (b), and 16 mJ (c) with particle size distributions (insets), InN nanocrystal (20 nm in diameter) produced with 16-mJ laser pulse energy and its crystal spacing of 2.7 A° (c, inset) [50]. TEM images and size distribution values of InN-NCs generated with (d and g) 3-, (e and h) 4- and (f and i) 5-mJ laser pulse energies. The average sizes for InN-NC were 10.84 ± 0.71, 16.27 ± 0.68, and 16.06 ± 0.59 nm for 3-, 4-, and 5-mJ laser energies, respectively. Laser energy of 3 mJ led to the smallest nanoparticles and narrower size distribution [51].

In the second study, to obtain ultra-small InN-NCs (<5 nm in diameter), the starting material was determined as InN powder target rather than HPCVD-grown InN thin film. It was shown that lower laser energy and lower ablation duration led to smaller nanoparticles. When ablation time is increased and the smaller nanoparticles are produced, they tended to aggregate. Hexagonal InN-NCs smaller than 5 nm in diameter were successfully produced in ethanol (**Figure 6d–f**) [51].

As a result, it was determined that the starting material, either a bulk target or a suspension, affects the final nanoparticle size in PLAL. In the literature, it was previously demonstrated that the ablation of suspensions leads to smaller nanoparticle dimensions compared to the ablation of bulk target. The particles in the suspension were suggested to absorb all the laser energy since they encounter a number of pulses during the whole ablation process as a result of continuous stirring. By contrast, in case of ablation of the bulk target, laser energy is absorbed only by a specific point on the target leading to larger nanoparticles. Moreover, the nanoparticle size at the end of the PLAL depends on the initial size of the suspended particles [52, 53].

We showed that pure InN-NCs can be successfully produced with PLAL method. Other nitrides in the same group might also be successfully produced with PLAL. Moreover, low decomposition temperatures might be overcome for this kind of material. Therefore, this technique might be successfully applied for the production of nanoparticles from other nitrides in the same group. As an outlook, femtosecond laser ablation might be performed to overcome the low decomposition temperature disadvantage of InN material.

2.2.5. MoS_2 nanoparticles

The different geometries of MoS_2 nanocrystal generation have been demonstrated by PLAL method in different liquids [54]. 3D MoS_2 nanocrystals are generated in water, and 2D MoS_2 nanosheets are generated in methanol as shown in **Figure 7a**. The optical microscope images of the MoS_2 nanosheets are represented in **Figure 7b**. XRD analysis shows that the MoS_2 nanosheets have a hexagonal crystal structure. The E1 2g and A1 g phonon modes are

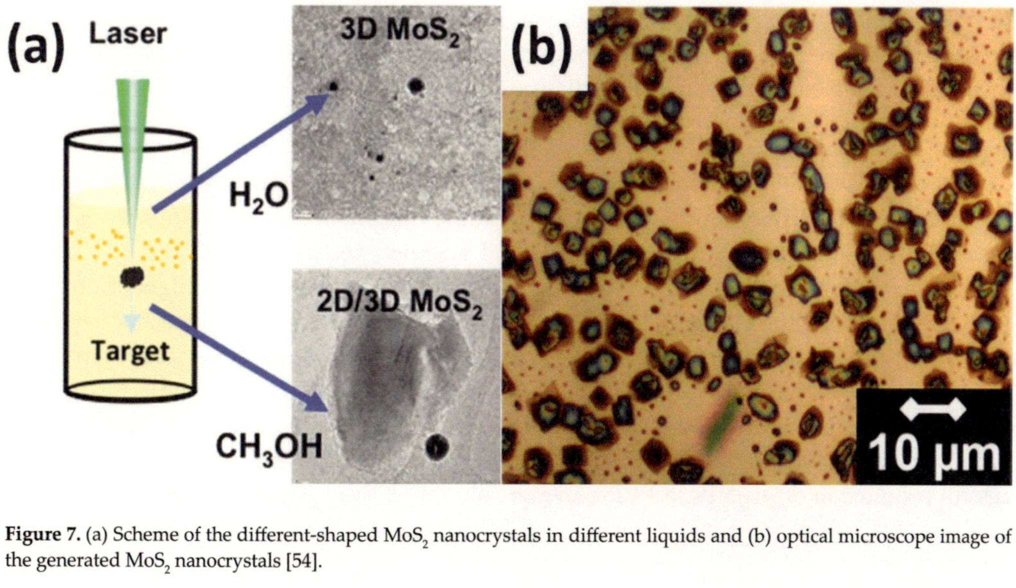

Figure 7. (a) Scheme of the different-shaped MoS_2 nanocrystals in different liquids and (b) optical microscope image of the generated MoS_2 nanocrystals [54].

determined by the peaks at 380.33 and 405.79 cm^{-1}, respectively, by Raman spectroscopy. This behavior of the generated nanocrystals in different liquids is associated with the chemical properties of the liquid where the laser ablation occurs. The ab initio calculation is used to examine the reason for the formation of 2D MoS_2 in methanol and 3D MoS_2 nanocrystals in water which are represented to correlate with vacancy formation.

2.3. Other types of nanoparticles produced by PLAL

2.3.1. Generation of nanoparticles from biological samples

Silica is useful for many plants, and a number of plants such as grasses, rice, sugar beet, and horse-tail contain high levels of biogenic silica [55]. Especially, sugar beet is an important biogenic silica source since the silica content of this plant is mostly found in bagasse, which is produced in significant amounts as a by-product in agricultural industry. On the other hand, silica has a number of application fields, in particular in drugs, cosmetics, printer toners, varnishes and food preservatives [53, 54]. Moreover, silica NPs are used in biomedical and biotechnological fields [56]. Thus, an environmentally friendly process for the production of silica NPs is crucial. A number of chemical methods were described in the literature for silica nanoparticle production [57, 58]. However, these chemical methods have various disadvantages such as being energy intensive, expensive, need for high temperatures, strong acidities and high pressures [59]. PLAL might be suggested as a suitable silica nanoparticle production method due to its versatility and low cost. Moreover, PLAL allows the synthesis of nanoparticles in large quantities. Our group demonstrated the production of silica nanoparticles by using the sugar beet bagasse as the target in PLAL since it contains large amounts of silica inherently. At the same time, chemical silica nanoparticle production method was utilized for the comparison of the end products. The successful silica nanoparticle production by PLAL was demonstrated with several characterization methods such as SEM, TEM, and dynamic light scattering (DLS). Silica nanoparticles with an average size of 74 nm were obtained with PLAL while chemical method resulted in a size range of 531–825 nm (**Figure 8**). Besides, silica nanoparticles produced with the chemical method resulted in a decrease in the growth of model aquatic algae. On the other hand, silica nanoparticles produced with PLAL led to an increase in the growth of these organisms. Due to the purity of silica nanoparticles produced by PLAL, this method might also be accepted as an environmentally friendly process for the nanoparticle production [60].

2.3.2. Nanocomposite generation

Significantly high surface-to-volume ratio of electrospun nanofibers and nanoparticles incorporated into these nanofibers attract substantial attention recently. Functional nanofibrous structures with unique physical, chemical, optical, and catalytical properties might be obtained with this incorporation and they are used in different applications such as biotechnology, sensors, photonics, optoelectronics, energy, and so on [61, 62]. Although there are two different ways to obtain polymeric nanofibers containing NPs [63, 64], both of these approaches include a number of optimized process for every single material and require the utilization of toxic-reducing agents, stabilizers, protecting agents or surfactants. Our group demonstrated the successful production of Au-NPs in PVP polymer solution with PLAL method and PVP/Au-NP

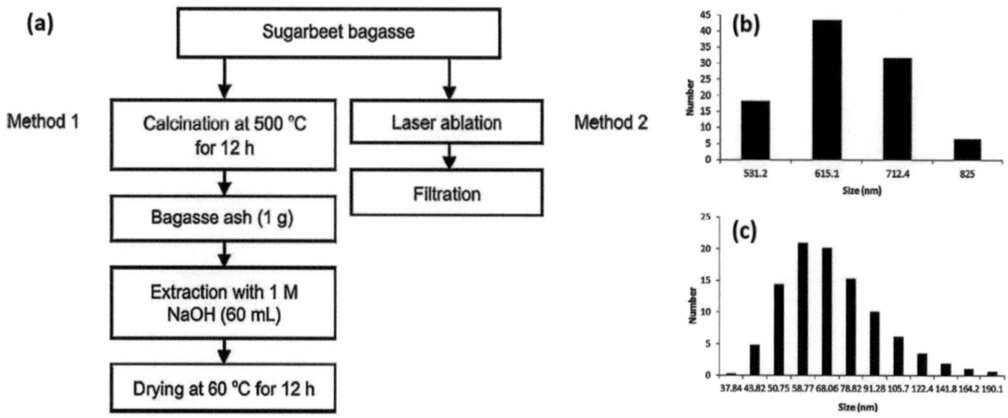

Figure 8. (a) The diagram demonstrating the chemical and PLAL procedures used to produce silica nanoparticles from sugar beet bagasse. The silica nanoparticle size distribution, (b) for the chemically produced nanoparticles, and (c) PLAL [60].

solution was further electrospun to obtain uniform nanofibers. Here, PLAL was a clean, chemically safe, and less time-consuming method for functional nanofibers containing nanoparticles. It can be suggested that other NPs can also be produced by PLAL in various polymer solutions. Thus, functional nanofibrous composites for different applications can be further developed depending on the NP and the polymer type [65].

3. Applications

3.1. Thin film MoS_2 nanocrystal-based ultraviolet photodetector

MoS_2 NCs-based photodetectors work well in the 300–400-nm UV range while omitting the response at the visible range due to MoS_2 NC absorption properties [66]. The UV range response of MoS_2 NCs generated by PLAL photodetectors gives a rise in the field of photonics and optoelectronics due to their structural and optical properties. The absorption of MoS_2 NCs synthesized by PLAL and the photodetectors based on them are very similar and their photoresponsivity has a good agreement. MoS_2 NC-based photodetectors could be more advantageous compared to silicon- and germanium-based photodetectors because of their low oxidation tendency [67–70].

3.2. A plasmonic-enhanced photodetector based on silicon nanocrystals obtained through laser ablation

A photodetector that works at visible range is fabricated and its electrical properties are enhanced by adding Ag-NCs layer which improves the plasmonic effects where Si-NCs are synthesized by PLAL method and Ag NCs are generated by atomic layer deposition (ALD)-annealing technique [71]. The SEM image of the Ag-NCs is represented in **Figure 9a** and the inset shows the size distribution of the Ag-NCs. In **Figure 9b**, the schematic representation

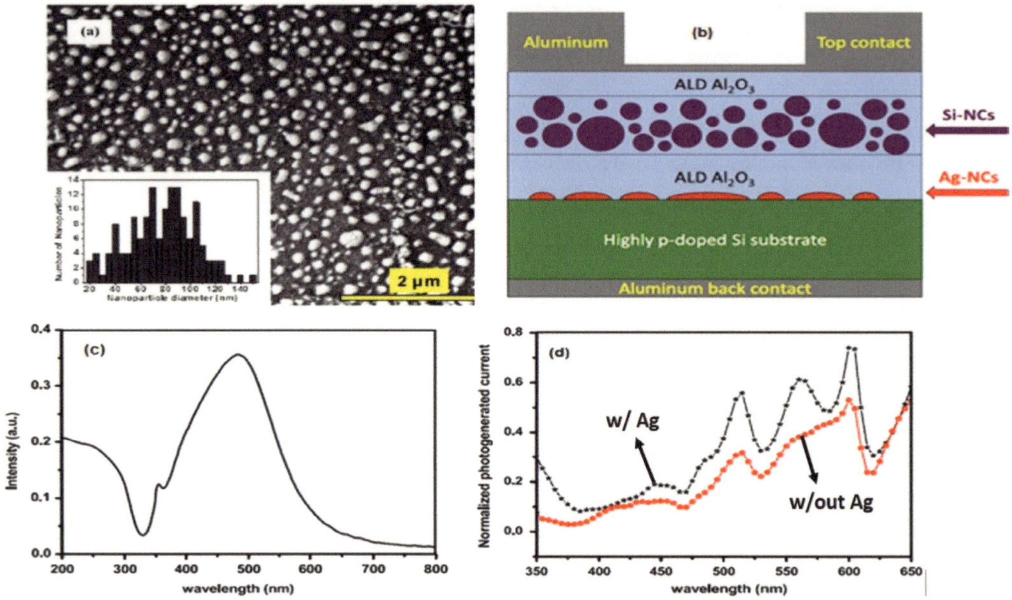

Figure 9. (a) SEM image of Ag-NCs; inset: size distribution of them, (b) schematic representation of the photodetector, (c) UV–vis absorption of Ag-NCs on quartz, (d) photo-generated current versus wavelength spectrum [71].

of the fabricated photodetector is shown where the active films (Si-NC and Ag-NCs) are sandwiched between insulating Al_2O_3 films. The absorption graph of the Ag nano islands is at the visible range and shown in **Figure 9c**. The photo-generated currents are measured for both cases; first one is without the Ag-NC film, and the second case is with the Ag-NC film. In **Figure 1d**, the measure photo-generated current versus wavelength is represented. The visible response is enhanced by adding Ag-NC film due to its plasmonic effects. Si-NC-based photodetector is widely used for visible range and its electronic and optical properties could be improved by Ag-NC film because of its unique plasmonic effects.

3.3. A near-infrared range photodetector based on indium nitride nanocrystals obtained through laser ablation

Plasmonic properties of nanomaterials enable the optical tuning of near-infrared range plasmonic devices and they are frequently studied by many areas such as nanophotonics and telecommunication [72]. The large portion of solar spectrum consists of NIR range and this increases the importance of plasmonic-enhanced devices that work at this range [73]. Indium nitride nanocrystals (InN-NCs) have a NIR plasmonic resonance property [74, 75]. They also have a low band-gap energy, which is in the range of 0.7–0.9 eV, and they have unique electronic features such as high electron mobility, which allows fabricating high speed and high frequency electronic devices [76]. A near-infrared ranged photodetector has been demonstrated by using InN-NCs [77]. InN-NCs are generated by PLAL method from chemical vapor-deposited thin film. When 1-V bias is applied to the device, 600–1100-nm wavelength range photosensitivity is recorded. In **Figure 10a**, TEM images of InN-NCs synthesized by PLAL method are represented. In **Figure 10b**, 3D scheme of photodetector is shown, and in **Figure 10c**, the photoresponsivity versus wavelength graph is shown.

3.4. Enhancement in c-Si solar cells using 16-nm InN nanoparticles

To increase the performance of Si solar cells, InN-NCs are used. Higher performance depends on the low reflection that causes light scattering which allows enhancement between 400 and 620 nm [78]. This study was related with improving the anti-reflective coating (ARC) layers such as texturing for light trapping and layer designing [79–83]. Due to InN-NC coating, 9.67 and 16.42% improvements are obtained for short circuit current density and efficiency, respectively.

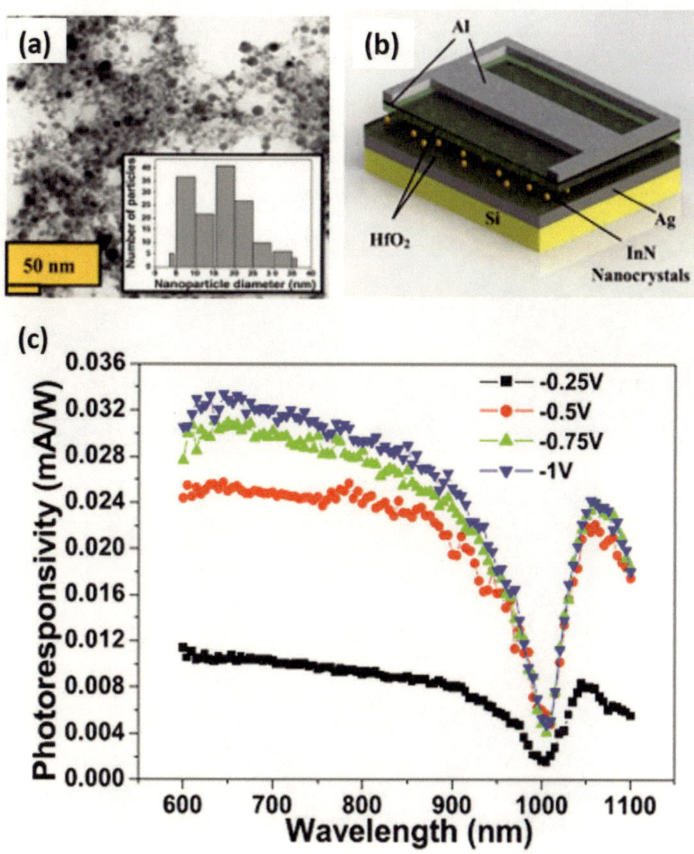

Figure 10. (a) TEM image of InN-NCs, (b) schematic of photodetector, and (c) photosensitivity versus wavelength graph [77].

3.5. Enhanced memory effect via quantum confinement in 16-nm InN nanoparticles embedded in ZnO charge-trapping layer

The fabrication of charge-trapping memory cells with InN-NCs synthesized by PLAL method has been demonstrated with ZnO charge-trapping layer [84]. The observed memory hysteresis is very low without the InN-NCs. After InN-NCs enhancement, a 5-V memory window is achieved.

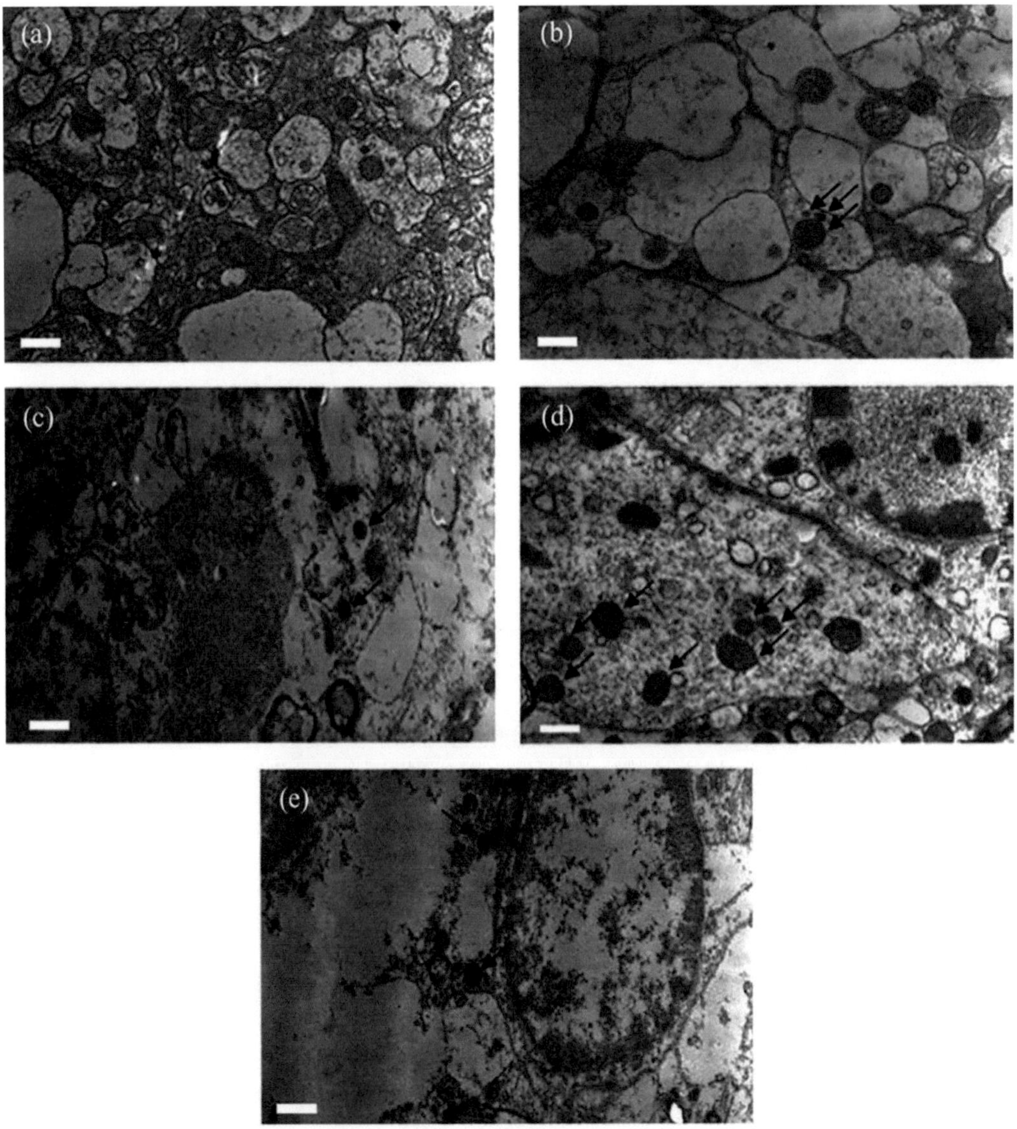

Figure 11. TEM images demonstrate the Ag-NP distribution in isolated rat hippocampal slice incubated with Ag-NPs, (a) control; (b) and (c) Ag-NPs in the extracellular region; (d) Ag-NPs in the intracellular region, localized in the cytoplasm; (e) phagocytosis of Ag-NPs. Arrows show the Ag-NPs. Scale bar for each figure is 500 nm [21].

3.6. Biological applications

Pure uncoated Ag-NPs were used for the evaluation of Ag-NP toxicity in an aquatic macrophyte and in rat hippocampus cells. It was shown that Ag-NPs led to a dose-dependent toxicity in lower nanoparticle concentrations, which implies that accidental contamination of water sources with Ag-NP at low doses might result in significant ecological toxicity [19]. In the other toxicity study, the purity of Ag-NPs produced by PLAL method led to the understanding of the importance of surface chemistry and the size for the uptake mechanism of Ag-NPs into isolated rat hippocampal cells. It was demonstrated that phagocytosis was endocytotic pathway type for the entry of larger Ag-NPs into the hippocampal neurons (**Figure 11**) [21].

4. Conclusion

Nanoparticle production is a crucial part of the nanoscience which leads to several applications in physics, electronics, medicine, biology and industry. As the scale goes down to nano, materials exhibit unique properties compared to their macroscale bulk counterparts. To reveal such properties, reaching nanoscale material production has become an issue. PLAL is a fast and cheap method to obtain pure nanomaterials in organic solutions. Also, it is easy to apply to different materials since it does not require high temperature and high-pressure conditions. The presented results demonstrate both nanomaterial generations and their applications are successfully achieved with various materials. To sum up, PLAL is a powerful nanomaterial generation method that is already demonstrated with different types of materials and applications successfully. Additionally, PLAL is a promising technique to become mass nanomaterial production with its unique properties.

Acknowledgements

The authors thank the Turkish Academy of Science—Outstanding Young Scientist Award Program (TUBA-GEBIP).

Author details

Bülend Ortaç*, Elif Uzcengiz Şimşek and Canan Kurşungöz

*Address all correspondence to: ortac@unam.bilkent.edu.tr

National Nanotechnology Research Center (UNAM), Institute of Materials Science and Nanotechnology, Bilkent University, Ankara, Turkey

References

[1] Oberdörster G, Oberdörster E, Oberdörster J. Review nanotoxicology: An emerging discipline evolving from studies of ultrafine particles. Environmental Health Perspectives. 2005;**113**(7):823-839

[2] Baeuerle D. Laser Processing and Chemistry. Linz, Austria: Springer Science & Business Media; 2000

[3] Ghofraniha N, Conti C, Ruocco G, Trillo S. Shocks in nonlocal media. Physical Review Letters. 2007;**99**(4):43903

[4] Staudt W, Borneis S, Pippert K-D. TFT annealing with excimer laser. Technology and market outlook. Physica Status Solidi. 1998;**166**(2):743-749

[5] Chrisey, Douglas B, Hubler GK. Pulsed Laser Deposition of Thin Films. New York: Wiley; 1994. p. 55

[6] Voloshko A, Itina TE. Nanoparticle formation by laser ablation and by spark discharges-properties, mechanisms, and control possibilities. In: Nanoparticles Technology. InTech; 2015

[7] Xu X, Duan G, Li Y, Liu G, Wang J, Zhang H, Dai Z, Cai W. Fabrication of gold nanoparticles by laser ablation in liquid and their application for simultaneous electrochemical detection of Cd2+, Pb2+, Cu2+, Hg2+. ACS Applied Materials & Interfaces. 2013;**6**(1):65-71

[8] Correard F, Maximova K, Estève MA, Villard C, Roy M, Al-Kattan A, Sentis M, Gingras M, Kabashin AV, Braguer D. Gold nanoparticles prepared by laser ablation in aqueous biocompatible solutions: Assessment of safety and biological identity for nanomedicine applications. International Journal of Nanomedicine. 2014;**9**:5415

[9] Burzhuev S, Dâna A, Ortac B. Laser synthesized gold nanoparticles for high sensitive strain gauges. Sensors and Actuators A: Physical. 2013;**203**:131-136

[10] Zamborini FP, Leopold MC, Hicks JF, Kulesza PJ, Malik MA, Murray RW, et al. Electron hopping conductivity and vapor sensing properties of flexible network polymer films of metal nanoparticles. Journal of the American Chemical Society. 2002;**124**(30):8958-8964

[11] Herrmann J, Müller KH, Reda T, Baxter GR, Raguse BD, De Groot GJJB, Chai R, Roberts M, Wieczorek L. Nanoparticle films as sensitive strain gauges. Applied Physics Letters. 2007;**91**(18):183105

[12] Wiesner MR, Lowry GV, Alvarez P, Dionysiou D, Biswas P. Assessing the risks of manufactured nanomaterials. Environmental Science & Technology. 2006;4336-4345

[13] Navarro E, Piccapietra F, Wagner B, Marconi F, Kaegi R, Odzak N, et al. Toxicity of silver nanoparticles to *Chlamydomonas reinhardtii*. Environmental Science & Technology. 2008;**42**(23):8959-8964

[14] Chen X, Schluesener HJ. Nanosilver: A nanoproduct in medical application. Toxicology Letters. 2008;**176**(1):1-12

[15] Evanoff DD, Chumanov G. Synthesis and optical properties of silver nanoparticles and arrays. ChemPhysChem. 2005;**6**(7):1221-1231

[16] Pingali KC, Rockstraw DA, Deng S, Pingali KC, Rockstraw DA, Deng S. Aerosol science and technology silver nanoparticles from ultrasonic spray pyrolysis of aqueous silver nitrate. Aerosol Science and Technology. 2005;**39**(10):1010-1014

[17] Barcikowski S, Hahn A, Walter J. Health risks of nanoparticulate emissions during femtosecond and picosecond pulsed laser machining. In: Proc SPIE 7201, Laser Applications in Microelectronic and Optoelectronic Manufacturing VII. 2009. p. 720109

[18] Akman E, Oztoprak BG, Gunes M, Kacar E, Demir A. Effect of femtosecond Ti : Sapphire laser wavelengths on plasmonic behaviour and size evolution of silver nanoparticles. Photonics and Nanostructures-Fundamentals and Applications. 2011;**9**(3):276-286

[19] Üçüncü E, Özkan AD, Kurşungöz C, Ülger ZE, Ölmez TT, Tekinay T, et al. Effects of laser ablated silver nanoparticles on *Lemna minor*. Chemosphere. 2014;**108**:251-257

[20] Bruchez M, Moronne M, Gin P, Weiss S, Alivisatos AP. Semiconductor nanocrystals as fluorescent biological labels. Science (80-). 1998;**281**(5385):2013-2016

[21] Kursungoz C, Taş ST, Sargon MF, Sara Y, Ortaç B. Toxicity of internalized laser generated pure silver nanoparticles to the isolated rat hippocampus cells. Toxicology & Industrial Health. 2017;1-9

[22] Torchynska TV, Vasilievitch YV. Nanocrystals and Quantum Dots of Group IV Semiconductors. American Scientific Publ.; 2010

[23] Trindade T, Brien PO, Pickett NL. Nanocrystalline semiconductors: synthesis, properties, and perspectives. Chemistry of Materials. 2001;**13**(11):3843-3858

[24] Hu MS, Wang WM, Chen TT, Hong LS, Chen CW, Chen CC, et al. Sharp infrared emission from single-crystalline indium nitride nanobelts prepared using guided-stream thermal chemical vapor deposition. Advanced Functional Materials. 2006;**16**(4):537-541

[25] Lambert TN, Andrews NL, Gerung H, Boyle TJ, Oliver JM, Wilson BS, et al. Water-soluble germanium(0) nanocrystals: Cell recognition and near-infrared photothermal conversion properties. Small. 2007;**3**(4):691-699

[26] Prabakar S, Shiohara A, Hanada S, Fujioka K, Yamamoto K, Tilley RD. Size controlled synthesis of germanium nanocrystals by hydride reducing agents and their biological applications. Chemistry of Materials. 2009;**22**(2):482-486

[27] Cheng K, Anthony R, Kortshagen UR, Holmes RJ. High-efficiency silicon nanocrystal light-emitting devices. Nano Letters. 2011

[28] Henderson EJ, Seino M, Puzzo DP, Ozin GA. Colloidally stable germanium nanocrystals for photonic applications. ACS Nano. 2010;**4**(12):7683-7691

[29] Danek M, Jensen KF, Murray CB, Bawendi MG. Synthesis of luminescent thin-film CdSe/ZnSe quantum dot composites using CdSe quantum dots passivated with an overlayer of ZnSe. Chemistry of Materials. 1996;**8**(3):173-180

[30] Holman ZC, Liu CY, Kortshagen UR. Germanium and silicon nanocrystal thin-film field-effect transistors from solution. Nano Letters. 2010;**10**(7):2661-2666

[31] Guo Z, Zhang D, Wei S, Wang Z, Karki AB, Li Y, et al. Effects of iron oxide nanoparticles on polyvinyl alcohol: Interfacial layer and bulk nanocomposites thin film. Journal of Nanoparticle Research. 2010;**12**(7):2415-2426

[32] Wender H, Andreazza ML, Correia RRB, Teixeira SR, Dupont J, Schmid G, et al. Synthesis of gold nanoparticles by laser ablation of an Au foil inside and outside ionic liquids. Nanoscale. 2011;**3**(3):1240

[33] Ortaç B, Kayaci F, Vural HA, Deniz AE, Uyar T. Photoluminescent electrospun polymeric nanofibers incorporating germanium nanocrystals. Reactive and Functional Polymers. 2013;**73**(9):1262-1267

[34] Švrček V, Sasaki T, Shimizu Y, Koshizaki N. Blue luminescent silicon nanocrystals prepared by ns pulsed laser ablation in water. Applied Physics Letters. 2006;**89**(21)

[35] Yang S, Cai W, Zeng H, Li Z. Polycrystalline Si nanoparticles and their strong aging enhancement of blue photoluminescence. Journal of Applied Physics. 2008;**104**(2)

[36] Abderrafi K, Raul GC, Gongalsky MB, Suarez I, Abarques R, Chirvony VS, et al. Silicon nanocrystals produced by nanosecond laser ablation in an organic liquid. Journal of Physical Chemistry C. 2011;**115**:5147-5151

[37] Alkis S, Okyay AK, Ortaç B. Post-treatment of silicon nanocrystals produced by ultra-short pulsed laser ablation in liquid: Toward blue luminescent nanocrystal generation. Journal of Physical Chemistry C. 2012;**116**(5):3432-3436

[38] Akasaki I, Amano H. Crystal growth and conductivity control of group III nitride application to short wavelength light emitters semiconductors and their. Japanese Journal of Applied Physics. 1997;**36**(9R):5393

[39] Kazlauskas K, Tamulaitis G, Žukauskas A, Khan MA, Yang JW, Zhang J, et al. Exciton and carrier motion in quaternary AlInGaN. Applied Physics Letters. 2003;**82**(25):4501-4503

[40] Son KA, Liao A, Lung G, Gallegos M, Hatake T, Harris RD, Scheick LZ, Smythe W. GaN-based high temperature and radiation-hard electronics for harsh environments. Nanoscience and Nanotechnology Letters. 2010;**2**(2):89-95

[41] Borsella E, Garcia MA, Mattei G, Maurizio C, Mazzoldi P, Cattaruzza E, et al. Synthesis of GaN quantum dots by ion implantation in dielectrics. Journal of Applied Physics. 2001;**90**(9):4467-4473

[42] Miyamura M, Tachibana K, Arakawa Y. High-density and size-controlled GaN self-assembled quantum dots grown by metalorganic chemical vapor deposition. Applied Physics Letters. 2002;**80**(21):3937-3939

[43] Gyger F, Bockstaller P, Gröger H, Gerthsen D, Feldmann C. Quantum-confined GaN nanoparticles synthesized via liquid-ammonia-in-oil-microemulsions. Chemical Communications. 2014;**50**(22):2939

[44] Petersen S, Barcikowski S. In situ bioconjugation: Single step approach to tailored nanoparticle-bioconjugates by ultrashort pulsed laser ablation. Advanced Functional Materials. 2009;**19**(8):1167-1172

[45] Wu H, Yang R, Song B, Han Q, Li J, Zhang Y, et al. Biocompatible inorganic fullerene-like molybdenum disulfide nanoparticles produced by pulsed laser ablation in water. ACS Nano. 2011;**5**(2):1276-1281

[46] Demirel A, Öztaş T, Kurşungöz C, Yılmaz İ, Ortaç B. Synthesis of blue-shifted luminescent colloidal GaN nanocrystals through femtosecond pulsed laser ablation in organic solution. Journal of Nanoparticle Research. 2016;**18**(5):1-10

[47] Mohammad S, Noor HM. Progress and prospects of group-III nitride semiconductors. Progress in Quantum Electronics. 1996;**20**(5-6):361-525

[48] Wu C, Li T, Lei L, Hu S, Liu Y, Xie Y, et al. Indium nitride from indium iodide at low temperatures: Synthesis and their optical properties. New Journal of Chemistry. 2005;**29**:1610-1615

[49] Chen Z, Li Y, Cao C, Zhao S, Fathololoumi S, Mi Z, et al. Large-scale cubic InN nanocrystals by a combined solution- and vapor-phase method under silica confinement. Journal of the American Chemical Society. 2012;**134**(2):780-783

[50] Alkis S, Alevli M, Burzhuev S, Vural HA, Okyay AK, Ortaç B. Generation of InN nanocrystals in organic solution through laser ablation of high pressure chemical vapor deposition-grown InN thin film. Journal of Nanoparticle Research. 2012;**14**(8)

[51] Kurşungöz C, Şimşek EU, Tuzaklı R, Ortaç B. Generation of ultra-small InN nanocrystals by pulsed laser ablation of suspension in organic solution. Applied Physics A. 2017;**123**(3):209

[52] Kim D, Jang D. Synthesis of nanoparticles and suspensions by pulsed laser ablation of microparticles in liquid. Applied Surface Science. 2007;**253**(19):8045-8049

[53] Zeng H, Du XW, Singh SC, Kulinich SA, Yang S, He J, et al. Nanomaterials via laser ablation/irradiation in liquid: A review. Advanced Functional Materials. 2012;**22**(7):1333-1353

[54] Oztas T, Sen HS, Durgun E, Ortaç B. Synthesis of colloidal 2D/3D MoS2 nanostructures by pulsed laser ablation in organic liquid environment. Journal of Physical Chemistry C. 2014;118:30120

[55] Sun L, Carbide S, Sic P. Silicon-based materials from rice husks and their applications. Industrial & Engineering Chemistry Research. 2001;**40**(25):5861-5877

[56] Clément L, Zenerino A, Hurel C, Amigoni S, Taffin de Givenchy E, Guittard F, et al. Toxicity assessment of silica nanoparticles, functionalised silica nanoparticles, and HASE-grafted silica nanoparticles. Science of the Total Environment. 2013;**450-451**:120-128

[57] Cadby AJ, Tolbert SH. Controlling optical properties and interchain interactions in semiconducting polymers by encapsulation in periodic nanoporous silicas with different pore sizes. Journal of Physical Chemistry B. 2005;**109**(38):17879-17886

[58] Molenkamp WC, Watanabe M, Miyata H, Tolbert SH. Highly polarized luminescence from optical quality films of a semiconducting polymer aligned within oriented mesoporous silica. Journal of the American Chemical Society. 2004;**126**(14):4476-4477

[59] Bansal V, Ahmad A, Sastry M. Fungus-mediated biotransformation of amorphous silica in rice husk to nanocrystalline silica. Journal of the American Chemical Society. 2006;**128**(43):14059-14066

[60] San NO, Kurşungöz C, Tümtaş Y, Yaşa Ö, Ortaç B, Tekinay T. Novel one-step synthesis of silica nanoparticles from sugarbeet bagasse by laser ablation and their effects on the growth of freshwater algae culture. Particuology. 2014;**17**:29-35

[61] Ramakrishna S, Fujihara K, Teo WE, Yong T, Ma Z, Ramaseshan R. Electrospun nanofibers: Solving global issues. Mater Today. 2006;**9**(3):40-50

[62] Teo WE, Ramakrishna S. Electrospun nanofibers as a platform for multifunctional, hierarchically organized nanocomposite. Vol. 69, Composites Science and Technology. Elsevier Ltd; 2009. pp. 1804-1817

[63] Wang Y, Yang Q, Shan G, Wang C, Du J, Wang S, et al. Preparation of silver nanoparticles dispersed in polyacrylonitrile nanofiber film spun by electrospinning. Materials Letters. 2005;**59**(24-25):3046-3049

[64] He D, Hu B, Yao QF, Wang K, Yu SH. Large-scale synthesis of flexible free-standing SERS substrates with high sensitivity: electrospun PVA nanofibers embedded with controlled alignment of silver nanoparticles. ACS Nano. 2009;**3**(12):3993-4002

[65] Deniz AE, Vural HA, Ortaç B, Uyar T. Gold nanoparticle/polymer nanofibrous composites by laser ablation and electrospinning. Materials Letters. 2011;**65**(19-20):2941-2943

[66] Alkis S, Öztaş T, Aygün LE, Bozkurt F, Okyay AK, Ortaç B. Thin film MoS2 nanocrystal based ultraviolet photodetector. Optics Express. 2012;**20**(19):21815-21820

[67] Li Q, Newberg JT, Walter EC, Hemminger JC, Penner RM. Polycrystalline molybdenum disulfide (2H-MoS2) nano-and microribbons by electrochemical/chemical synthesis. Nano Letters. 2004;**4**(2):277-281

[68] Shanmugam M, Bansal T, Durcan CA, Yu B. Molybdenum disulphide/titanium dioxide nanocomposite-poly 3-hexylthiophene bulk heterojunction solar cell. Applied Physics Letters. 2012;**100**(15):153901

[69] Yoon Y, Ganapathi K, Salahuddin S. How good can monolayer MoS2 transistors be? Nano Letters. 2011;**11**(9):3768-3773

[70] Ross S, Sussman A. Surface oxidation of molybdenum disulfide. Journal of Physical Chemistry. 1955;**59**(9):889-892

[71] Alkis S, Oruç FB, Ortaç B, Koşger AC, Okyay AK. A plasmonic enhanced photodetector based on silicon nanocrystals obtained through laser ablation. Journal of Optics. 2012;**14**(12):125001

[72] Palomaki PKB, Miller EM, Neale NR. Control of plasmonic and interband transitions in colloidal indium nitride nanocrystals. Journal of the American Chemical Society. 2013;**135**(38):14142-14150

[73] Paz-Soldan D, Lee A, Thon SM, Adachi MM, Dong H, Maraghechi P, et al. Jointly tuned plasmonic-excitonic photovoltaics using nanoshells. Nano Letters. 2013;**13**(4):1502-1508

[74] Sardar K, Deepak FL, Govindaraj A, Seikh MM, Rao CNR. InN nanocrystals, nanowires, and nanotubes. Small. 2005;**1**(1):91-94

[75] Hsieh JC, Yun DS, Belcher AM. Ambient pressure, low-temperature synthesis and characterization of colloidal InN nanocrystals. Journal of Materials Chemistry. 2010; **20**(8):1435-1437

[76] Mohammad SN, Morkoç H. Progress and prospects of group-III nitride semiconductors. Progress in Quantum Electronics. 1996;**20**(5-6):361-525

[77] Tekcan B, Alkis S, Alevli M, Dietz N, Ortac B, Biyikli N, et al. A near-infrared range photodetector based on indium nitride nanocrystals obtained through laser ablation. IEEE Electron Device Letters. 2014;**35**(9):936-938

[78] Chowdhury FI, Alnuaimi A, Alkis S, Ortaç B, Aktürk S, Alevli M, et al. Enhancement in c-Si solar cells using 16nm InN nanoparticles. Materials Research Express. 2016;56202

[79] Wang Y, Zhang X, Bai L, Huang Q WZY. Effective light trapping in thin film silicon solar cells from textured Al doped ZnO substrates with broad surface feature distributions. Applied Physics Letters. 2012;263508

[80] Tark SJ, Kang MG, Park S, Jang JH, Lee JC, Kim WM, Lee JS, Kim D. Development of surface-textured hydrogenated ZnO:Al thin-films for μc-Si solar cells. Current Applied Physics. 2009;**9**:1318-1322

[81] Anna Selvan JA, Delahoy AE, Guo S, Li Y-M. A new light trapping TCO for nc-Si:H solar cells. Solar Energy Materials and Solar Cells. 2006;**90**:3371-3376

[82] Meiners B-M, Holinski S, Schäfer P, Hohage S, Borchert D. Investigation of anti-reflection coating stacks for silicon heterojunction solar cells. In: 29th European PV Solar Energy Conf and Exhibition. 2014

[83] Chowdhury FI, Blaine T, Gougam AB. Optical transmission enhancement of fluorine doped tin oxide (FTO) on glass for thin film photovoltaic applications. Energy Procedia. 2013;**42**:660-669

[84] Nazek El-Atab AN, Cimen F, Alkis S, Ortaç B, Alevli M, Dietz N, Okyay AK. Enhanced memory effect via quantum confinement in 16 nm InN nanoparticles embedded in ZnO charge trapping layer. Applied Physics Letters. 2014;**104**:253106

Section 3

Applications of Laser Ablation

Chapter 8

Matrix-Assisted Pulsed Laser Evaporation of Organic Thin Films: Applications in Biology and Chemical Sensors

Alexandra Palla Papavlu, Valentina Dinca,
Mihaela Filipescu and Maria Dinescu

Additional information is available at the end of the chapter

http://dx.doi.org/10.5772/intechopen.70676

Abstract

Polymer and biomolecule processing for medical and electronics applications, i.e. the fabrication of sensors and biosensors, microarrays, or lab on chip devices is a cornerstone field which shows great promise. Laser based thin film deposition techniques such as pulsed laser deposition or matrix-assisted pulsed laser evaporation (MAPLE) are competing with conventional methods for integrating new materials with tailored properties for novel technological developments. Successful polymer and protein thin film deposition requires several key elements for depositing viable and functional thin films, i.e. the characteristics of the laser depositing system, the choice of targets and receiver substrates, etc. This chapter reviews the following topics: brief presentation of the MAPLE process including several examples of polymer materials deposited by MAPLE, thus illustrating the potential of the technique as a gentle laser-assisted deposition method. In particular, the "synthesis" of new materials, their analysis and correlation of the bulk and interface properties to its bio-environment shall be discussed as a method to tackle some bioengineering issues. We will also focus on recent breakthroughs of the MAPLE technique for the fabrication of functional devices, i.e. sensor devices based either on chemoresponsive polymers or on proteins.

Keywords: maple, polyethylenimine, polyepichlorohydrin, polyisobutylene, lactoferrin, odorant-binding proteins, saw, hydroxyapatite nanoparticles, ha, polyethylene glycol-co-polycaprolactone methyl ether, PEG-PCL-me

1. Introduction

A little more than 50 years ago, on 16 May 1960, *LASER* light was first generated, at the time being described as *"a solution looking for a problem."* Today, this acronym for *Light Amplification by Stimulated Emission Radiation* has become a common part of our vocabulary. The properties of emitted light make lasers useful devices in many applications where monochromatic, coherent light is required. The use of lasers today is so far reaching and widespread that a comprehensive enumeration is impossible. Lasers cross many areas of manufacturing, such as medicine, surgery, telecommunications, homeland security, lighting, displays, and nanotechnology, just to name a few.

New tools bring new capabilities to research. In materials science, laser techniques revolutionized our understanding of materials by making it possible to design and integrate new materials with tailored properties for novel technology developments. Shortly after the discovery of lasers, researchers began irradiating every possible target material and phase.

Three years after the discovery of laser, Breech and Cross studied the laser vaporization and excitation of atoms from solid surfaces, while in 1965, Smith and Turner deposited the first thin films using a ruby laser. Starting with that moment, laser ablation became more and more popular for the deposition of various materials as thin films.

Pulsed laser deposition (PLD) is a growth technique that has been extensively used over the years, in different configurations. The common principle of these different approaches relies on irradiating a target (solid or liquid) with a pulsed laser source that is powerful enough to produce the ablation of the respective target. As most PLD experiments are carried out in vacuum, in a specialized deposition chamber, the ablated species are ejected from the target in the form of plasma, commonly referred to as "ablation plume." The high kinetic energy of the plume species (up to 100 eV) ensures their fast transport to a collecting substrate, usually positioned few centimeters away from the target. Ultimately, the ablated material gradually accumulates on the substrate in order to form a thin film, whose properties can be tuned by an appropriate control of the various experimental conditions.

Although the PLD technique is simple in concept, the flexibility in tuning the different parameters that affect the properties of the resulting films gives it a high degree of sophistication, which results in many advantages: (i) the laser source is positioned exterior to the deposition chamber, making it easy to operate changes to the ablation geometry and to deposition parameters involving ablation wavelength, energy per pulse, or the use of multiple ablation sources; (ii) most known solids and liquids have ablation thresholds well within the capabilities of present-day lasers dedicated to ablation procedures; (iii) for a given set of experimental parameters, the ablation rate of a target can be determined, which allows for a precise control of the film thickness down to a single pulse; (iv) a selective ablation of the target is achieved, limited to the area irradiated by the laser; (v) stoichiometric transfer between target material and substrate is achievable under appropriate conditions; (vi) the high kinetic energy of the ablated species provides them increased mobility, with respect to non-ablation techniques, thereby decreasing the substrate temperature that is otherwise necessary for the formation of a specific compound; (vii) PLD allows the obtaining of peculiar materials and metastable states that cannot be reproduced by other techniques [1, 2].

The ablation threshold of most target materials is well below that of their decomposition. However, there are several types of polymers, biopolymers, and proteins for which this condition does not hold true. These materials are very sensitive to pulse energies commonly delivered by ablation laser systems, and their direct irradiation, even at low fluence, can result in photochemical or thermal decomposition. Therefore, in order to avoid permanent structural damage, a softer laser ablation technique has been tailored around the specificities of organic and polymeric materials. This novel technique is referred to as "matrix-assisted pulsed laser evaporation" (MAPLE) and has been shown to be a powerful tool for the obtaining of organic thin films [1, 3]. In principle, it is a variation of the PLD concept, but its peculiar target preparation and handling procedures set it apart from a technical point of view. In MAPLE, the material of interest (e.g., a polymer or biomolecule) is diluted or dispersed in an inert solvent (i.e., matrix), and its weight concentration in the resulting solution is typically in the range of 0.1–5%. The solution is homogenized by means of magnetic stirring, and then frozen on a liquid nitrogen-cooled support, thus resulting in a solid target. When the target is ablated by the laser pulses, the solvent evaporates and is pumped out, whereas the material of interest (polymer, biomaterial) acquires sufficient kinetic energy to be transported and collected on the substrate. This approach provides a mechanism for thin-film growth that avoids structural damage of the dissolved material, due to the fact that the energy of the laser pulse is absorbed selectively by the solvent.

Two critical conditions must be fulfilled in order to achieve successful deposition by MAPLE:

1. The matrix should exhibit a strong absorption profile at the laser wavelength, whereas the guest material should show as little absorption as possible.
2. There must be no photochemical-mediated interaction between the solvent and the material of interest.

The most important advantages of MAPLE are (i) ultra-high vacuum is not required (a base pressure of 10^{-5}–10^{-6} mbar is enough) and (ii) this method is very flexible for the selection of the solvents. If we have a material that can be dissolved in a solvent, we can try to apply MAPLE to produce thin films, but we have to consider the two conditions mentioned above.

Previous studies have shown that by a careful choice of the experimental conditions, that is, choice of laser wavelength, pulse length, laser fluence, substrate temperature, and background atmosphere, it is possible to deposit a wide range of organic and biological compounds, without any chemical or structural alterations. Here, we only present a reduced list of such materials: polymers (polyethylene glycol (PEG) [4, 5], poly(D,L-lactide) [6], polyalkylthiophene [7], polyaniline [8]), polymer blends [9], active proteins (lysozyme [10], lactoferrin [11], mussel protein [12]), nanoparticles (TiO_2 and SnO_2 [13, 14]), and polymer-carbon nanotube composites [15, 16].

Moreover, MAPLE has been successfully used in sensor applications. Several groups have successfully deposited different polymers for sensor applications. For example, in [17], Pique et al. prove the suitability of MAPLE for the deposition of chemoselective polymers such as a fluoroalcoholpolysiloxane polymer (SXFA), and [18] demonstrated the deposition of polysiloxane thin films with applications in chemical sensors.

Although in the above-mentioned cases MAPLE has demonstrated its suitability for obtaining organic thin films, the deposition mechanism is still not completely elucidated. In [19, 20], molecular dynamics (MD) studies are performed for molecular systems consisting of polymer molecules dissolved in a volatile molecular matrix. The MD studies revealed a significant influence of the polymer molecules on the ablation process, even at relatively low concentrations in the range of several weight percent. Characteristic features detected on the surface of the MAPLE-deposited thin films such as elongated viscous droplets and "molecular balloons" composed of polymer layers which enclose matrix material were explained by these MD studies.

This chapter reviews the MAPLE process for the deposition of different types of materials (in particular polymers and proteins) as thin films for applications in biology and also their integration in sensor devices. Specifically, the following topics will be discussed: the experimental part in the MAPLE process for the deposition of polymer thin films; several examples of polymer materials deposited by MAPLE, illustrating the potential of this technique as a gentle laser-assisted deposition method. We will also focus on recent breakthroughs of the MAPLE technique for the fabrication of functional devices, that is, sensor devices based either on chemoresponsive polymers or on proteins, and biomedical applications of polymer thin films, that is, multifunctional active and responsive biointerfaces, coatings for orthopedic applications, biodegradable and non-biodegradable coatings as antitumoral systems.

2. MAPLE of polymers and proteins for sensor and biosensor applications

In the following section, we will address couple of recent breakthroughs of the MAPLE technique for the fabrication of functional devices, that is, sensor devices based either on chemoresponsive polymers or on proteins.

The detection of the analyte is based on the interaction between the polymer and target molecules which depends on the formation of a weak hydrogen bonding. The functional polymer group comprises a hydrophilic group, that is, a hydroxyl (OH) unit (or may be an NH_2 unit) and a hydrophobic group (e.g., fluorinated (CF_n) or siloxanes groups -Si-C-O-Si- or -Si-O-Si-C) which repel water due to the unpolar nature. The (OH) unit binds a single oxygen atom within the target analyte molecule, and a weak hydrogen bond is thus formed.

The first example of functional sensors presented in this work is based on the works published in [21–23]. Three polymers have been chosen, that is, polyepichlorohydrin (PECH), polyisobutylene (PIB), and polyethylenimine (PEI) to be used as chemical interactive membranes for the fabrication of surface acoustic wave (SAW) sensors. These polymers have the ability to specifically and selectively identify target gases immediately, and in addition, they can be dissolved in common solvents, that is, PECH in acetone, PIB in toluene, and PEI in ethanol, which makes the fabrication of MAPLE targets rather easy.

In order to carry out the MAPLE experiments, the polymers (three independent experiments) were dissolved in the appropriate solvents at concentrations between 0.1 and 2 wt%, and the as-obtained solutions were flash frozen in liquid nitrogen. The polymer-containing targets were irradiated with a laser beam from a "Surelite II" pulsed Nd:YAG laser (Continuum Company, 266 nm wavelength). When laser light irradiated this target, the solvent evaporated and the polymer material was collected on the Si(100) substrate placed parallel and at 4 cm distance from the frozen target. All substrates were cleaned prior to any deposition by dipping them in different solvents, that is, first in acetone, followed by ethanol and finally in the ultrasonic bath with ultrapure water. The last step was to blow dry them in a nitrogen flow.

The laser fluence was varied between 0.08 and 0.8 J cm^{-2}. The substrates were kept at ambient temperature during the deposition. The number of pulses was varied from 6000 pulses to 54,000, resulting in thin polymer films (of PEI, PIB, and PECH) with variable thicknesses (between 40 and 300 nm). A frozen target during irradiation in a MAPLE experiment is shown in **Figure 1**.

Prior to their use in SAW sensor applications, the polymer films produced by MAPLE have been investigated from the surface morphology and chemical structure point of view. Atomic force microscopy (AFM) has been used to analyze the roughness and thickness of the deposited polymer films. AFM (XE 100 AFM setup from Park) measurements were carried out to analyze the films surface roughness on several different areas and dimensions. The chemical structure has been tested with Fourier transform infrared spectroscopy (FTIR). FTIR is a chemical analysis method that detects the characteristic vibrations of functional groups in a sample. The infrared spectrum of the native molecule was measured and compared with the thin-film spectra. The FTIR measurements were carried out with a Jasco

Figure 1. Picture taken during a typical MAPLE deposition.

FT/IR-6300 type A spectrometer in the range 500–7000 cm^{-1}. All spectra were obtained by accumulating 128 scans and CO_2/H_2O correction.

Investigating the morphology of the deposited polymer (PIB, PECH, and PEI) films, it has been noticed that their quality is in general affected by the laser fluence, while the thickness of the deposited layers is related to the number of pulses. As general comments, it has been seen that there is an optimum range of laser fluences for which the surface of the thin polymer films deposited by MAPLE is uniform, with a low density of droplets and cracks. For example, the PIB thin films deposited by MAPLE on silicon substrates were generally rough, the best depositions were achieved for fluences between 0.1 and 0.3 J/cm^2, with 0.08 J/cm^2 being the threshold laser fluence. For SAW sensor applications, polymer layers with low roughness are required, due to the fact that SAW scattering and diffraction can be minimized with a less rough active surface. This was achieved by using targets with polymer concentrations of 1 wt% and laser fluences of 0.1 J/cm^2 (in the case of all three polymers) for evaporation. 2D topographical AFM images of the three polymers deposited by MAPLE from targets containing 1 wt% polymer in the solution are shown in **Figure 2**. The laser fluence applied for the deposition was 0.1 J/cm^2.

Further on, FTIR analysis was applied to investigate the chemical structure of polymer films. It has been found that for the laser fluences applied, where the polymer films are uniform and exhibit the lowest roughness (i.e., 0.1 J/cm^2), the important infrared active bands are very well reproduced (see **Figure 3** and **Figure 4** as examples).

In conclusion, MAPLE was successfully used to grow uniform and continuous polymer (PEI, PIB, and PECH) thin films which maintain their chemical structure similar to that in bulk. In order to prove their feasibility in SAW sensors, the polymer films were deposited onto SAW sensors and tested to evaluate the performances with respect to sensitivity, resolution, and response time. The responses of the PEI-, PIB-, and PECH-coated devices to different concentrations of dimethyl methylphosphonate (DMMP) were evaluated. The frequency shift of the bare and PIB-coated SAW sensor prior to being tested for DMMP analyte is shown in **Figure 5** (left). The response curve of the SAW sensor coated with PECH, PIB, and PEI polymer exposed to different concentrations of DMMP vapor in N_2 is shown in **Figure 5**.

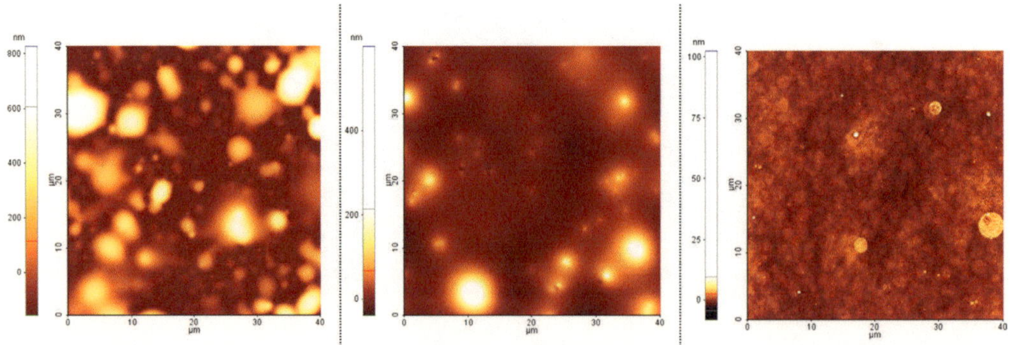

Figure 2. AFM images on a 40 × 40 μm areas of (left)PIB, (middle) PECH, and (right) PEI polymers.

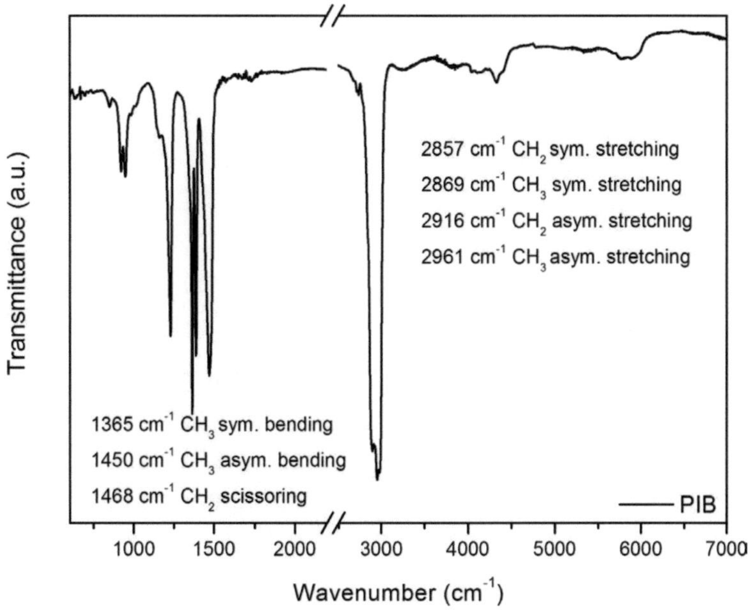

Figure 3. FTIR spectra of the PIB polymer layer deposited at a laser fluence of 0.1 J/cm² (for band assignment, see [24]).

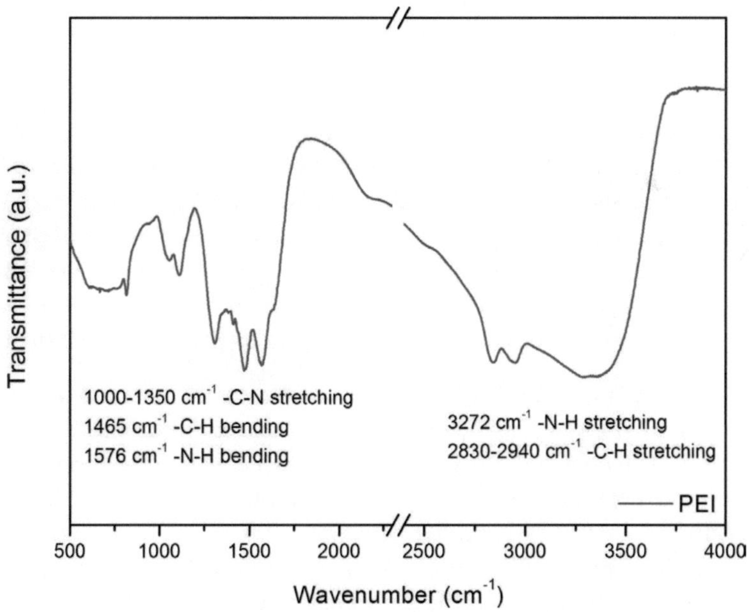

Figure 4. FTIR spectra of the PEI polymer layer deposited at a laser fluence of 0.1 J/cm².

The tests carried out with the SAW sensors showed good performances of the sensors fabricated by MAPLE, in particular the response curve behaviors of the three sensors demonstrated that PIB polymer has a higher sensitivity to DMMP vapor in comparison to PEI and PECH polymers. Therefore, a higher resolution was obtained by using PIB coating.

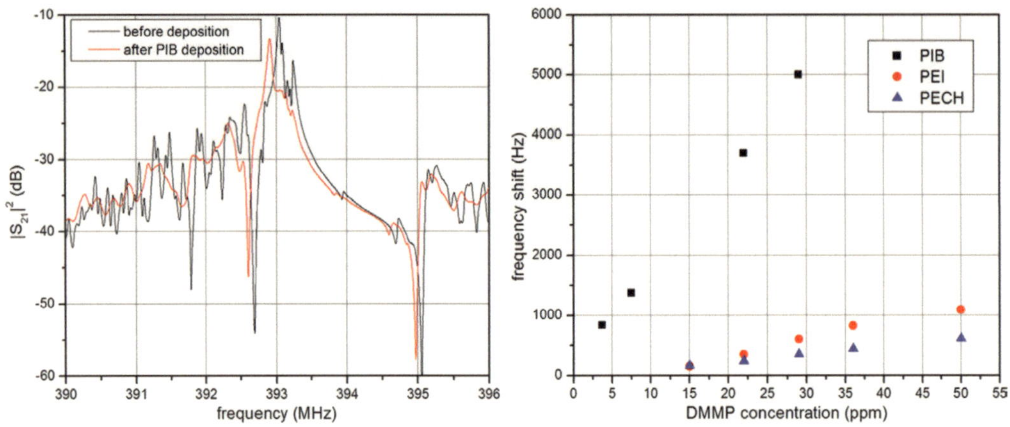

Figure 5. (left) frequency responses of an uncoated SAW sensor and a PIB-coated SAW sensor. (right) PIB-, PEI-, and PECH-coated SAW sensor responses to various concentrations of DMMP (adopted from [21]).

In addition to polymers, proteins and other biomolecules are of great interest in the development of novel sensors and biosensors, and microarray chip devices. Proteins, and in particular odorant-binding proteins (OBPs), show tremendous perspective for integration in devices aimed at the detection of contaminants in food, essential to avoid risks for humans. OBPs are small extracellular proteins which belong to the lipocalin super-family [25, 26]. They have an important role in odor detection by carrying, deactivating, and/or selecting the odorant molecules [27]. In recent studies [28], the possibility of depositing OBPs thin films through MAPLE onto the active area of a SAW device for the development of a biosensor was presented. The sensing system proposed by the authors exploited the high sensitivity and fast response time typical of SAW-based sensors in combination with the adaptable selectivity of the OBPs [29]. The biosensor fabricated in [28] was based on SAW resonators coated through MAPLE with wild-type OBP from bovine (wtbOBP), characterized by different binding specificity, plus an uncoated SAW device used as reference. To demonstrate the functionality of the biosensor, the SAW devices coated by MAPLE were exposed to different concentrations of octanol and carvone, two odorant compounds largely used in the food industry.

The typical coatings of morphology together with the frequency response of the SAW device before and after the MAPLE deposition of wtbOBP (320 mJ/cm^2 and 46,000 pulses) are shown in **Figure 6a** and **b**. The obtained sensitivities are proportional to the surface density of the wtbOBP coating and, hence, correlated to the laser parameters. Based on the known molecular weight of wtbOBP (37,000 Da) and the frequency shift of 502 kHz (**Figure 6b**) (adapted from [28]), the obtained surface density was found to be 25.86 × 10^{-6} kg/m^2, corresponding to 42 × 104 molecules/μm^2. In particular, the SAW biosensor showed a higher sensitivity to the carvone odorant (**Figure 6c**, adapted from [28]).

These results demonstrate that MAPLE is a powerful technique to fabricate biosensors with ultimate applications in assessment of food contamination by molds or for the evaluation of indoor air quality in buildings.

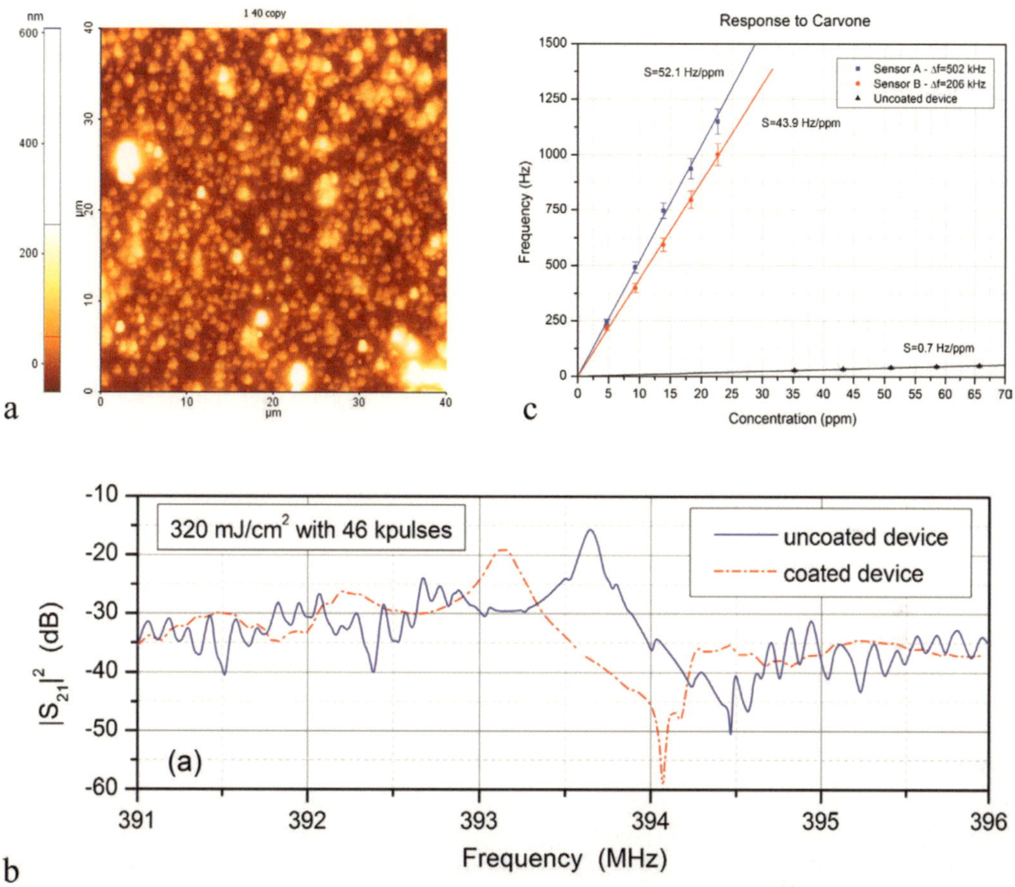

Figure 6. (a) AFM images of wtOBP deposited at 320 mJ/cm² with 46,000 pulses on SAW resonators; (b) the frequency response (amplitude of S_{21}) of SAW devices before and after wtOBP deposition; (c) the response curves for wtOBP-based SAW biosensors and for an uncoated device upon exposure to different concentrations of carvone.

3. Multifunctional active and responsive biointerfaces

Most of the applications related to the biomedical field imply the use of well-controlled biocompatible biointerfaces for medical implants, drug delivery, and lab-on-a-chip devices. Although bulk characteristics of materials are important, due to the direct contact and interaction with the bioenvironment, the surface characteristics, that is, its physical and chemical modification, are directly responsible for its biological outcome and response (e.g., cells ingrowth, protein adsorption, etc.). Therefore, the surface engineering and modifications of the materials used for such applications must take into consideration specific characteristics such as cytocompatibility, mechanical and chemical properties, adhesion with the substrate, controlled morphology and roughness, the feasibility of working with various materials using the same technique, behavior under physiological medium for short and long periods of time, and corrosion.

The ability to deposit or transfer both small and complex, large molecular-mass organic compounds with no or minimum photo-thermal decomposition and preservation of the chemical structure and functionality makes MAPLE an ideal technique to functionalize any type of 2D or 3D surfaces.

It was shown that MAPLE was used not only for obtaining one element coatings (e.g., nanoparticles, proteins, and polymers), but also as a single-step process for various coatings embedding drugs, natural proteins within biodegradable polymeric matrices for bone-related or antitumoral coatings. For example, MAPLE was successfully used for the deposition of hybrid and complex coatings such as multi layers composed of natural and synthetic compounds, for example lactoferrin (Lf), hydroxyapatite nanoparticles (HA), biodegradable polymer (polyethylene glycol-co-polycaprolactone methyl ether (PEG-PCL Me)) [30-34].

Therefore, by combining various specific characteristics of the compounds of interest with the laser parameters in order to tune the interface characteristics of a bio surface, the potential of MAPLE for surface modification and engineering is of crucial importance.

4. Coatings for orthopedic applications

Nowadays, smart devices unifying multiple functionalities, with a high level of integration in a patient body, are the future targets in the medical implant field. The challenges in the medium- and long-term clinical use and performance of orthopedic metallic implants are considering the tailoring and improvement of traditional implant engineering materials, and are related to stable anchorage in the bone tissue by rapid osseointegration and low inflammatory response. Therefore, multifunctional hybrid smart coatings having bone-matching properties, combined osseo-inductive, osseo-conductive and infection-preventing abilities, low inflammatory response as well as highly adhesion on the substrate surface and flexibility in tailoring the composition, appropriate surface chemistry, and architecture are highly desirable.

MAPLE is crucial for the realization of joining process between metallic alloy substrate and the biomimetic coatings as it confers high adhesion, thickness, and morphology control. Obtaining of optimal performance of the multifunctional biomimetic coatings (bioactive layers mimicking the bone bio mineralization process and which jointly mimic the biological osseo-integration processes, bone bonding, and prevent infections, having a controlled release of bioactive factor) was approached by combining the specific properties of the different compounds:

- Biodegradable and bioresorbable copolymer PEG-PCL-Me, as multifunctional coating and matrix for embedding bioactive factor, is mainly used as protein repellent; bone bonding polymer for better adhesion to metallic implant substrate; matrix for entrapping and controlled release of bone induction factors/bioactive factors.

- HA for osteogenic potential and for improving mechanical properties of the coating;

- Lactoferrin, as natural protein with antibacterial properties and rich chemistry for minimizing host inflammatory reactivity. In addition, it can be used as bone induction regulator, maintenance, and repair and for further binding of active factors and as bone growth factor by promoting proliferation, differentiation, and survival of osteoblasts.

Although the mentioned materials have distinctive chemistry and characteristics, the realization of joins between them, as heterostructures or composites and hybrids, is possible by applying a modified MAPLE technique as shown in **Figure 7**.

The high potential of the MAPLE technique in embedding multiple bioactive factors, into a biodegradable synthetic polymeric thin film (PEG-PCL Me) in a single step and under vacuum conditions, with minimum influence of solvents or deposition conditions on the functionality of Lf or HA, is the main reason for choosing it as flexible method for a controlled functionalization of implant surface in terms of quality, quantity, and morphology for promoting survival, proliferation, and differentiation of murine MC3T3-E1 pre-osteoblasts. The initial attachment of cells to the biomaterial is dependent on surface properties, and detectable changes in cell shape and cytoskeleton organization were remarked, as shown in **Figure 8** (after 2 h) and **Figure 9** (after 24 h).

In vitro biological assessment of all analyzed biomaterials indicated that they differentially promoted cell adhesion and proliferation and supported different degrees of extracellular matrix mineralization with improved initial bioresponse for the PEG-PCL-Me-HA.

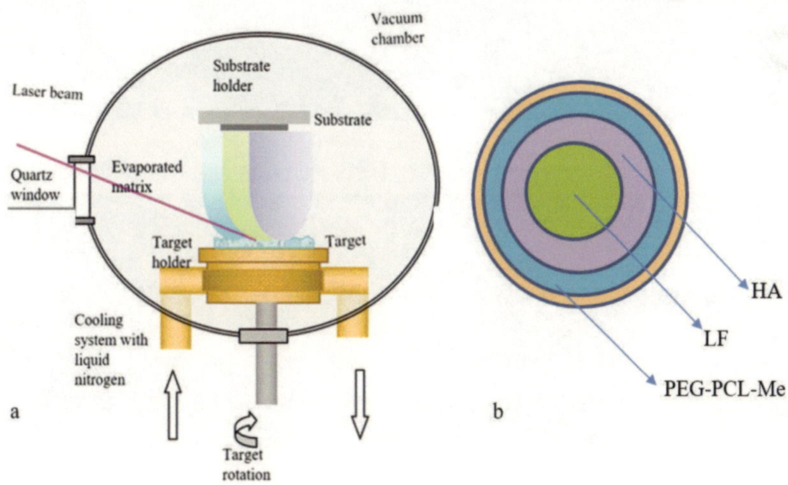

Figure 7. Experimental setup (a) and target modification (b) from [34].

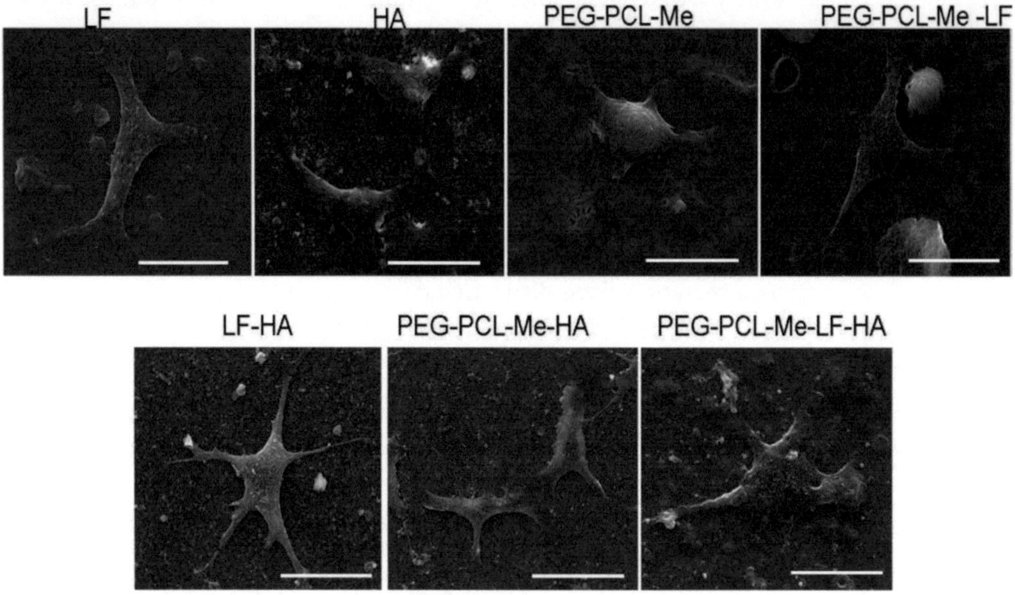

Figure 8. Early biological response of MC3T3-E1 cells to the analyzed biomaterials: The ability of pre-osteoblasts to adhere on the analyzed substrates after 2 h of culture. Scale bar is 20 μm in all images.

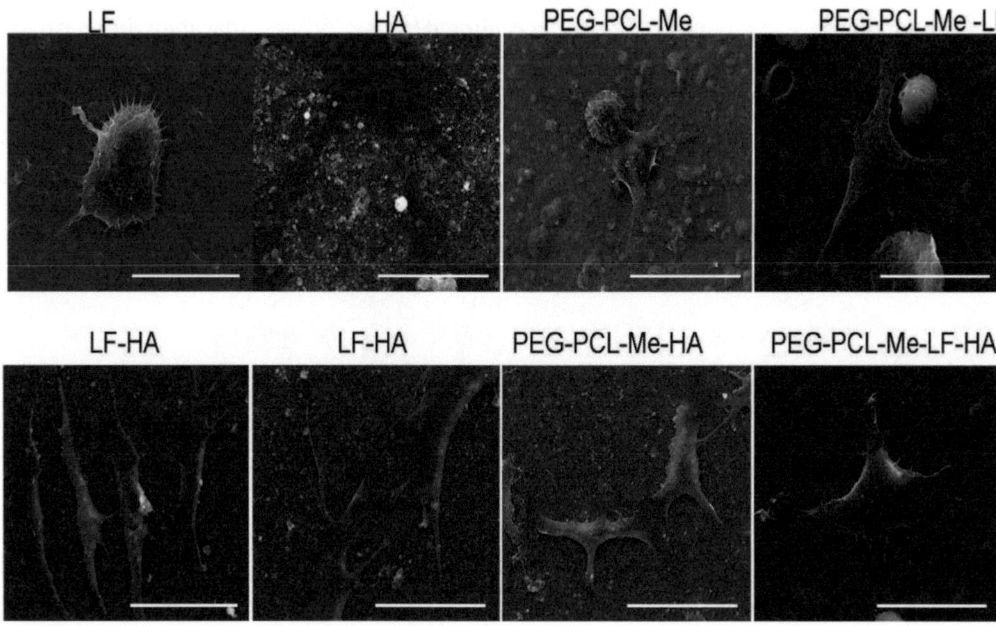

Figure 9. Cellular morphological features at 24 h post-seeding. Scale bar is 20 μm in all images.

5. Biodegradable and non-biodegradable coatings as antitumoral systems

The use of bio platforms for the short- and long-term study of the cancer cells exposed to various drugs represents one research direction of interest. Given the fact that tumoral cell lines

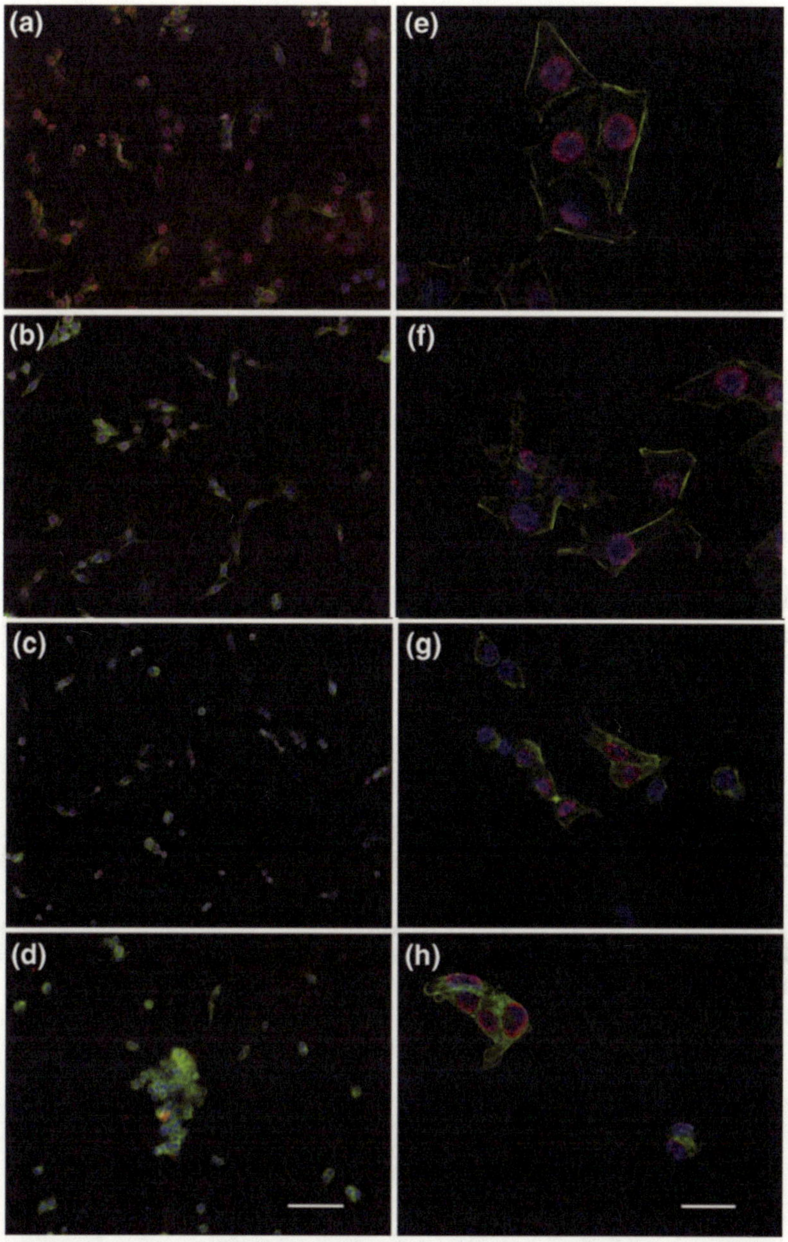

Figure 10. Proliferation and adhesion of B16-F10 melanoma cells after 24 h, onto different substrates: PCL (a, e), PCL_Cis (b, f), PCL_Apo-rLf (c, g), and PCL_Apo-rLf_Cis (d, h). Immune-fluorescence microscopy reveals Ki67 for proliferation and actin filaments for adhesion. Scale bar is 200 μm (a–d) and 50 μm (e–h). (adopted from [28].).

such as malignant melanoma are resistant to most of the chemotherapeutic agents, finding an optimal solution by synergetic effect of drugs with natural proteins is a requirement. In addition to resistance to the chemotherapeutic effect, drugs such as cisplatin (Cis) can induce a nephrotoxic effect on cells. By using both recombinant iron-free lactoferrin [(Apo-rLf)] and Cis embedded within a biodegradable polycaprolactone coating, the effect on the murine melanoma B16-F10 cells' morphology and proliferation was investigated revealing a decreased viability and proliferation, in the case of melanoma cells cultured on both Apo-rLf and Cis thin films (**Figure 10**).

Furthermore, since its initial synthesis, the temperature-responsive surfaces of poly (N-isopropylacrylamide) (pNIPAM) were studied for biomedical applications related to sensors, drug delivery, or tissue engineering applications [35, 36].

For example, MAPLE was used to demonstrate that pNIPAM coatings could be tailored in terms of thickness and roughness as thermo-responsive surfaces for L929 fibroblast cell line adhesion and single-cell detachment studies. It was found that the cells did not change their shape or viability, indicating that the control of cell attachment-detachment by changing temperature is a reversible and reproducible process [37].

By combining pNIPAM with dacarbazine drug, the effect on HT29 tumoral cell line was evaluated. There were no restraining in cell development on the PNIPAM coatings while embedding dacarbazine led to significant changes in both cellular numbers and shape **Figures 11** and **12**.

By laser surface engineering and/or functionalization, a wide range of structural, chemical, and morphological characteristics could be achieved, with optimization steps applied

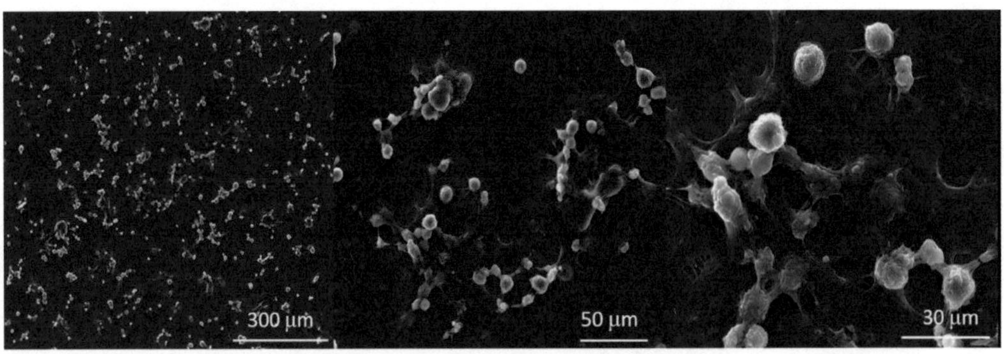

Figure 11. SEM images of HT29 cells onto the pNIPAM coatings obtained by MAPLE at 36,000 pulses and 400 mJ/cm^2 laser fluence.

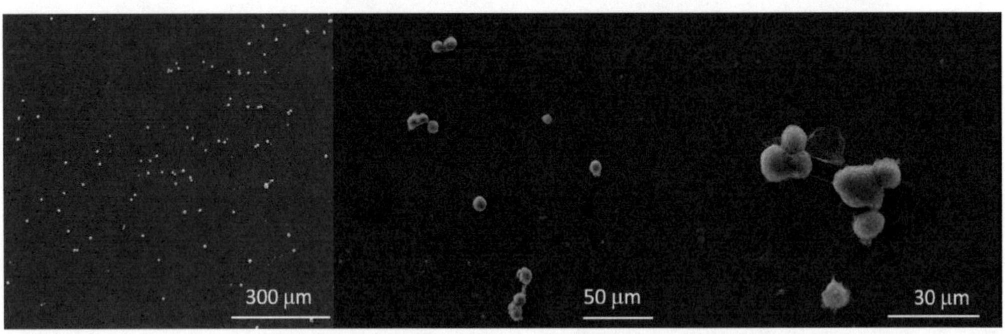

Figure 12. SEM images of HT29 cells onto the pNIPAM-dacarbazine coatings obtained by MAPLE at 36,000 pulses and 400 mJ/cm^2 laser fluence.

http://dx.doi.org/10.5772/intechopen.70676

accordingly to the specific application. Therefore, a scientific interdisciplinary approach involving the "synthesis" of new materials, analysis, and correlation of the materials bulk and interface characteristics to its bioenvironment represents a method to tackle the future of bioengineering.

6. Conclusions

This chapter summarizes some applications and developments in the field of matrix-assisted pulsed laser evaporation (MAPLE) of polymers and biomolecules, that is, protein thin film deposition. The data shown suggest that MAPLE is a promising approach for depositing proteins (lactoferrin, odorant-binding proteins) and polymers (polyisobutylene, polyethylenimine, and polyepichlorohydrin) as thin films that can be utilized for the fabrication of novel biosensors, or smart devices with multiple functionalities. Furthermore, one might combine different specific characteristics of the compounds of interest with the laser processing parameters, and is therefore being able to tune the interface characteristics of a bio surface, thus enhancing the potential of MAPLE for surface modification and engineering.

In addition, it has been shown that the deposition process of the polymer and protein materials can be optimized by modifying the process parameters, that is, laser wavelength, laser fluence, active material concentration in the target, etc. Considering the above-mentioned advantages of the MAPLE technique on depositing materials with very high reproducibility, one could envision that this approach could provide a new strategy to engineer and/or functionalize new materials to be used for regenerative biomedicine, tissue engineering studies, medical implants, or biosensors.Conflict of interestThe authors declare no competing interests.

Acknowledgements

Financial support from the Romanian Nucleus Program: contract 4 N/2016 and from the Romanian National Authority for Scientific Research, CNCS: UEFISCDI project number PN-II-PT-PCCA-2013-4-1992 (SOLE), PN-II-PT-PCCA-2013-4-1870 (ARCOLAS), PN-II-PT-PCCA-2013-4:199 (ORTHOBIOMIM), PN-II-RU-TE-2014-4-2334, and PN-II-RU-TE-2014-4-2311 (FLEXSENS), is gratefully acknowledged.

Author details

Alexandra Palla Papavlu*, Valentina Dinca, Mihaela Filipescu and Maria Dinescu

*Address all correspondence to: alexandrapalla@yahoo.co.uk

Lasers Department, National Institute for Lasers, Plasma, and Radiation Physics, Magurele, Romania

References

[1] Gammino S, Mezzasalma AM, Neri F, Torrisi L, editors. Plasma Production by Laser Ablation, PPLA 2003. Messina and Catania, Italy, World Scientific; 2004. p. 41-46. DOI: 10.1142/9789812702555_0005D. Book chapter: Pulsed laser deposition of oxide thin films, authors: Brodoceanu, N. D. Scarisoreanu, M. (Morar) Filipescu, G. N. Epurescu, D. G. Matei, P. Verardi, F. Craciun, and M. Dinescu

[2] Chrisey DB, Piqué A, RA MG, Horwitz JS, Ringeisen BR, Bubb DM, Wu PK. Laser deposition of polymer and biomaterial films. Chemical Reviews. 2003;**103**(2):553-576. DOI: 10.1021/cr010428w

[3] Schou J. Physical aspects of the pulsed laser deposition technique: The stoichiometric transfer of material from target to film. Applied Surface Science. 2009;**255**:5191. DOI: 10.1016/j.apsusc.2008.10.101

[4] Toftmann B, Rodrigo K, Schou J, Pedrys R. High laser fluence deposition of organic materials in water ice matrices by MAPLE. Applied Surface Science. 2005;**247**:211-216. DOI: 10.1016/j.apsusc.2005.01.175

[5] Purice A, Schou J, Dinescu M. Particle emission from polymer-doped water ice matrices induced by non-linear absorption of laser light at 1064 nm. Chemical Physics Letters. 2006;**427**:251-254. DOI: 10.1016/j.cplett.2006.06.118

[6] Cristescu R, Doraiswamy A, Patz T, Socol G, Grigorescu S, Axente E, Sima F, Narayan RJ, Mihaiescu D, Moldovan A, Stamatin I, Mihailescu IN, Chisholm B, Chrisey DB. Matrix assisted pulsed laser evaporation of poly(D,L-lactide) thin films for controlled-release drug systems. Applied Surface Science. 2007;**253**:7702-7706. DOI: 10.1016/j.apsusc.2007.02.042

[7] Gutierrez-Llorente A, Horowitz G, Perez-Casero R, Perriere J, Fave JL, Yassar A, Sant C. Growth of polyalkylthiophene films by matrix assisted pulsed laser evaporation. Organic Electronics. 2004;**5**:29-34. DOI: 10.1016/j.orgel.2003.11.003

[8] Constantinescu C, Scarisoreanu N, Moldovan A, Dinescu M, Vasiliu C. Thin films of polyaniline deposited by MAPLE technique. Applied Surface Science. 2007;**253**:7711-7714. DOI: 10.1016/j.apsusc.2007.02.057

[9] Paun IA, Ion V, Moldovan A, Dinescu M. Thin films of polymer blends for controlled drug delivery deposited by matrix-assisted pulsed laser evaporation. Applied Surface Science. 2010;**96**:243702. DOI: 10.1063/1.3453756

[10] Purice A, Schou J, Kingshott P, Pryds N, Dinescu M. Characterization of lysozyme films produced by matrix assisted pulsed laser evaporation (MAPLE). Applied Surface Science. 2007;**253**:6451-6455. DOI: 10.1016/j.apsusc.2007.01.066

[11] Constantinescu C, Palla-Papavlu A, Rotaru A, Florian P, Chelu F, Icriverzi M, Nedelcea A, Dinca V, Roseanu A, Dinescu M. Multifunctional thin films of lactoferrin for biochemical use deposited by MAPLE technique. Applied Surface Science. 2009;**255**:5491-5495. DOI: 10.1016/j.apsusc.2008.07.163

[12] Cristescu R, Patz T, Narayan RJ, Menegazzo N, Mizaikoff B, Mihaiescu DE, Messersmith PB, Stamatin I, Mihailescu IN, Chrisey DB. Processing of mussel adhesive protein analog thin films by matrix assisted pulsed laser evaporation. Applied Surface Science. 2005;**247**:217-224. DOI: 10.1016/j.apsusc.2005.01.066

[13] Rella R, Spadavecchia J, Manera MG, Capone S, Taurino A, Martino M, Caricato AP, Tunno T. Acetone and ethanol solid-state gas sensors based on TiO2 nanoparticles thin film deposited by matrix assisted pulsed laser evaporation. Sensors and Actuators B. 2007;**127**:426-431. DOI: 10.1016/j.snb.2007.04.048

[14] Caricato AP, Luches A, Rella R. Nanoparticle thin films for gas sensors prepared by matrix assisted pulsed laser evaporation. Sensors 2009;**9**(4):2682-2696. DOI: 10.3390/s90402682

[15] Wu PK, Fitz-Gerald J, Piqué A, Chrisey DB, McGill RA. Laser-solid interactions for materials processing. In: Kumar D, Norton DP, Lee CB, Ebihara K, Xi XX, editors. MRS Symposia Proceedings No. 617. Warrendale, PA: MRS; 2000 (MRS, Warrendale, PA, 2000); J2-3

[16] Sellinger AT, Leveugle EM, Gogick K, Zhigilei LV, Fitz-Gerald JM. Laser processing of polymer nanocomposite thin films. Journal of Vacuum Science & Technology A: Vacuum, Surfaces, and Films. 2006;**24**:1618-1622. DOI: 10.1116/1.2167980

[17] Pique A, RCY A, Stepnowski JL, Weir DW, Arnold CB, RA MG, Chrisey DB. Laser processing of polymer thin films for chemical sensor applications. Surface and Coatings Technology. 2003;**163**:293-299. DOI: 10.1016/S0257-8972(02)00606-0

[18] Houser EJ, Chrisey DB, Bercu M, Scarisoreanu ND, Purice A, Colceag D, Constantinescu C, Moldovan A, Dinescu M. Functionalized polysiloxane thin films deposited by matrix-assisted pulsed laser evaporation for advanced chemical sensor applications. Applied Surface Science. 2006;**252**:4871-4876. DOI: 10.1016/j.apsusc.2005.07.159

[19] Leveugle E, Zhigilei LV. Molecular dynamics simulation study of the ejection and transport of polymer molecules in matrix-assisted pulsed laser evaporation. Journal of Applied Physics. 2007;**102**:074914. DOI: 10.1063/1.2783898

[20] Claude R. Phipps; Editor. Proc. of SPIE Vol. 7005, High-Power Laser Ablation VII; 2008. 700517. DOI: 10.1117/12.782524

[21] Palla-Papavlu A, Dinca V, Dinescu M, Di Pietrantonio F, Cannatà D, Benetti M, Verona E. Matrix-assisted pulsed laser evaporation of chemoselective polymers. Applied Physics A: Materials Science & Processing. 2011;**105**:651-659. DOI: 10.1007/s00339-011-6624-5

[22] Dinca V, Palla Papavlu A, Matei A, Luculescu C, Dinescu M, Lippert T, Di Pietrantonio F, Cannata D, Benetti M, Verona E. A comparative study of DRL-lift and lift on integrated polyisobutylene polymer matrices. Applied Physics A. 2010;**101**:429-434. DOI: 10.1007/s00339-010-5826-6

[23] Dinca V, Palla-Papavlu A, Dinescu M, Shaw Stewart J, Lippert T, Di Pietrantonio F, Cannata D, Benetti M, Verona E. Polymer pixel enhancement by laser-induced forward transfer for sensor applications. Applied Physics A. 2010;**101**:559-565. DOI: 10.1007/s00339-010-5898-3

[24] Palla-Papavlu A, Constantinescu C, Dinca V, Matei A, Moldovan A, Mitu B, Dinescu M. Polyisobutylene thin films obtained by matrix assisted pulsed laser evaporation for sensors. Sensor Letters. 2010;**8**:502-506. DOI: 10.1166/sl.2010.1302

[25] Lobel D, Jacob M, Volkner M, Breer H. Odorants of different chemical classes interact with distinct odorant binding protein subtypes. Chemical Senses. 2002;**27**:39-44. DOI: 10.1093/chemse/27.1.39

[26] Spinelli S, Ramoni R, Grolli S, Bonicel J, Cambillau C, Tegoni M. The structure of the monomeric porcine odorant binding protein sheds light on the domain swapping mechanism. Biochemistry. 1998;**37**:7913-7918. DOI: 10.1021/bi980179e

[27] Tegoni M, Ramoni R, Bignetti E, Spinelli S, Cambillau C. Domain swapping creates a third putative combining site in bovine odorant binding protein dimer. Nature Structural Biology. 1996;**3**(10):863-867. DOI: 10.1038/nsb1096-863

[28] Di Pietrantonio F, Benetti MF, Dinca V, Cannatà D, Verona E, D'Auria S, Dinescu M. Tailoring odorant-binding protein coatings characteristics for surface acoustic wave biosensor development. Applied Surface Science. 2014;**302**:250 -255. DOI: 10.1016/j.apsusc.2013.10.112

[29] Ramoni R, Bellucci S, Grycznyski I, Grycznyski Z, Grolli S, Staiano M, De Bellis G, Micciulla F, Pastore R, Tiberia A. The protein scaffold of the lipocalin odorant-binding protein is suitable for the design of new biosensors for the detection of explosive components. Journal of Physics: Condensed Matter. 2007;**19**:395012. DOI: 10.1088/0953-8984/19/39/395012

[30] Dinescu M, Matei A, Dinca V, Palla Papavlu A, Di Pietrantonio F, Cannata D, Verona E, Lippert T. Laser processing of organic materials: Applications in tissue engineering and chemical sensing. Romanian Reports in Physics. 2013;**65**:1019-1031 WOS:000325599200038

[31] Palla-Papavlu A, Rusen L, Dinca V, Filipescu M, Lippert T, Dinescu M. Characterization of ethylcellulose and hydroxypropyl methylcellulose thin films deposited by matrix-assisted pulsed laser evaporation. Applied Surface Science. 2014;**302**:87-91. DOI: 10.1016/j.apsusc.2013.11.058

[32] Dinca V, Florian PE, Sima LE. MAPLE-based method to obtain biodegradable hybrid polymeric thin films with embedded antitumoral agents. Biomedical Microdevices. 2014;**16**:11-21. DOI: 10.1007/s10544-013-9801-9

[33] Rusen L, Neacsu P, Cimpean A, Ion V, Brajnicov S, Dumitrescu LN, Banita J, Dinca V, Dinescu M. In vitro evaluation of poly(ethylene glycol)-block-poly(ε-caprolactone) methyl ether copolymer coating effects on cells adhesion and proliferation. Applied Surface Science. 2016;**374**:23-30. DOI: 10.1016/j.apsusc.2015.08.214

[34] Rusen L, Brajnicov S, Neacsu P, Marascu V, Bonciu A, Dinescu M, Dinca V, Cimpean A. Novel degradable biointerfacing nanocomposite coatings for modulating the osteoblast response. Surface and Coatings Technology. 2017; **325**:397-409. https://doi.org/10.1016/j.surfcoat.2017.06.045; 0257-8972 (accepted)

[35] Cole MA, Voelcker NH, Thissen H, Griesser HJ. Stimuli-responsive interfaces and systems for the control of protein–surface and cell–surface interactions. Biomaterials. 2009;**30**(9): 1827-1850. DOI: 10.1016/j.biomaterials.2008.12.026

[36] Yamato M, Akiyama Y, Kobayashi J, Yang J, Kikuchi A, Okano T. Temperature responsive cell culture surfaces for regenerative medicine with cell sheet engineering. Progress in Polymer Science. 2007;**32**:1123-1133. DOI: 10.1016/j.progpolymsci.2007.06.002

[37] Rusen L, Dinca V, Mitu B, Mustaciosu C, Dinescu M. Temperature responsive functional polymeric thin films obtained by matrix assisted pulsed laser evaporation for cells attachment-detachment study. Applied Surface Science. 2014;**302**:134-140. DOI: 10.1016/j.apsusc.2013.09.122

Chapter 9

Micro- and Nano-Structuring of Materials via Ultrashort Pulsed Laser Ablation

Chung-Wei Cheng and Jinn-Kuen Chen

Additional information is available at the end of the chapter

http://dx.doi.org/10.5772/intechopen.70454

Abstract

Laser material processing has been demonstrated as an effective means for machining almost every solid material. The quality of laser machining depends on the processing parameters that dictate material ablation mechanisms. The understanding of the complex physics associated with ultrashort pulsed laser (USPL) material interaction and ablation has advanced significantly owing to a great many theoretical and experimental studies in the past 20 years. To date, USPLs have been considered as a novel tool for micro- and nano-machining of bulk or thin film materials and for internal modification of transparent materials via multi-photon absorption in a tiny focal volume. Moreover, USPL material processing is now gaining interest in other applications, such as in sensors, electronics and medical device industries.

Keywords: pulsed laser, femtosecond laser, laser material ablation, two-temperature model, LIPSS

1. Introduction

Since the first laser was invented in 1960, many different types of lasers have been developed due to rapidly increasing areas of applications. In the selection of a laser for a particular application, the following parameters are usually considered: pulse duration, wavelength, fluence, pulse repetition rate, beam uniformity and stability, bandwidth, weight and size of laser system, lifetime and reliability, hardware design, and cost. Among them, the first five are more crucial for material processing. Sufficient laser pulse energy and proper repetition rate (if a pulse laser is selected) are needed for processing the material, wavelength that determines laser energy absorption, and pulse duration that controls the heat affected zone. The interaction among the above four effects, together with the beam quality, controls the material process quality.

Ultrashort pulsed lasers (USPLs) are a particular class of lasers whose pulse duration (t_p) is less than a few picoseconds (ps). This classification is based on the fact that for most solids the time to establish thermal equilibrium between electrons and phonons is of the order of picoseconds. For a 100 femtoseconds (fs) and 0.1 J laser pulse, for example, the peak power is one terawatt (1.0×10^{12} W). If the laser pulse is focused onto an area of 0.1 mm^2, its peak intensity can reach 10^{15} W/cm^2. In view of the extremely short pulse duration and high intensity, USPLs have been explored and demonstrated great potential for a large variety of applications in physics, chemistry, life science, materials, and engineering science in the past 20 years.

Ultrashort time and high peak power are the two unique features of USPLs such that a thin layer of material can be ablated before the absorbed laser energy diffuses into the surrounding bulk. Because of that, a material can be processed very precisely with minimal or even without collateral damage. Such concepts and applications to micro/nano precision processing have been demonstrated in laboratories since the mid-1990s. Furthermore, with the possibility of reaching an extremely high intensity by focusing ultrashort pulses in a tiny volume or area, multi-photon absorption is another niche that makes USPLs well suited for 3D micro and nano processing and fabrication. The advantages of USPL material processing include: (i) high precision [1, 2], (ii) minimal collateral damage, (iii) capability of processing practically any material, (iv) 3D fabrication [3–5], and (v) a single step for creating different surface structures, such as laser-induced periodic surface structures (LIPSS) [6]. Larger material removal rates than those with conventional pulsed lasers by ultrafast bursts of pulses have also been reported recently [7].

In this chapter, three areas of laser material processing are considered: (a) mechanisms of laser-material interaction and ablation, (b) modeling of laser material ablation, and (c) applications. For the mechanisms, the characteristic times of materials and the difference in material interactions with long, short and ultrashort laser pulses are discussed. The modeling is limited to laser-material interaction for metal and semiconductor materials, including ablation for metals and damage for semiconductors. Examples of the USPL application focus on modification of material properties below the surface of transparent materials and LIPSS for sensors, precision molds, and medical devices.

2. Mechanisms of laser-material interaction and ablation

The mechanisms of laser-material interaction and ablation mainly depend on the characteristic times and thermophysical properties of materials as well as laser parameters such as wavelength, pulse duration and fluence. This section will focus on laser interactions with industrial materials.

2.1. Laser interaction with metals

Laser irradiation to a metal solid is a two-step heating process [8]. The incident photons collide with free electrons that are confined within the laser light propagation path. The absorbed laser energy is then converted to kinetic energy of the excited electrons, allowing them to travel relatively long distance (tens or hundreds of nanometer (nm), depending on the materials) before collisions with other electrons. This is referred to as ballistic electron transport in metals. After the electron-electron collisions, their kinetic energy is spread among the electrons into the Fermi-Dirac distribution. This phase is referred to as thermalization, and is completed in a few tens of femtoseconds. The electron temperature (T_e) is measureable at this time. Afterwards, the electron thermal energy diffuses, through electrons, into a deeper part of the material. In the meantime, a part of the electron thermal energy is transferred to lattice (phonons) through electron-phonon collision. It takes few picoseconds (ps) for the excited phonons to become thermalized so that the lattice temperature (T_l) can be measured. The energy exchange between the electrons and the lattice lasts for tens or hundreds of picoseconds, leading to thermal equilibrium ($T_e = T_l$) between the two subsystems (electrons and lattice), then common thermal diffusion drives the heat dissipation in the bulk.

During pulsed laser irradiation, since the heat capacity of electrons is about two orders of magnitude lower than that of the metal lattice, the electron temperature can shoot up to a very high temperature (e.g., 10^3–10^4 K) while the lattice mainly remains in low temperature state. Hence, the electron and lattice temperatures can be quite different before thermal equilibrium is established. This two-step heating is particularly true for laser pulses that are shorter than 100 ps [8, 9]. **Figure 1** shows the time histories of the calculated electron and lattice temperatures at the irradiated surface of a copper foil by a single femtosecond laser pulse (duration 120 fs, wavelength 800 nm) of two fluences (0.5 J/cm^2 and 3.0 J/cm^2). The smooth evolutions of both T_e and T_l with the time shown in the case of a laser fluence of 0.5 J/cm^2 are the same as

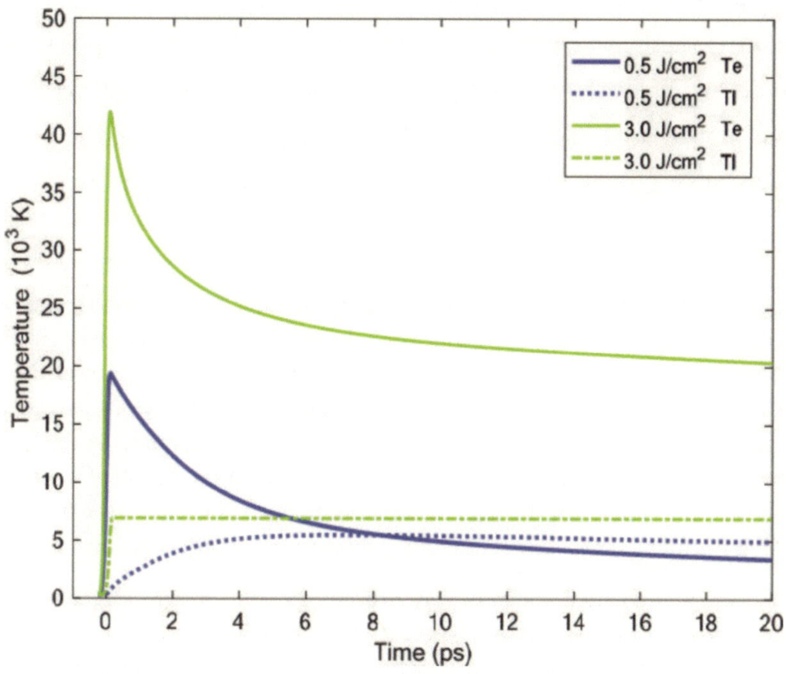

Figure 1. Calculated electron and lattice temperatures as functions of time.

most simulation results of USPL heating reported previously. For heating by nanosecond (ns) or longer laser pulses, since the characteristic times of energy transfer between the excited electrons and lattice are much shorter than the laser pulse duration, thermal equilibrium between the two subsystems is nearly established while lasing. This is particularly true for laser pulses longer than 10 ns [10]. Therefore, one-step heating ($T_e = T_l$) is usually considered for the case of conventional, long-pulse laser heating.

2.2. Laser ablation of metals

The laser energy is primarily deposited within the optical penetration depth (a few tens of nm for metals) with an exponential decay. Because the penetration depth is so shallow, the gradient of the induced electron temperature during and shortly after USPL irradiation can be immense. This non-uniform electron temperature can generate a considerably large hot-electron blast force that could cause severe deformation in the cold lattice [9–12]. The non-uniform lattice temperature that develops later is another cause for the ultrafast deformation. Due to the small size of the affected thermal zone, the gradient of lattice temperature could be very sharp. As a result, severe thermal stresses are induced in the lattice. In the case of a relatively low laser fluence, the hot-electron blast force and thermal stress are the vital power that destroys the bulk material underneath the irradiated surface before the thermal energy is significantly conducted into the bulk [11, 13]. For high laser fluences, the lattice, on the other hand, can be superheated and subsequently undergo a metastable phase transformation from solid to liquid and then, further to vapor if the fluence is sufficiently high [14–17].

Using a simple criteria for material removal, Momma et al. [2] and Nolte et al. [18] deduced two well-known logarithmic functions for the USPL material ablation depth: $\delta = \eta \ln(F_o / F_{th}^{\eta})$ for low fluences and $\delta = \zeta \ln(F_o / F_{th}^{\zeta})$ for high fluences, where η is the optical penetration depth; ζ is the effective electron heat diffusion length; F_o is the laser fluence; and F_{th}^{η} and F_{th}^{ζ} are the ablation threshold for low and high laser fluences, respectively. The failure of these two simple functions to describe the ablation rate for higher fluences [19] suggests that more mechanisms may be involved in the ablation process.

For USPL heating at laser fluences slightly exceeding the ablation threshold, only a thin layer of material is ablated by a single pulse due to the effects of the hot-electron blast and/or thermal expansion. Material ablation can take place during the lasing or after the laser pulse is off but before the next pulse arrives. When the material is removed, its associated thermal energy is also eradicated. Because very little or no molten liquid remains in the bulk material, hydrodynamic motion is negligible or never even occurs. As a result of a very thin layer ablated with little or no thermal damage to the bulk material by each laser pulse, the hole or crater resulting from this non-thermal ablation process can be very clean and precise [2].

Although non-thermal ablation can precisely process materials with minimal damage, it suffers from its very slow processing due to insufficient laser energy. This, therefore, creates a need for developing a more efficient means of processing. One potential approach to efficiently process a material target, for example, is to use a laser beam that irradiates high-fluence pulses (for high ablation rate) and then, is followed by relatively low-fluence pulses (for high precision). One can consider the low-fluence finishing as an "integrated" processing step, which results in high quality without the need for a separate post-processing. **Figure 2** shows the excellent quality of a hole produced by a femtosecond laser beam of 120 fs and 800 nm with low-fluence finishing.

For laser fluences well exceeding the ablation threshold, non-thermal ablation only occurs in the early time of the USPL heating [11]. Because the lattice temperature rises so drastically, the irradiated material rapidly undergoes phase transformation. Due to the fact that time is too short to allow necessary heterogeneous nuclei to form, the melted material is unable to boil [14, 20]. Instead, it is superheated past the normal boiling point to more or less the thermodynamic equilibrium critical temperature (T_{tc}). At that state, the tensile strength of the superheated liquid falls to zero and fluctuation in volume becomes dramatic, leading to a tremendous number of homogeneous nuclei being formed. As a result of the bubbles formed at such an extremely high rate, the subsurface layer of metastable liquid relaxes explosively into a mixture of vapor and equilibrium liquid droplets, which are immediately ejected from the bulk material. This thermal ablation mechanism is referred to as phase explosion. The phase explosion and material removal of metals caused by nanosecond [15] and USPL [16, 17, 21–23] heating have widely been investigated experimentally and theoretically. However, most studies investigated the thermal ablation within a low laser fluence regime (e.g., <10 J/cm^2), and a high laser fluence regime was rarely investigated.

As discussed above, high stress (non-thermal) and phase explosion (thermal) are the two main mechanisms for metal ablation. Non-thermal ablation provides a precision controllability which is crucial to micro- and nano-scale material processing. Another advantage is that little or no post processing is needed. Therefore, USPLs with a fluence slightly exceeding the ablation threshold are a good choice for microfabrication for which precision and small sizes are required. On the other hand, USPLs with fluences well beyond the ablation threshold may not offer an advantage for precise material processing due to the strong thermal (superheating and bubble forming) and mechanical (fluid ejection) effects that accompany them. However, there could be a trade-off between precision and efficiency of laser material processing [18].

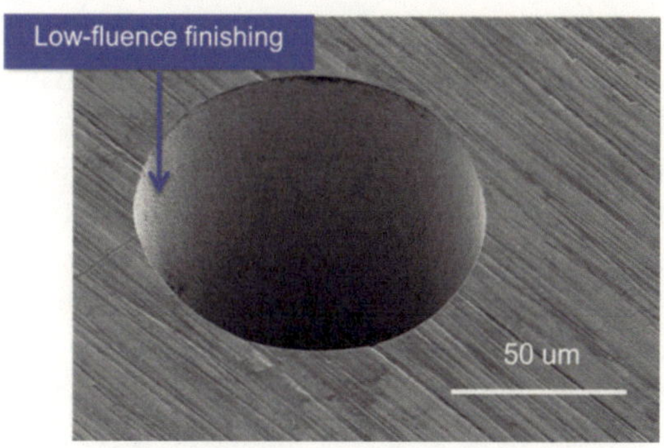

Figure 2. Drilling of aluminum by a femtosecond laser beam of 120 fs and 800 nm with low-fluence finishing.

The LIPSS created on a metal surface by multiple ultrashort laser pulses with a fluence near the ablation threshold are another interesting result of laser matter interaction [6]. Formation of micro- and nano-sized periodic ripples and column arrays are dependent on the number of laser pulses, pulse duration, fluence, wavelength, radiation polarization, and the ambient conditions. The formation of periodic-like surface structures on metal can be attributed to the interference between an incident femtosecond laser beam and the surface scattered wave [24–27].

2.3. Laser interaction with semiconductor and dielectric materials

The difference between semiconductor and dielectric materials is the number of free electrons in the conduction band. When a semiconductor or dielectric material is irradiated by a laser, the electrons in the valence band absorb the photon energy and then transit to the conduction band via single- or multi-photon absorption, depending on the photon energy (hv, where h is the Planck constant and v is the laser-light frequency) and the band-gap energy (E_g) of the material. The interband transition of the electrons creates holes in the valence band. The excess energy of the created electron-hole pairs, $khv - E_g$ ($k = 1$ for single-photon absorption and $k > 1$ for k-photon absorption), is the kinetic energy of the excited carriers that determines the carrier temperature. As the electrons and holes undergo temporal and spatial evolution, some of them recombine, for example, through the three-body Auger process. Meanwhile, additional electron-hole pairs could be generated via impact ionization from those free carriers with kinetic energy that is equal to or greater than the band gap. The excited electron-hole pairs thermalize to the Fermi-Dirac distribution via carrier-carrier collisions on a <100 fs timescale after photon absorption. In the meantime, thermalization between the carriers and phonons proceeds until the thermal equilibrium state is inevitably established.

2.4. Laser ablation of semiconductor and dielectric materials

Three main factors that affect damage and ablation of semiconductor and dielectric materials by lasers are: (a) free-electron density, (b) free-electron kinetic energy, and (c) carrier and phonon temperature. Increasing of the carrier and phonon temperature leads to the hot-electron blast force, thermal expansion, superheating, and phase explosion that are basically the same as those in metal materials described previously.

At the very beginning of USPL excitation, the density rise of the electron-hole pairs generated grows exponentially after a very short transient [28]. Depending on the peak laser intensity, for example, 12 TW/cm^2, this could result in a high density (10^{21}–10^{22} cm^{-3}) of the electron-hole pairs. When the kinetic energy of an excited electron exceeds the surface barrier and the momentum component normal to the surface of the material is positive, the excited electrons can escape from the irradiated material into the surrounding air or vacuum. Consequently, the irradiated surface gains highly positive charges, leading to a repulsive force. If the repulsion force between ions is greater than the lattice binding strength, the atomic bonds are broken and a subsurface layer of the material is disintegrated. Damage and removal of material by this repulsive force is referred to as Coulomb explosion [29]. This ablation mechanism is non-thermal because Coulomb explosion occurs prior to significant heating of the phonon subsystem. Like the non-thermal ablation in metals, this is the mechanism that can lead to high precision in processing semiconductor or dielectric materials with USPLs.

It is quite natural to use focused USPLs for 3D micromachining of semiconductor and dielectric materials because of their extremely high peak intensity. With the advantages of superior localization and great optical penetration depth (longer laser wavelength), multi-photon absorption, in general, is a better mechanism for 3D micromachining, especially for processing inside bulk material. One-photon absorption may only be suitable for surface processing due to linear absorption and smaller penetration depth (shorter laser wavelength).

When a USPL beam is tightly focused inside a transparent material, multi-photon ionization, tunneling ionization, and avalanche ionization could occur in the focal volume. Like metals, the excited electrons transfer their energy to the ions via collision. With some time delay, the electrons and ions eventually reach thermal equilibrium. When the laser intensity is below a certain threshold, electrons recombine with holes in a non-radiative way. For laser intensities exceeding the certain level (referred to as plasma threshold), a plasma spark, resulting from the recombination of high-density electron plasma with holes, can be observed. The spark is associated with an optical breakdown of the transparent materials, having a characteristic duration of the order of 10 ns. After recombination, the thermal energy diffuses away from the focal volume on a microsecond timescale. At even higher intensities, laser-induced permanent modification of material properties is observed [4]. If the laser intensity is above the modification threshold, the hot electron-hole plasma and ions explosively expand from the focal volume into the surrounding material.

With tight focusing of ultrashort laser pulses, different kinds of permanent structure changes in a transparent material can be made by adjusting the incident pulse energy, in principle. Fabrication/machining/processing can be performed on the surface or within the bulk of the material by moving the laser focus along the desired paths. More detailed information can be found in the papers cited in the review paper [30].

3. Modeling of laser material ablation

In this section, we focus on two-step heating models for ultrafast thermal transport, ablation models for metals, and damage models for semiconductor materials.

3.1. Two-temperature model for metals

The two-temperature (2T) model was pioneered by Anisimov et al. [31] in 1974 to describe a two-step heating process of metals subjected to short-pulse laser irradiation. It was not until the early 1990s that USPL material interactions received considerable attention. Since then, numerous modified versions of 2T have been proposed [9, 32]. Based on the Boltzmann transport approximation, a semi-classical 2T model is derived for the dynamics of electron concentration (n), mean (drift) velocity ($\overline{v}_e$), and average energy for metal materials [9]:

$$\frac{\partial n}{\partial t}+\nabla\cdot\left(n\bar{v}_e\right)=0 \tag{1}$$

$$m_e\frac{\partial \bar{v}_e}{\partial t}+m_e\bar{v}_e\cdot\nabla\bar{v}_e+\left[k_B\left(1+\frac{T_e}{C_e}\frac{\partial C_e}{\partial T_e}\right)-e\beta_e\right]\nabla T_e=-\frac{eT_e\bar{v}_e}{\mu_o T_l} \tag{2}$$

$$C_e\left(\frac{\partial T_e}{\partial t}+\bar{v}_e\cdot\nabla T_e+\frac{2}{3}T_e\nabla\cdot\bar{v}_e\right)+\nabla\cdot\bar{Q}_e=-G\left(T_e-T_l\right)+S \tag{3}$$

In the above equations, m_e is the electron mass, $\bar{Q}_e$ is the heat flux vector in the electron subsystem, k_B is the Boltzmann constant, e is the electron charge, μ_o represents the mobility of the electrons, β_e is the temperature-dependent parameter for free electrons, C_e is the heat capacity of the electron, G represents the electron-phonon coupling factor, S is the volumetric laser heat source, and ∇ is the divergence operator. Since the electron relaxation time of metals is around ten femtoseconds, the constitutive relation for $\bar{Q}_e$ and T_e is given by [32]:

$$\tau_e\frac{\partial \bar{Q}_e}{\partial t}+\bar{Q}_e=-K_e\nabla T_e \tag{4}$$

where τ_e is the electron relaxation time (the mean time for electrons to change their states), and K_e is the electron thermal conductivity.

For the lattice subsystem, the thermal transport equation includes an energy exchange with the electrons and a thermal relaxation effect in a general case [33]:

$$C_l\frac{\partial T_l}{\partial t}=-\nabla\cdot\bar{Q}_l+G\left(T_e-T_l\right) \tag{5}$$

$$\tau_l\frac{\partial \bar{Q}_l}{\partial t}+\bar{Q}_l=-K_l\nabla T_l \tag{6}$$

In Eqs. (5) and (6), $\bar{Q}_l$ is the heat flux vector in the lattice, τ_l is the relaxation time in phonon collisions, and C_l and K_l are the heat capacity and thermal conductivity of lattice, respectively.

By neglecting the electron drift velocity, energy equation (3) is simplified to:

$$C_e\frac{\partial T_e}{\partial t}=-\nabla\cdot\bar{Q}_e-G\left(T_e-T_l\right)+S \tag{7}$$

Thus, the semi-classical 2T model, seen in Eqs. (1)–(6), is reduced to the dual-hyperbolic 2T model (HH2T) [33].

For pure metals, K_l is much smaller than K_e. The hyperbolic 2T model (H2T) [32] can be retrieved from the HH2T model by neglecting the heat conduction in the lattice, i.e., $\nabla\cdot\bar{Q}_l=0$ in Eq. (5). Further neglecting the electron relaxation effect in the H2T model leads to the parabolic 2T model (P2T) [8, 34].

After the thermal equilibrium, $T_e = T_l = T$, $\overline{Q}_e + \overline{Q}_l = \overline{Q}$, $C_e + C_l = C$, and $K_e + K_l = K$. Combining Eqs. (5) and (7) yields the macroscopic energy balance equation:

$$C\frac{\partial T}{\partial t} = -\nabla \cdot \overline{Q} + S \tag{8}$$

The above energy equation (8) together with Fourier's heat conduction law is a well-known, parabolic one-temperature heat conduction model (P1T). When the Cattaneo equation [35] is considered, the result becomes a hyperbolic (thermal wave) one-temperature model (H1T). The above 1T models represent one-step heating and have been widely used for long-pulse laser heating since the electrons and lattice are assumed to be in thermal equilibrium instantaneously when a medium is heated.

The thermophysical properties, C, K and G, control thermal response in laser-irradiated material. The expressions for these three thermophysical properties can be found in Refs. [10, 36]. Recently, Lin et al. [37] presented data C_e and G for different metals over a wide range of electron temperatures from room temperature to 5×10^4 K. A formula for K_e up to Fermi temperature was given by Anisimov and Rethfeld [38].

The mechanical stresses caused by the hot-electron blast force and the thermal stresses induced by non-uniform temperatures in the lattice can be solved by the equation of motion of lattice:

$$\rho_l \frac{\partial^2 u_i}{\partial t^2} = \sigma_{ij,j} + \frac{2}{3}(C_e T_e)_{,i} \tag{9}$$

where u_i ($i = x$, y, and z) is the displacement vector of a material (lattice) point, σ_{ij} is the stress tensor at the point, and the subscript "," denotes the spatial derivative. The last term on the right-hand side of Eq. (9) is the hot-electron blast force. The thermal strains resulting from a non-uniform lattice temperature are included in the stress-strain relations. The deformation can be linearly elastic or non-linearly plastic.

The exchange of thermal and mechanical energy should not be neglected due to the extremely high strain rate (~10^9 s^{-1}). Thus, the energy equation of lattice, Eq. (5), is re-written as:

$$C_l \frac{\partial T_l}{\partial t} = -\nabla \cdot \overline{Q}_l + G(T_e - T_l) - (3\lambda + 2\mu)\varphi T_l \frac{\partial \varepsilon_{kk}}{\partial t} \tag{10}$$

where λ is the Lamé constant, μ is the shear modulus, φ is the thermal expansion coefficient, and ε_{kk} $(= \nabla \cdot \overline{u})$ is the volume dilatation. In addition, the strain rate effects on plastic deformation and failure could be substantial, and the existing stress-strain relations and fracture strength may be insufficient for ultrafast deformation. To accurately predict non-thermal material ablation by stresses, the strain rate effects on material behavior should be investigated further and quantified.

3.2. Laser heat source

The most popular laser beams are Gaussian in both space and time. For simplicity, let us consider an incident laser beam that is normal to the material target. The volumetric laser heat source S in Eqs. (3) and (7) is given as [39]:

$$S(r,z;t)=\sqrt{\frac{\beta}{\pi}}\frac{[1-R(r,0;t)]F_o}{t_p}\alpha(r,z;t)\exp\left[-2(\frac{r}{r_o})^2-\int_0^z\alpha dz-\beta(\frac{t-mt_p}{t_p})^2\right] \quad (11)$$

where r is the radial distance from the center of the beam, z is the coordinate in the direction of laser beam propagation, $F_o = J_o/(\pi r_o^2/2)$ with J_o denoting the pulse energy, r_o being the beam radius defined at the e^{-2} distance, $R(r, 0; t)$ is the surface reflectivity of the material, α is the absorptivity coefficient, t_p is the full width half maximum (FWHM) of the Gaussian temporal pulse, m is an integer, and $\beta = 4\ln(2)$. The lasing, assumed to start at $t = 0$, reaches its peak power at $t = mt_p$, and ends at $t = 2mt_p$. The value of m can be set at 2 or 3 since the laser energy outside this period of time is inconsequential. If the effects of the hot electrons ballistic motion [40] are taken into account, the laser heat source is modified to:

$$S(r,z;t)=\sqrt{\frac{\beta}{\pi}}\frac{[1-R(r,0;t)]F_o}{t_p}\frac{1}{\delta(r,z;t)+\delta_b(r,z;t)}\exp\left[-2(\frac{r}{r_o})-\int_0^z\frac{1}{\delta+\delta_b}dz-\beta(\frac{t-mt_p}{t_p})^2\right] \quad (12)$$

where $\delta = 1/\alpha$ is the temperature-dependent optical penetration depth and δ_b is the ballistic electron penetration depth. The temperature-dependent optical properties, R and α, in Eqs. (11) and (12) can be determined from optical properties and the Fresnel function.

Recently, an extended Drude model [41] was proposed to characterize the temperature-dependent R and α, and a critical point model with three Lorentzian terms for interband transition [42] for copper. The numerical results show that when a metal is irradiated by a USPL at high fluence, the dynamic changes of R and α during the laser irradiation could significantly alter laser energy absorption and the distribution of laser heat density. Although it is only true for low electron temperatures, the constant R and α at room temperature have been widely employed in 2T modeling. In this case, the integral term becomes a constant, z/δ in Eq. (11) and $z/(\delta + \delta_b)$ in Eq. (12).

3.3. Comparison of different thermal models

Figure 3 compares the predictions of four thermal models, HH2T, H2T, H1T, and P1T, for lattice temperatures at the front surface of a 200 nm gold film heated by a laser pulse of F_o = 0.5 J/cm^2 and t_p = 100 ps and 1 ns [10]. Evidently, both of the one-temperature models significantly overestimated the bulk temperature for the 100 ps pulse; hence, they are inadequate for simulating USPLs heating. The results, shown in **Figure 3(b)**, confirm that the Fourier heat conduction model is sufficient for long-pulse laser heating. In fact, the difference among the four models becomes indistinct for a 10 ns laser pulse [10].

3.4. Ablation models for metals

The 2T models discussed above only characterize thermal transport in a solid metal. For material ablation, other physical processes such as phase transitions and material removal need to be considered. Different approaches have been proposed to simulate laser material ablation, including ultrafast thermoelasticity [11], dynamics of thermal ablation [17], hydrodynamic modeling [43], molecular dynamics [22], etc. Although those numerical analyses [17]

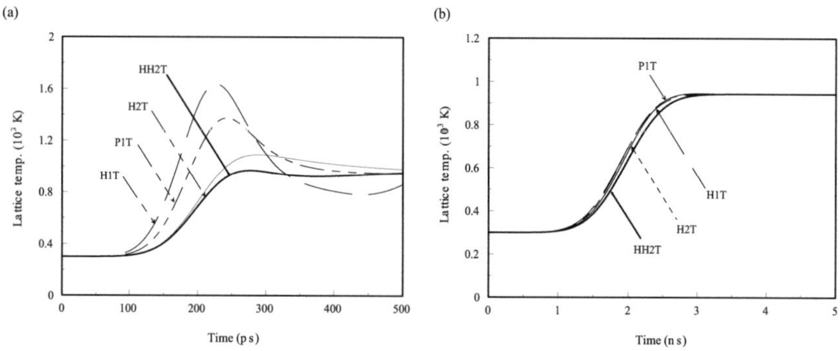

Figure 3. Comparison of the four thermal model predictions for lattice temperature at the front surface of a 200 nm gold film heated by a laser pulse of F_o = 0.5 J/cm^2 and (a) t_p = 100 ps and (b) t_p = 1 ns. Reproduced with permission from publisher [10].

show a certain degree of success in comparison with experimental measurements, none of the aforementioned models have been validated thoroughly with experimental data for USPL ablation over a wide pulse duration and fluence.

A comprehensive USPL ablation model is needed to be able to accurately describe the entire ablation process and predict the ablation depth and rate. It should be developed based on universal regulations and methods that allow for the description of non-equilibrium responses and cover the complex physical phenomena for the entire laser process. Such phenomena include photon-electron interaction, laser-pulse propagation and ionization, phase transitions, superheating through a metastable liquid phase, rapid nuclei formation, explosion and dynamics of homogeneous nuclei (bubbles), generation of recoil pressure, hydrodynamic motion, formation and dynamics of the plasma plume, laser-plasma interaction, radiation from the resulting plasmas, condensation and re-solidification of the liquid/vapor, etc.

To accurately simulate thermal ablation by USPLs, a semi-classical 2T model integrated with models of phase transformations for ultrafast melting, evaporation and re-solidification and a phase explosion model for ejection of metastable liquid and vapor was attempted recently [44]. When superheated liquid temperature reaches $0.9T_{tc}$, phase explosion is assumed to take place [20] and that material, including both electrons and lattice, is then removed under the assumption of phase explosion. Once phase explosion no longer occurs, vaporization could continue until the lattice temperature drops significantly. **Figure 4** shows the time history of the ablation depth for copper foil irradiated by a single femtosecond laser pulse (duration 120 fs, wavelength 800 nm) of different fluences. The steep occurrences of material ablation result mainly from phase explosion, while the sloping parts result from vaporization. As shown in **Figure 4 (b)** for the case of laser fluence 6.1 J/cm^2, the ablation depth by phase explosion and vaporization are 258.4 nm (up to 310 ps) and 5.9 nm (from 310 to 600 ps), respectively. The results shown here indicate that for high fluences, phase explosion is the dominating material ablation mechanism in USPL material ablation. It is noted that this ablation model is not yet fully comprehensive.

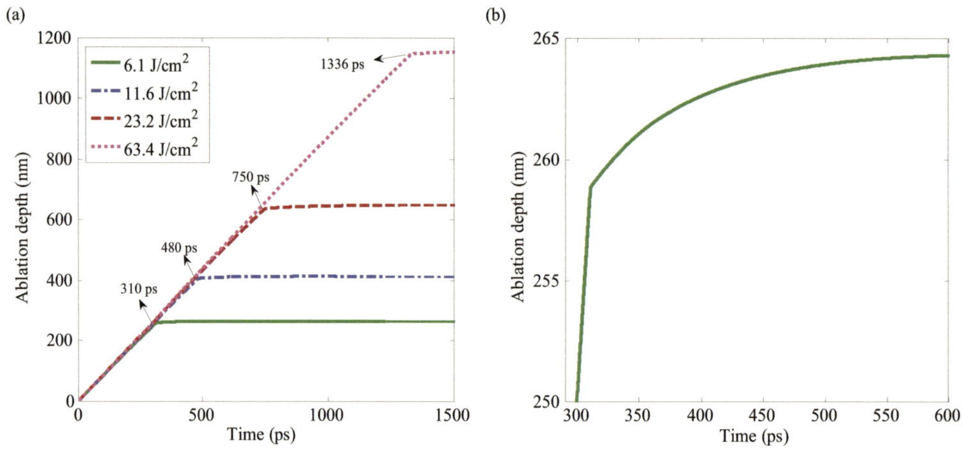

Figure 4. (a) Time histories of the ablation depths of a copper foil by a single femtosecond laser pulse (duration 120 fs, wavelength 800 nm) of different fluences, (b) magnification of the fluence case of 6.1 J/cm^2. Reproduced with permission from publisher [45].

3.5. Ultrafast transport dynamics models for semiconductors

To model the transport process of a large number of hot electrons, holes, and phonons in a semiconductor, the formalism must be based on the principle of statistical mechanics. A self-consistent model for transport dynamics in semiconductors caused by USPL heating can be found in [46]. Based on the relaxation-time approximation of the Boltzmann equation, the rate equations are derived for the dynamics of electron-hole carrier number density and current, ambipolar energy current, carrier energy, and lattice energy.

3.5.1. Rate equation for carrier pairs

The balance equation for the electron-hole pairs number density generated by a laser pulse for two-photon absorption is:

$$\frac{\partial n}{\partial t} = \frac{\beta_1 I(x,t)}{h\nu} + \frac{\beta_2 I^2(x,t)}{2h\nu} - \gamma n^3 + \theta n - \nabla \cdot J \tag{13}$$

where β_1 is the avalanche coefficient, β_2 is the 2-photon absorption coefficient, γ is the Auger recombination coefficient, θ is the impact ionization coefficient, and J is the carrier current. The magnitude of β_1 is much greater than that of β_k ($k > 1$). For silicon at room temperature, for instance, $\beta_1 = 1.0 \times 10^4\ cm^{-1}$ and $\beta_2 = 2.0 \times 10^{-9}\ cm/W$. The last three terms on the right-hand side of Eq. (13) represent Auger recombination, impact ionization, and the loss due to carrier current, respectively. The Auger recombination is described as the time when an electron-hole pair recombines, giving up its energy to a third electron in the conduction band; this reduces the carrier number density. The reverse effect is impact ionization. Due to the nonlinearity nature (I^k), multi-photon absorption could be more efficient and occur faster than single-photon absorption when laser intensities are sufficiently high.

3.5.2. Rate equation for carrier energy

The balance equation for the electron-hole pairs energy produced by a laser pulse for two-photon absorption is:

$$C_{e-h}\frac{\partial T_e}{\partial t} = (\beta_1 + \Theta n)I(x,t) + \beta_2 I^2(x,t) - \nabla\cdot W - \frac{C_{e-h}}{\tau_e}(T_e - T_l)$$
$$-\frac{\partial n}{\partial t}\left\{E_g + \frac{3}{2}k_B T_e\right\} - n\left(\frac{\partial E_g}{\partial n}\frac{\partial n}{\partial t} + \frac{\partial E_g}{\partial T_l}\frac{\partial T_l}{\partial t}\right) \quad (14)$$

where C_{e-h} is the heat capacity of electron-hole pairs, Θ is the free carrier absorption cross-section, and W is the ambipolar energy current which is the sum of the carrier energy currents in electrons and holes.

3.5.3. Rate equation for lattice energy

For semiconductor materials the thermal conductivity of the lattice is comparable with the bulk value. It is necessary to consider the thermal transfer in the lattice. The lattice energy rate is expressed as:

$$C_l\frac{\partial T_l}{\partial t} = \nabla\cdot(k_l \nabla T_l) + \frac{C_{e-h}}{\tau_e}(T_e - T_l) \quad (15)$$

The three rate equations (13)–(15), together with the constitutive equations for the carrier pair current and the ambipolar energy current [46], compose a complete self-consistent model for ultrafast transport dynamics in semiconductors subjected to USPL irradiation.

Figure 5 shows the time histories of the density and temperature of carriers and the lattice temperature at the incident surface of a silicon sample irradiated by a 500 fs laser pulse with fluence 0.005 J/cm^2 [46]. The peak laser power is at t = 1.5 ps (m = 3). As shown in **Figure 5**, the carrier temperature reaches its maximum at about 0.68 ps and the number density reaches its maximum at about 2.21 ps. It should be noted that when the peak carrier temperature occurs, the laser power is only at about 0.06% of its maximum. The contradictory intuition of this high temperature is attributed to a very small amount of electron-hole pairs that are created during this early time. Consequently, the carrier heat capacity is very small, thereby leading to a rapid, noticeable rise in the carrier temperature although the net carrier thermal energy is quite low. As time is prolonged, the carrier density increases drastically since much more laser energy has been absorbed. As a result, the carrier heat capacity becomes greater and greater. Meanwhile, the energy loss, due to the change in carrier energy density (the second last term on the right hand side of Eq. (14)), becomes pronounced, and so does the loss due to thermal transfer from the carriers to the lattice. Those changes make the rate of change of the net energy unable to remain positive at some point and thereafter, even though considerable laser energy is absorbed. This explains why the carrier temperature quickly reaches its peak very early in the laser irradiation and then falls. The numerical result in Ref. [46] also shows that one-photon absorption and Auger recombination are two crucial factors that alter the carrier's number density.

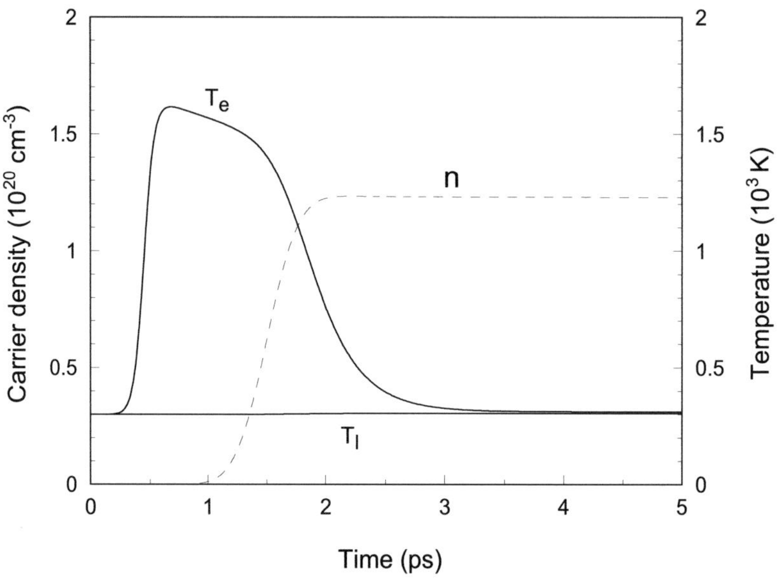

Figure 5. Time evolution of carrier density (n), carrier temperature (T_e), and lattice temperature (T_l) at the front surface of a 20 μm silicon sample heated by a 500 fs, 775 nm, 0.005 J/cm^2 laser pulse. Reproduced with permission from publisher [46].

3.6. Damage models for semiconductors

For pulse durations longer than the carrier–lattice energy relaxation time (a few picoseconds), it has generally been accepted that the solid–liquid phase transition by high laser fluences is a thermal (melting) process. On the other hand, experiments with femtosecond laser pulses of high fluence have demonstrated an ultrafast phase transformation on a subpicosecond time scale. The mechanism of this ultrafast phase transformation is nonthermal melting, differing from the above thermal melting.

Among the continuum models, there are two approaches that are frequently used in evaluation of the damage threshold for semiconductor materials. One approach employs a single rate equation to evaluate electron density in the conduction band [47]. The equation includes ionization rates for single and multi-photon absorption, avalanche ionization, and other relaxation terms. Damage is assumed when the calculated electron density exceeds a critical value, which is often determined semi-empirically by matching with the experimental data. In most cases, the critical value used differs from that of the critical density of electron-hole plasma that makes the plasma opaque (optical breakdown). The other approach uses a self-consistent model in which the rate equation for the carrier number density is coupled to the energy balance equations for both the carriers and phonons [46]. The onset of damage is determined by which condition, carrier number density, or lattice temperature is first met. **Figure 6** shows the comparison between the theoretical damage fluence thresholds with measured values for Si [9, 21]. It appears that the self-consistent model agrees fairly well with the experimental data for laser pulse durations up to several picoseconds; on the other hand, the carrier density model is only in good agreement for laser pulse durations of subpicoseconds [46]. In view of

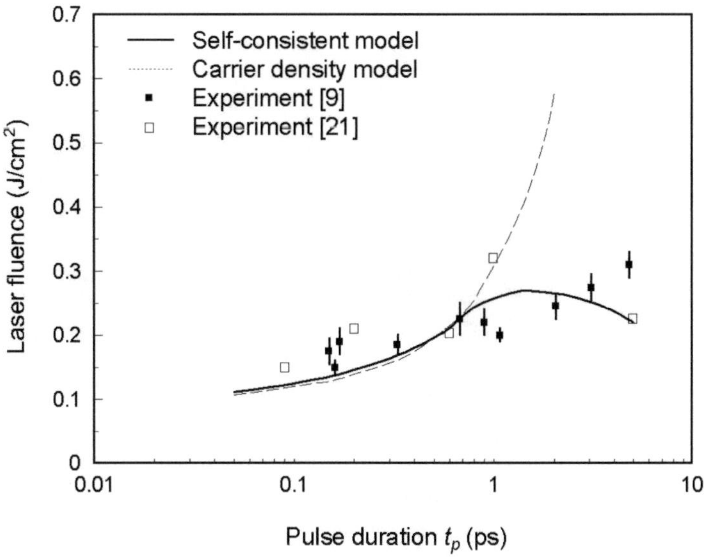

Figure 6. Theoretical and experimental damage thresholds versus laser pulse duration for Si. Reproduced with permission from publisher [46].

both approaches failing to predict the damage fluence thresholds for longer laser pulses, a more robust model or approach is suggested.

Modeling of the non-thermal ablation of semiconductor materials via Coulomb explosion can be found in the study by Stoian et al. [29]. On the other hand, a comprehensive modeling of thermal ablation via phase explosion, to the authors' knowledge, has not yet been reported in open literature.

4. Applications

As aforementioned, the great applications of USPLs are attributed to two unique features: ultrashort pulse duration and extremely high intensity. The benefits of USPL material processing include high precision with a minimal heat affected zone, the ability to locally modify below the surface of transparent materials, the single step for creating different surface structures, etc. A great number of applications of USPLs have been proposed, for example, to industrial technologies and bio/medicine [48, 49]. Recently, applications of LIPSS have been of particular interest.

The ability to locally modify material properties below the surface of transparent materials is an important area of USPL application. One interesting example is the repair of a packaged.

TFT-LCD panel. In this repair, a femtosecond laser beam passes through the polarizer, glass substrate, and then, focuses on the color filter (color photoresist) corresponding to the hot pixels such that the phenomenon of nonlinear multi-photon absorption can be induced within

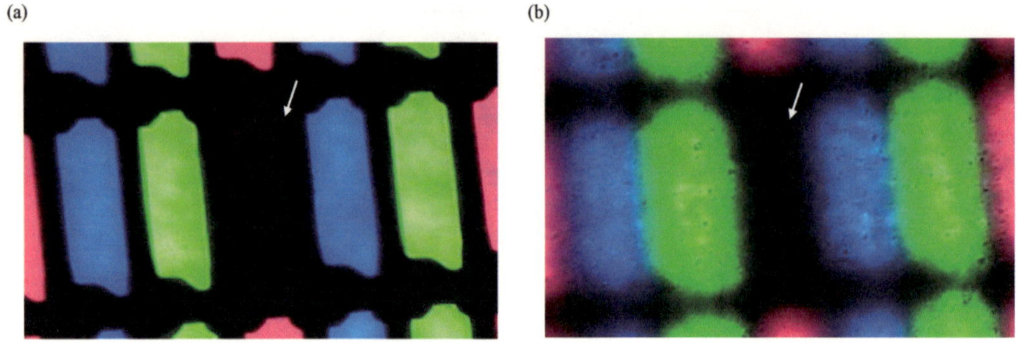

Figure 7. A red color filter after laser repair: (a) image focused at the photoresist and (b) image focused at the polarizer.

the localized area. Then, the property of the color photoresist layer is changed. The hot pixels become dead pixels, and any defective pixels inside the panel of the packaged LCD can be repaired directly. **Figure 7** shows a red colored filter after laser repair. The image is captured from different focal positions. As shown in **Figure 7(a)**, the red colored filter (marked by an arrow) has been blackened. In **Figure 7(b)**, however, the polarizer corresponding to the red colored filter (marked by an arrow) is not damaged after laser irradiation. Accordingly, the defective pixel is eliminated by the transformation of the bright one into a dark one.

LIPSS by USPLs allows for a new range of functionalized surface processing that permits novel applications for sensors, medical devices, precision molds, etc. For sensor applications, fabrication of Ag nanostructure-covered surface structures can be used for surface-enhanced Raman scattering (SERS) application. The SERS intensities of rhodamine 6G (R6G) at LIPSS-treated Ag substrates are 15 times greater than those at non-treated Ag substrates [50].

For medical devices applications, LIPSS by USPLs can be employed to fabricate electrosurgical blades [51, 52]. The average temperatures of tissue cut by conventional and LIPSS-treated electrosurgical blades are 145°C and 116°C, respectively. This corresponds to an improvement of about 29°C in temperature. As a result, there is less damage to tissue that is cut by the proposed electrosurgical blade.

LIPSS by USPLs is also employed to fabricate medical stainless steel [53]. As shown in **Figure 8**, the attachment assay with 72 h in a NIH3T3 culture demonstrates that significantly more cells are attached to the LIPSS-treated sample. This means that the cells spread over the treated surface more rapidly than over the non-treated surface. LIPSS-treated stainless steel is believed to possess better biocompatibility than stainless steel without surface modification.

For precision mold applications, for example, LIPSS by USPLs can be employed to fabricate structured molds and fast replication of large-area hierarchical micro/nano structures (lotus-leaf-like patterns) on a plastic part by injection molding [54]. Compared with an ordinary plastic surface, the contact angle of the structured plastic parts is increased by 38%, from 97 to 134°.

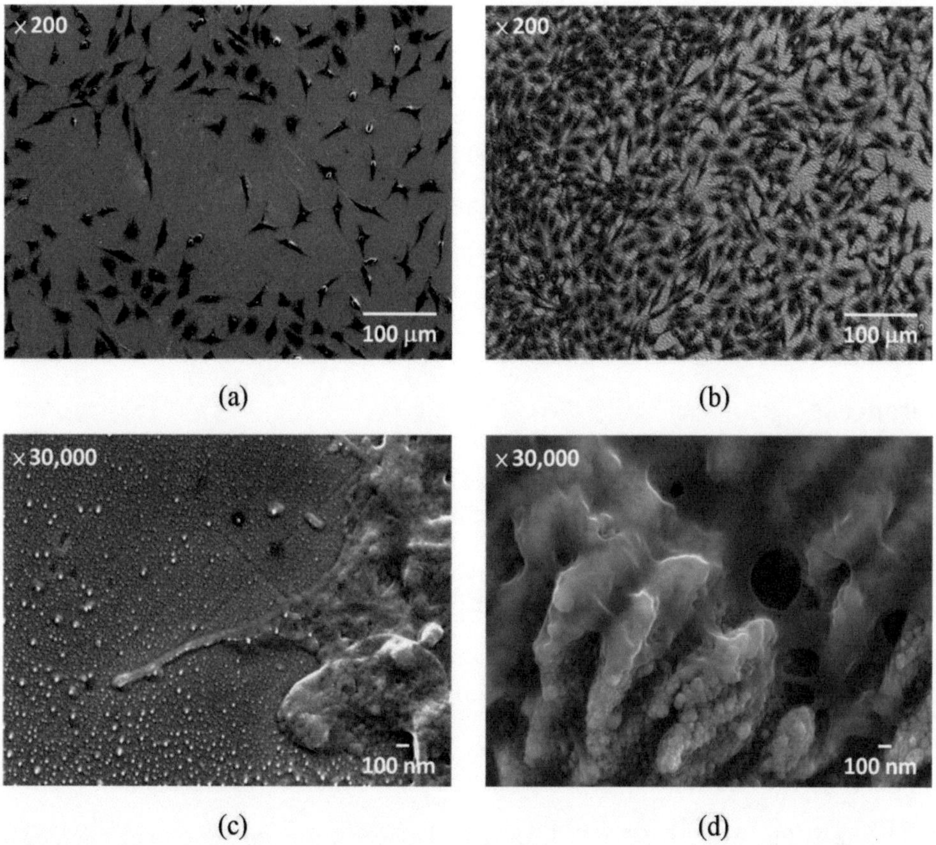

Figure 8. SEM images of an NIH3T3 cell after being in culture for 72 h: (a) control sample, (b) LIPSS-treated sample, (c) magnification of (a), and (d) magnification of (b). Reproduced with permission from publisher [53].

5. Conclusion

In reviewing laser micro- and nano-structuring of materials through pulse laser ablation, three main areas have been covered, including laser system development, experimentation and modeling, and applications. In this single chapter, we focused on laser material ablation mechanisms, modeling, and some applications of material properties modification and LIPSS that have been developed or demonstrated. Four areas in the modeling of laser matter interaction have been presented, including temperature response with temperature-dependent material properties and optical properties, ablation models for metals, ultrafast transport dynamics models for semiconductors, and damage models for semiconductors. Understanding the mechanisms and modeling of laser material interaction with materials can provide an insight for optimizing the processing parameters for precisely machining a variety of materials and can allow the development of high-value and innovative laser process techniques for a variety of applications.

Author details

Chung-Wei Cheng[1]* and Jinn-Kuen Chen[2]

*Address all correspondence to: weicheng@nctu.edu.tw

1 Department of Mechanical Engineering, National Chiao Tung University, Hsinchu, Taiwan

2 Department of Mechanical and Aerospace Engineering, University of Missouri, Columbia, USA

References

[1] Pronko P, Dutta S, Squier J, Rudd J, Du D, Mourou G. Machining of sub-micron holes using a femtosecond laser at 800 nm. Optics Communications. 1995;**114**:106-110

[2] Momma C, Chichkov BN, Nolte S, von Alvensleben F, Tünnermann A, Welling H, et al. Short-pulse laser ablation of solid targets. Optics Communications. 1996;**129**:134-142

[3] Davis KM, Miura K, Sugimoto N, Hirao K. Writing waveguides in glass with a femtosecond laser. Optics Letters. 1996;**21**:1729-1731

[4] Glezer E, Milosavljevic M, Huang L, Finlay R, Her T-H, Callan JP, et al. Three-dimensional optical storage inside transparent materials. Optics Letters. 1996;**21**:2023-2025

[5] Kawata S, Sun H-B, Tanaka T, Takada K. Finer features for functional microdevices. Nature. 2001;**412**:697-698

[6] Vorobyev AY, Guo C. Direct femtosecond laser surface nano/microstructuring and its applications. Laser & Photonics Reviews. 2013;**7**:385-407

[7] Kerse C, Kalaycıoğlu H, Elahi P, Çetin B, Kesim DK, Akçaalan Ö, et al. Ablation-cooled material removal with ultrafast bursts of pulses. Nature. 2016;**537**:84-88

[8] Anisimov SI, Kapeliovich BL, Perel'man TL. Electron emission from metal surfaces exposed to ultrashort laser pulses. Journal of Experimental and Theoretical Physics. 1974;**39**:375-377

[9] Chen J, Tzou D, Beraun J. A semiclassical two-temperature model for ultrafast laser heating. International Journal of Heat and Mass Transfer. 2006;**49**:307-316

[10] Chen J, Beraun J, Tham C. Investigation of thermal response caused by pulse laser heating. Numerical Heat Transfer: Part A: Applications. 2003;**44**:705-722

[11] Chen J, Beraun J, Grimes L, Tzou D. Modeling of femtosecond laser-induced non-equilibrium deformation in metal films. International Journal of Solids and Structures. 2002;**39**:3199-3216

[12] Chen J, Beraun J, Tham C. Ultrafast thermoelasticity for short-pulse laser heating. International Journal of Engineering Science. 2004;**42**:793-807

[13] Falkovsky L, Mishchenko E. Electron-lattice kinetics of metals heated by ultrashort laser pulses. Journal of Experimental and Theoretical Physics. 1999;**88**:84-88

[14] Miotello A, Kelly R. Laser-induced phase explosion: New physical problems when a condensed phase approaches the thermodynamic critical temperature. Applied Physics A: Materials Science & Processing. 1999;**69**:S67-S73

[15] Bulgakova N, Bulgakov A. Pulsed laser ablation of solids: Transition from normal vaporization to phase explosion. Applied Physics A: Materials Science & Processing. 2001;**73**: 199-208

[16] Dömer H, Bostanjoglo O. Phase explosion in laser-pulsed metal films. Applied Surface Science. 2003;**208**:442-446

[17] Chen J, Beraun J. Modelling of ultrashort laser ablation of gold films in vacuum. Journal of Optics A: Pure and Applied Optics. 2003;**5**:168

[18] Nolte S, Momma C, Jacobs H, Tunnermann A, Chichkov BN, Wellegehausen B, et al. Ablation of metals by ultrashort laser pulses. Journal of the Optical Society of America B-Optical Physics. Oct 1997;**14**:2716-2722

[19] Furusawa K, Takahashi K, Kumagai H, Midorikawa K, Obara M. Ablation characteristics of Au, Ag, and Cu metals using a femtosecond Ti: Sapphire laser. Applied Physics A. 1999;**69**:S359-S366

[20] Kelly R, Miotello A. Comments on explosive mechanisms of laser sputtering. Applied Surface Science. 1996;**96–98**:205-215

[21] Bulgakova NM, Bourakov IM. Phase explosion under ultrashort pulsed laser ablation: Modeling with analysis of metastable state of melt. Applied Surface Science. 2002;**197**:41-44

[22] Schäfer C, Urbassek HM, Zhigilei LV. Metal ablation by picosecond laser pulses: A hybrid simulation. Physical Review B. 2002;**66**:115404

[23] Leveugle E, Zhigilei L. Microscopic mechanisms of short pulse laser spallation of molecular solids. Applied Physics A. 2004;**79**:753-756

[24] Huang M, Zhao F, Cheng Y, Xu N, Xu Z. Origin of laser-induced near-subwavelength ripples: Interference between surface plasmons and incident laser. ACS Nano. Dec 2009;**3**:4062-4070

[25] Bonse J, Rosenfeld A, Krueger J. On the role of surface plasmon polaritons in the formation of laser-induced periodic surface structures upon irradiation of silicon by femtosecond-laser pulses. Journal of Applied Physics. Nov 15 2009;**106**:104910

[26] Hashida M, Namba S, Okamuro K, Tokita S, Sakabe S. Ion emission from a metal surface through a multiphoton process and optical field ionization. Physical Review B. 2010;**81**: 115442

[27] Huang M, Zhao F, Cheng Y, Xu N, Xu Z. The morphological and optical characteristics of femtosecond laser-induced large-area micro/nanostructures on GaAs, Si, and brass. Optics Express. Nov 8 2010;**18**:A600-A619

[28] Stuart BC, Feit MD, Herman S, Rubenchik AM, Shore BW, Perry MD. Optical ablation by high-power short-pulse lasers. JOSA B. 1996;**13**:459-468

[29] Stoian R, Ashkenasi D, Rosenfeld A, Campbell E. Coulomb explosion in ultrashort pulsed laser ablation of Al_2O_3. Physical Review B. 2000;**62**:13167

[30] Gattass RR, Mazur E. Femtosecond laser micromachining in transparent materials. Nature Photonics. 2008;**2**:219-225

[31] Anisimov S, Kapeliovich B, Perelman T. Electron emission from metal surfaces exposed to ultrashort laser pulses. Zhurnal Eksperimental'noi i Teoreticheskoi Fiziki. 1974;**66**:375-377

[32] Tien C, Qiu T. Heat transfer mechanism during short pulse laser heating of metals. American Society of Mechanical Engineers Journal of Heat Transfer. 1993;**115**:835-841

[33] Chen JK, Beraun JE. Numerical study of ultrashort laser pulse interactions with metal films. Numerical Heat Transfer Part A-Applications. Jul 2001;**40**:1-20

[34] Fujimoto J, Liu J, Ippen E, Bloembergen N. Femtosecond laser interaction with metallic tungsten and nonequilibrium electron and lattice temperatures. Physical Review Letters. 1984;**53**:1837

[35] Cattaneo C. A form of heat conduction equation which eliminates the paradox of instantaneous propagation. Compte Rendus. 1958;**247**:431-433

[36] Chen J, Latham W, Beraun J. The role of electron–phonon coupling in ultrafast laser heating. Journal of Laser Applications. 2005;**17**:63-68

[37] Lin Z, Zhigilei LV, Celli V. Electron-phonon coupling and electron heat capacity of metals under conditions of strong electron-phonon nonequilibrium. Physical Review B. 2008;**77**: 075133

[38] Anisimov SI, Rethfeld B. Theory of ultrashort laser pulse interaction with a metal. Proc. SPIE 3093, 1997; 192-203

[39] Wang S, Ren Y, Chang K, Cheng C, Chen J, Tzou D. Ablation of copper by a single ultrashort laser pulse. Journal of Laser Micro Nanoengineering. 2014;**9**:88-92

[40] Byskov-Nielsen J, Savolainen J-M, Christensen MS, Balling P. Ultra-short pulse laser ablation of copper, silver and tungsten: Experimental data and two-temperature model simulations. Applied Physics A-Materials Science & Processing. May 2011;**103**:447-453

[41] Ren YP, Chen JK, Zhang YW, Huang J. Ultrashort laser pulse energy deposition in metal films with phase changes. Applied Physics Letters. May 2011;**98**:191105

[42] Ren Y, Chen JK, Zhang Y. Optical properties and thermal response of copper films induced by ultrashort-pulsed lasers. Journal of Applied Physics. Dec 1 2011;**110**:113102

[43] Colombier JP, Combis P, Bonneau F, Le Harzic R, Audouard E. Hydrodynamic simulations of metal ablation by femtosecond laser irradiation. Physical Review B. Apr 2005;**71**: 165406

[44] Ren Y, Chen JK, Zhang Y. Modeling of ultrafast phase changes in metal films irradiated by an ultrashort laser pulse using a semiclassical two-temperature model. International Journal of Heat and Mass Transfer. 2012;**55**:1260-1627

[45] Cheng CW, Wang SY, Chang KP, Chen JK. Femtosecond laser ablation of copper at high laser fluence: Modeling and experimental comparison. Applied Surface Science. 2016;**361**: 41-48

[46] Chen J, Tzou D, Beraun J. Numerical investigation of ultrashort laser damage in semiconductors. International Journal of Heat and Mass Transfer. 2005;**48**:501-509

[47] Sokolowski-Tinten K, von der Linde D. Generation of dense electron-hole plasmas in silicon. Physical Review B. 2000;**61**:2643

[48] Fermann ME, Galvanauskas A, Sucha G. Ultrafast Lasers: Technology and Applications. Vol. 80. CRC Press; 2002

[49] Dausinger F, Lichtner F, Lubatschowski H. Femtosecond Technology for Technical and Medical Applications. Vol. 96. Springer Science & Business Media; 2004

[50] Chang H-W, Tsai Y-C, Cheng C-W, Lin C-Y, Lin Y-W, Wu T-M. Nanostructured Ag surface fabricated by femtosecond laser for surface-enhanced Raman scattering. Journal of Colloid and Interface Science. Aug 1 2011;**360**:305-308

[51] Cheng CW, Lin CY, Tseng WP, Ou KL, Peng PW. Electrosurgical unit with micro/nano structure and the manufacturing method thereof. ed: Google Patents; 2013

[52] Cheng CW, Lin CY, Wu PH, Chang KP, Horng JB, Wu WT, et al. Novel applications by femtosecond laser in electronics and medical device industries. In: Presented at the Smart Laser Processing Conference Yokohama, Japan; 2014

[53] Lin CY, Cheng CW, Ou KL. Micro/nano-structuring of medical stainless steel using femtosecond laser pulses. Physics Procedia. 2012;**39**:661-668

[54] Wu PH, Cheng CW, Chang CP, Wu TM, Wang JK. Fabrication of large-area hydrophobic surfaces with femtosecond-laser-structured molds. Journal of Micromechanics and Microengineering. 2011;**21**:115032

Chapter 10

Multi-Beam Multi-Target Pulsed Laser Deposition of AZO Films with Polymer Nanoparticles for Thermoelectric Energy Harvesters

Abdalla M. Darwish, Sergey S. Sarkisov,
Paolo Mele and Shrikant Saini

Additional information is available at the end of the chapter

http://dx.doi.org/10.5772/intechopen.70678

Abstract

In comparison with metallic thermoelectric films, oxide films with artificial nanodefects have been seldom studied. And there has been no report on the incorporation of island-shaped organic nanoparticles. We describe a new approach to introduce nanometer-sized phonon scatterers in aluminum-doped ZnO (AZO) thermoelectric thin films–concurrent multi-beam multi-target-pulsed laser deposition and the matrix-assisted pulsed laser evaporation (MBMT-PLD/MAPLE). The approach was used to make nanocomposite thin films of AZO matrix with evenly dispersed poly(methyl methacrylate) (PMMA) nanoparticles. The introduction of the nanoparticles enhanced phonon scattering with consequent decrease of thermal conductivity by 20%. The electrical conductivity did not decrease after the addition of the second phase, as it would be predicted by Wiedemann-Franz law, but improved by 350% over pure AZO film. The thermoelectric figure of merit of the nanocomposite film became twice that of the pure AZO film. Taking advantage of room-temperature deposition, optimized AZO nanocomposite films are expected to be used in real applications, such as thin film modules deposited on flexible polymeric substrates for ubiquitous harvesting of the waste heat.

Keywords: laser ablation, pulsed laser deposition, matrix-assisted pulsed laser evaporation, nanocomposite films, AZO, polymer nanoparticles, thermos-electric energy harvesters

1. Introduction

ZnO is a well-known *n*-type semiconductor used in a variety of applications such as optical devices, piezoelectric transducers, transparent electrodes, and gas sensors [1]. Furthermore, interesting results have been published on Al-doped ZnO (AZO) as a sustainable thermoelectric material to replace toxic and expensive materials based on Se, Te, and Bi [2]. More recently, researchers have focused on the preparation of AZO in thin film form due to the possibility to control better the morphology and crystalline orientation of the films. In pure AZO films prepared by the conventional pulsed laser deposition (PLD) on single crystal substrates at relatively low temperatures (400–600°C), the thermal conductivity κ has been reduced significantly with respect to the bulk material due to the enhanced phonon scattering at the film substrate interface and at grain boundaries [3]. Ultimately, this approach has yielded a higher thermoelectric figure of merit, ZT [4]. Further improvement of ZT can be achieved by the introduction of multilayer structures and nanodefects acting as additional phonon scatterers reducing the thermal conductivity [5, 6].

In contrast to metallic thermoelectric films, oxide films with artificial nanodefects have been seldom studied: $SrTiO_3$/Nb-$SrTiO_3$ [7], AZO/hydroquinone [8], and AZO/Y_2O_3 [9]. There is still no report on oxide films with nanodefects in the form of polymer nanoparticles. This chapter describes nanocomposite thin films made of AZO matrix with evenly dispersed poly (methyl methacrylate) (PMMA) nanoparticles produced by the concurrent multi-beam multi-target pulsed laser deposition of AZO and matrix-assisted pulsed laser evaporation of the polymer (MBMT-PLD/MAPLE) [10–13].

A wide variety of nanocomposite materials based on polymers and inorganic substances can be classified as "polymer nanocomposites," where the polymer serves as a host for inorganic nanoparticles [14, 15]. In case, when the volume fraction of the polymer decreased below 90% (and the interaction between the inorganic inclusions became stronger, as illustrated in **Figures 1** and **2**), the resulting materials were often called as "organic-inorganic hybrid composites" [14]. In some occasions, one polymer was used as a host for a nanoparticles made of

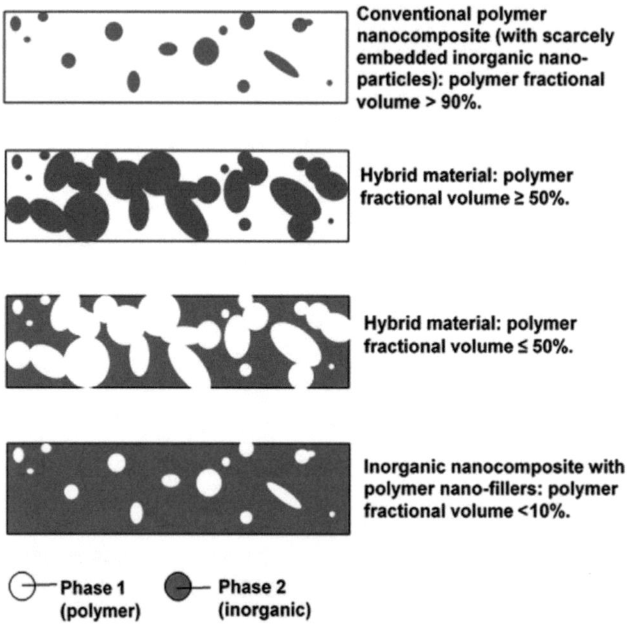

Figure 1. Two-phase polymer-inorganic nanocomposite films.

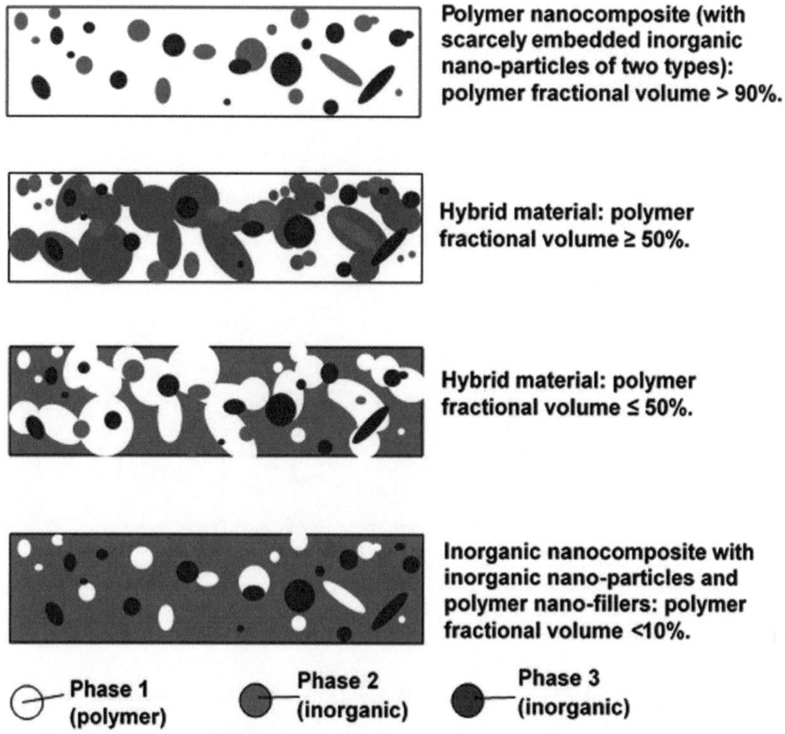

Figure 2. Three-phase polymer-inorganic nanocomposite films.

another polymer [16]. Making a uniformly mixed nanocomposite films with low fractional volume (<10%) of the polymer and at least one inorganic phase prevailing and serving as a host for the polymer nanoparticles (**Figures 1** and **2**) has remained an unachievable goal because the list of suitable technologies was limited mainly to the sol-gel process [17]. Hence, the two novelty features of this work are: (i) the use of MBMT-PLD/MAPLE to produce an oxide-polymer nanocomposite film at room temperature with the oxide acting as a host to the polymer nanoparticles and (ii) the use of the polymer nanoparticles as phonon scatterers. The description given below demonstrates how the approach improves the thermoelectric properties of AZO films.

2. Methods and materials

The MBMT-PLD/MAPLE system at Dillard University is schematically presented in **Figure 3**. Two pulsed laser beams ablated concurrently two targets and ejected the target materials in the plumes that propagated toward the surface of a substrate where they mixed and formed a nanocomposite film. Linear actuators tilted the targets and changed the directions of the plumes in order to secure uniform mixing of the target materials in the film. Plume 1 is formed by laser beam 1 ablating a conventional solid inorganic PLD target. The second target (to the right) was a polymer solution frozen by circulating liquid nitrogen (LN). Accordingly, the process of polymer deposition was the matrix-assisted pulsed laser evaporation (MAPLE) [10–13]. All the

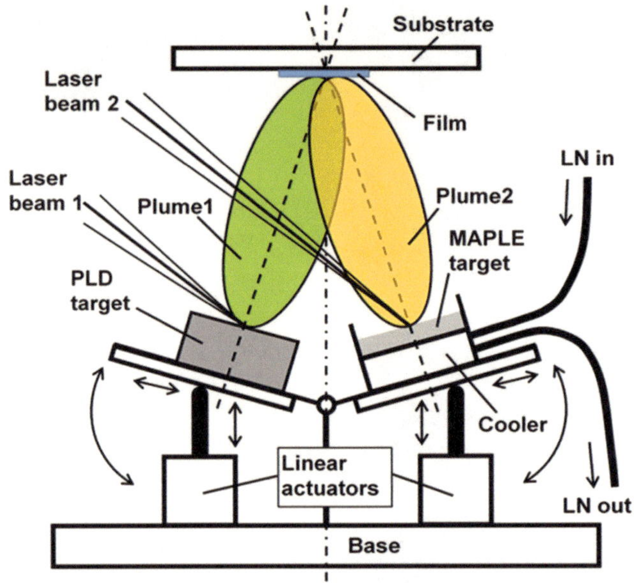

Figure 3. The schematic of the multi-beam multi-target-pulsed laser deposition system used to make the films.

components of the deposition system were placed in a vacuum chamber (not shown) with optical windows for the laser beams. The inorganic, AZO target was a pellet of $Zn_{0.98}Al_{0.02}O$ where the aluminum fraction was 2% of the total by weight as compared with zinc, not counting the oxygen. The AZO pellet had 20 mm in diameter and 3 mm in thickness. The pellet was prepared by the spark plasma sintering method as described elsewhere [18]. The MAPLE target was a solution of PMMA in chlorobenzene at a proportion of 1 g solids per 10 mL liquids filtered with 0.2 μm filter. The solution was poured in a copper cup of the MAPLE target assembly (**Figure 3**) and frozen in liquid nitrogen (LN). The laser source was a Spectra Physics Quanta Ray Nd:YAG Q-switched Pro-250-50 laser with a pulse repetition rate of 50 Hz, 750 mJ energy per pulse at the 1064 nm fundamental wavelength, and 400 mJ energy per pulse at the 532 nm second harmonic. The AZO target was ablated with the 532 nm beam. The fluence was tuned up between 0.8 and 1.0 J/cm^2 per pulse. The MAPLE target was evaporated with the 1064 nm beam. The fluence was ranging from 0.84 to 2.4 J/cm^2 per pulse to maintain a volume fraction of PMMA in the AZO matrix of about 5%. The deposition time was 3.0 min. The thickness of the deposited films was approximately 150 nm as measured with an atomic force microscope. The films were deposited on Al_2O_3 (sapphire) (100) single crystal substrates at room temperature in vacuum. The inorganic target was rotated during the irradiation of laser beam. The substrate-target distance was maintained about 35 mm. Deposition of pure AZO films in the same experimental conditions was carried out in order to have reference samples.

3. Results and discussion

X-ray diffraction (XRD) analysis of the as-grown AZO-PMMA films conducted with a Bruker D2 Phaser diffractometer (**Figure 4**) revealed the presence of polycrystalline AZO, while peaks

http://dx.doi.org/10.5772/intechopen.70678

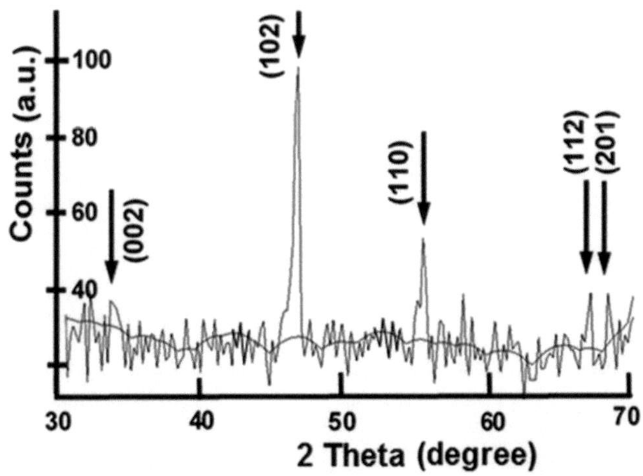

Figure 4. X-ray diffraction spectrum of the films (peaks are indexed with [*hkl*] reflections of AZO matrix).

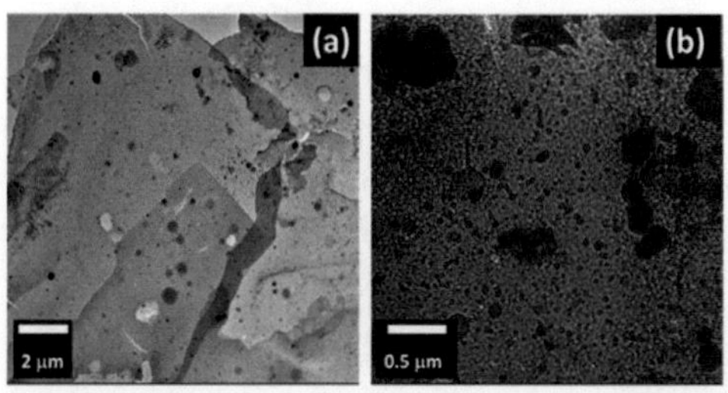

Figure 5. Top-view TEM images of the films deposited on Al_2O_3 substrate with (a) low and (b) high magnification.

of PMMA were absent. Polycrystallinity of AZO is due to the fact that the ablation in MBMT-PLD/MAPLE system was performed at room temperature to solidify PMMA after its condensation on the substrate (PMMA has been reported to melt at about 430 K [19]). As widely observed in literature, AZO films deposited at room temperature by conventional PLD are commonly polycrystalline [20–22].

Figure 5 (a) and **(b)** present TEM images of the film samples that were peeled off from the Al_2O_3 substrate. Typically, polymer nanoparticles-nanoclusters of a size of 10–50 nm are seen embedded in the AZO matrix. The formation of the nanoclusters is due to the heavy entanglement of the polymer molecules ruled by chemical interactions during their condensation on the substrates and chemical interaction with the ceramic host. Some polymer molecules could also self-assemble in individual fibers creating networks across the AZO matrix.

The electrical conductivity versus temperature ($\sigma - T$) characteristics of the films were measured by the conventional four-probe technique, and Seebeck coefficient was measured by a commercial system (from MMR technologies) in the temperature range from 300 to 600 K. Further, the thermal conductivity κ of the films was measured at 300 K using the time domain thermoreflectance (TDTR) technique [23, 24].

Sample	Electrical conductivity σ (S/cm) (300K/600K)		Seebeck coefficient S (μV/K) (300K/600K)		Power factor σS^2 (mW/(m.K^2)) (300K/600K)		Thermal conductivity κ (Wm^{-1}K^{-1}) 300K		ZT (300K/600K)	
	Pure AZO	AZO + PMMA	Pure AZO	AZO + PMMA	Pure AZO	AZO + PMMA	Pure AZO	AZO + PMMA	Pure AZO	AZO + PMMA
Film on Al_2O_3	433/539	1382/1630	−15/−30	−9/−20	0.01/0.05	0.01/0.07	7.4 ± 0.2	5.9 ± 0.3	0.005/0.04	0.0055/0.07
Bulk AZO	206/152		−132/−150		0.35/0.34		34		0.0035/0.014	

Table 1. Comparative performance of the nanocomposite thermoelectric AZO-PMMA films.

Electro-conductive and thermo-electrical characteristics of the AZO-PMMA nanocomposite films in the range of 300–600 K are presented in **Figures 6** through 11 and **Table 1**. Performance of pure AZO films deposited on the same substrate is also reported for comparison.

Electrical conductivity σ of AZO + PMMA films is higher than that of the AZO films over all the measured temperature range and increases with the increase of temperature (**Figure 6**). The highest value of the electrical conductivity is 1630 S/cm at 600 K, three times increase comparing to pure AZO films. 1 S/cm (siemens per centimeter) is 100 times 1 S/m, the unit of electrical conductivity in SI, 1 S/m = 1 (A^2.s^3)/(kg.m^3). The significant increase of σ can be explained as follows. At first, as long as PMMA remains unaffected by the temperatures applied during the measurement (below 430 K), the polymer can be assumed to increase the number of oxygen vacancies in AZO usually resulting in the increase of electrical conductivity in a hybrid system. Another scenario is that above 430 K, PMMA decomposes in CO_2 and by-products. Then, carbon dioxide converts into amorphous carbon, which is known to have good electrical conductivity (around 670 S/cm [25]). Since, virgin PMMA has very low electrical

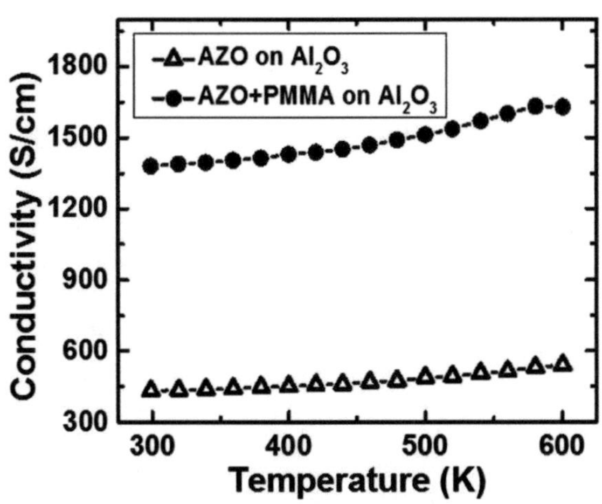

Figure 6. Temperature dependence of the electrical conductivity of AZO and AZO + PMMA films deposited on Al_2O_3 substrates.

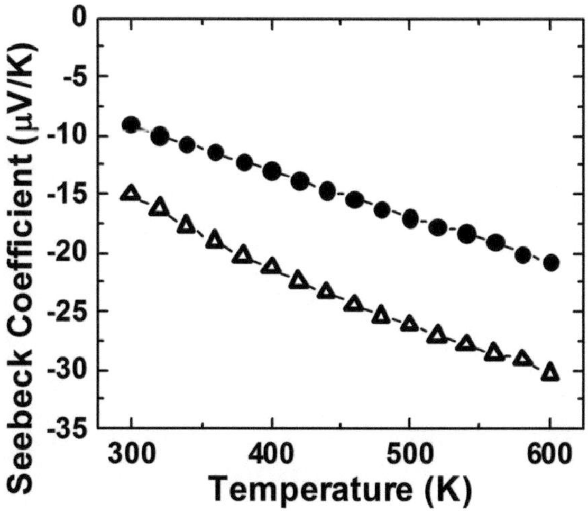

Figure 7. Temperature dependence of Seebeck coefficient of AZO and AZO + PMMA thin films on Al_2O_3 substrates. Symbols used to represent data points are the same as in **Figure 6**.

conductivity (of the order of 10^{-13} S/cm [26]), the degree of polymer conversion into carbon could be assumed rather substantially, and thus, affecting not only σ, but other physical properties that were becoming similar to those of an AZO-carbon nanocomposite.

Seebeck coefficient versus temperature plot is presented in **Figure 7**. The negative sign of Seebeck coefficient indicates electrons as the electric charge carriers and correspondingly the hybrid composite material being an *n*-type semiconductor. The magnitude of Seebeck coefficient increases with the temperature increase. Overall, the magnitude of Seebeck coefficient of AZO + PMMA hybrid films is lower than that of the pure AZO thin film due to the fact that it has more electric charge carriers. The highest absolute value of Seebeck coefficient is about −30 μV/K at 600 K for AZO thin films. Based on the measured electrical conductivity and Seebeck coefficient, power factor $PF = \sigma S^2$ was calculated. The value of *PF* for AZO + PMMA films (**Figure 8**) was found to be higher than that of the AZO thin films due to the higher electrical conductivity. The maximum value of *PF* is about 0.07 mW/(m.K^2) at 600 K.

Electrical conduction is governed by thermally activated hopping mechanism as is indicated by linear plot of $\ln(\sigma T)$ versus $1/T$ in **Figure 9**. The electrical conductivity follows relation $\sigma T = A\exp[-E_\sigma/(k_b T)]$ (where E_σ is the activation energy for electrical conductivity, A is the pre-exponential factor, k_B is the Boltzmann constant) [27]. Activation energy E_σ was calculated using the slope of the linear fit of the plots to be about 44 and 47 meV for AZO + PMMA and AZO films respectively. Slightly lower activation energy of the nanocomposite film indicates that the charge carriers with lower energy hopping between the grain boundaries also participate in electrical conduction. Due to this, AZO + PMMA film exhibits higher electrical conductivity. Further, the temperature dependence of Seebeck coefficient can be described as $S = (k_B/e)[E_S/(k_B T) + B]$, where E_S is the thermopower activation energy and B is the heat of transport taken as temperature independent [27]. Activation energy E_S was calculated from the slope of linear fitting of the plots of S versus $1/T$ (**Figure 10**) to be 7.0–9.2 meV for AZO + PMMA and

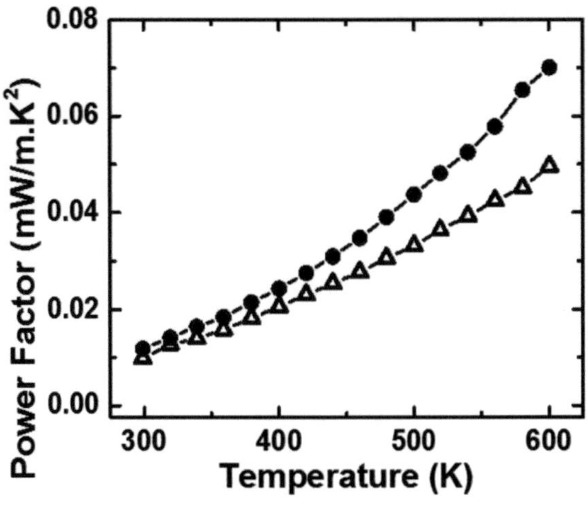

Figure 8. Temperature dependence of the power factor of AZO and AZO + PMMA thin films on Al_2O_3 substrates. Symbols used to represent data points are the same as in **Figure 6**.

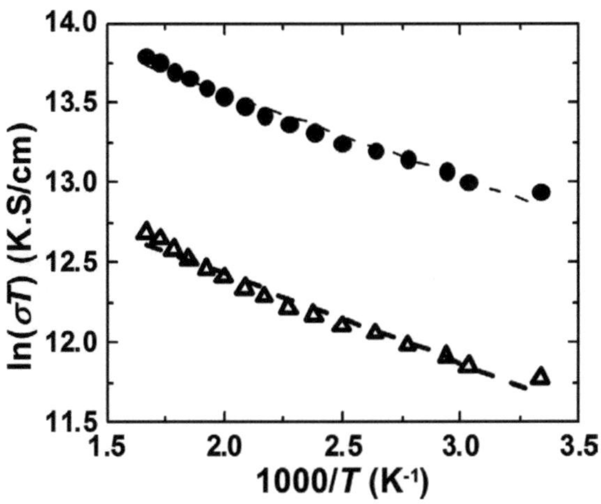

Figure 9. Inverse temperature dependence of ln(σT) of AZO and AZO + PMMA thin films on Al_2O_3 substrate. Dashed lines are the linear fit of the corresponding plots. Symbols used to represent data points for different samples as in **Figure 6**.

AZO films, respectively. The values of E_σ are about five to six times greater than E_S, which suggests that a part of the activation energy is arising from the activated mobility of the small polaron, which does not contribute to thermopower and the formation of which has been reported for a number of oxides [27, 28]. It is worth of mentioning that, besides the thermal hopping mechanisms, the temperature-independent tunneling electron transport mechanism can be sometimes observed in the granulated multi-component films prepared by PLD as it has been recently reported (for the films made of gold and silver nanogranules at a temperature below 60°C) in paper [29].

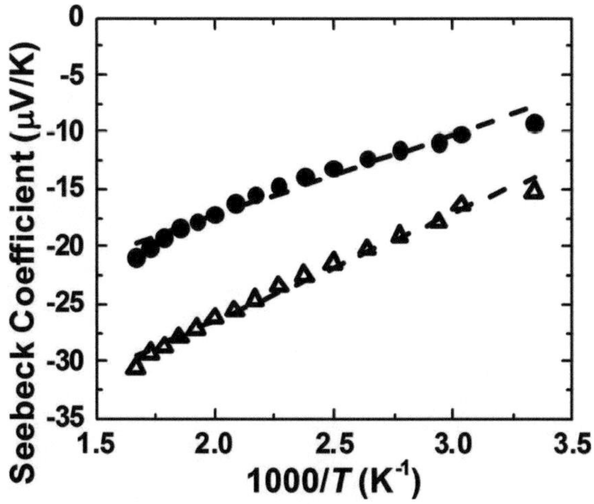

Figure 10. Inverse temperature dependence of Seebeck coefficient of AZO and AZO + PMMA films on Al_2O_3 substrates. Dashed lines are the linear fit of the corresponding plots. Symbols used to represent data points for different samples are the same as in **Figure 6**.

Thermal conductivity κ (at 300 K) was measured by the above-described TDTR method to be (5.9 ± 0.3) and (7.4 ± 0.2) W/m.K for AZO + PMMA and AZO film, respectively. Less κ for the hybrid film can be attributed to two factors: low thermal conductivity of PMMA (κ = 0.25 W/m.K at room temperature [30]) and the phonon scattering from uniformly dispersed PMMA nanoparticles (as shown in **Figure 3**). These out-of-plane film thermal conductivity values $\kappa(300K)$ measured at 300 K were further used in calculating the thermoelectric figure of merit $ZT = \sigma S^2 T/\kappa'(T)$ in the range of 300–600 K, where $\kappa'(T)$ is the film in-plane thermal conductivity at a given temperature T. For elevated temperatures (T > 300 K), the replacement of $\kappa'(T)$ by $\kappa(300K)$ can be validated by two facts: (i) in ZnO films, the in-plane thermal conductivity determined at different conditions was always higher than the out-of-plane one [31]; and (ii) the thermal conductivity of ZnO films was found to decrease with increasing temperature [32]. So, the replacement will not change the trend of ZT growth with T. **Figure 11** and **Table 1** show that ZT for AZO + PMMA film is always greater than that of the pure AZO film (twice as great at 600 K). This is due to the higher electrical conductivity and lower thermal conductivity of the nanocomposite film.

The electrical conductivity σ of AZO + PMMA films turned out to violate the Wiedemann-Franz law, $\kappa_{el}/\sigma = L_0 T$ (κ_{el} is the electronic component of the combined thermal conductivity $\kappa = \kappa_{ph} + \kappa_{el}$, κ_{ph} is the phonon component, L_0 is the Lorentz number) and got decoupled from the thermal conductivity, a common feature of complex and disordered media (so called "electron crystals and phonon glasses") [33]. The hypothetical structure of the composite AZO + PMMA medium can be schematically depicted in **Figure 12**. The polymer phase in the inorganic AZO matrix is assumed to be present in a variety of forms: nanoclusters (nanoparticles or nanodots), nanofibers, and fiber nanobundles. Nanoclusters are formed by the entangled long-chain polymer molecules. Nanofibers may be formed by straightened single polymer strands or the bundles of parallel polymer strands. All these formations (nanoclusters,

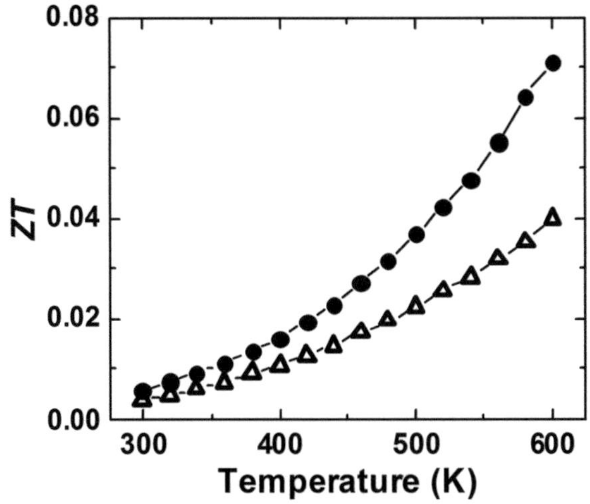

Figure 11. Temperature dependence of the thermoelectric figure of merit *ZT* of AZO and AZO + PMMA thin films on Al_2O_3 substrates. Symbols used to represent data points are the same as in **Figure 6**. *ZT* was calculated using the thermal conductivities of AZO and AZO + PMMA films on Al_2O_3 substrates measured at 300 K (see **Table 1**).

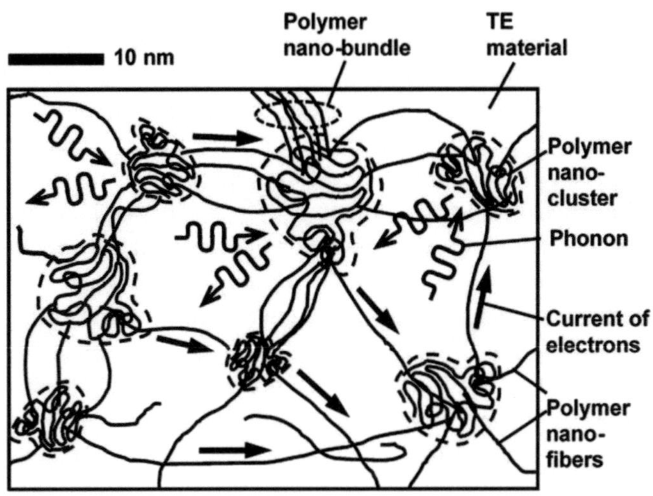

Figure 12. Schematic diagram of the AZO + PMMA nanocomposite film illustrating the effects of the polymer nanophase on the electrical and thermal conductivity of the AZO host.

nanofibers, and nanobundles) can be interconnected forming a network across the inorganic host. The polymer nanoclusters increased phonon scattering and reduced thermal conductivity of the nanocomposite films.

As was mentioned above, one of the mechanisms of significant (three times) increase of the electrical conductivity of AZO + PMMA films with respect to pure AZO could be the carbonization of the polymer phase. Carbonization occurs as a pyrolysis–a thermochemical decomposition of organic material at elevated temperatures without oxygen (or any halogen).

It involves the simultaneous change of chemical composition and physical phase. Pyrolysis is a type of thermolysis, and is most commonly observed in organic materials exposed to elevated temperatures. It is one of the processes involved in charring wood, starting at 200–300°C (390–570°F). It also occurs in fires where solid fuels are burning or when vegetation comes into contact with lava in volcanic eruptions. In general, pyrolysis of organic substances produces gas and liquid products and leaves a solid residue richer in carbon content, char. Pyrolysis leading to carbonization of polymers, such as PMMA, poly(vinylidene fluoride)—PVDF, poly (acrylonitrile)—PAN, and poly-N-vinylformamide—PNVF, during laser ablation and heat treatment has been reported before and used to modify the electrical conductivity of polymer coatings and fibers [34–36]. In case of AZO + PMMA films, carbonization could turn the network of the polymer nanoclusters and nanofibers into an additional efficient passage for the electric current (**Figure 12**) contributing to the overall increase of the electrical conductivity of the nanocomposite films and eventually to the increase of ZT.

4. Conclusions

The new concurrent MBMT-PLD/MAPLE deposition method has been successfully used to produce nanocomposite oxide-polymer thermoelectric films composed of AZO matrix with uniformly dispersed PMMA nanoparticles. The deposition was conducted at room temperature and on the cold (kept also at room temperature) substrate with no buffer gas. The volume fraction of the polymer material in the AZO matrix was chosen to be around 5%. These deposition conditions affected favorably the enhancement of the thermoelectric effect because the long-chain polymer molecules had a better chance to reach the substrate and mix with the inorganic matrix material without being decomposed. The chosen polymer fraction was small enough not to compromise substantially the electrical conductivity of the matrix. The PMMA nanoparticles allowed enhancing the phonon scattering with consequent decrease of thermal conductivity as compared to pure AZO film. The electrical conductivity did not drop after the addition of the second phase, as it would be expected from Wiedemann-Franz law, but grew three times higher over pure AZO film. This was assumingly due to the increase of the number of oxygen vacancies in AZO caused by PMMA or/and by the polymer carbonization at elevated temperatures. The thermoelectric figure of merit was improved by a factor of two. Further experiments would be necessary to establish the optimal proportion of the polymer nanoadditive in AZO matrix and other deposition parameters that maximize the thermoelectric performance of the nanocomposite films. Taking advantage of room-temperature deposition, optimized AZO nanocomposite films are expected to be used in real applications such as thin film modules deposited on flexible polymer substrates for ubiquitous harvesting of the waste heat.

Acknowledgements

A.D. appreciates the financial support from US Army Grant No. W911NF-15-1-0446.

Author details

Abdalla M. Darwish[1]*, Sergey S. Sarkisov[2], Paolo Mele[3] and Shrikant Saini[3]

*Address all correspondence to: adarwish@bellsouth.net

1 Dillard University, New Orleans, USA

2 SSS Optical Technologies LLC, Huntsville, USA

3 Muroran Institute of Technology, Muroran, Japan

References

[1] Ozgür U, Alivov YI, Liu C, Teke A, Reschikov MA, Dögan S, et al. A comprehensive review of ZnO materials and devices. Journal of Applied Physics. 2005;**98**(4):103/041301. DOI: 10.1063/1.1992666

[2] Ohtaki M, Tsubota T, Eguchi K, Arai H. High-temperature thermoelectric properties of $(Zn_{1-x}Al_x)O$. Journal of Applied Physics. 1996;**79**:1816

[3] Mele P, Saini S, Honda H, Matsumoto K, Miyazaki K, Hagino H, et al. Effect of substrate on thermoelectric properties of Al-doped ZnO thin films. Applied Physics Letters. 2013;**102**:253903

[4] Saini S, Mele P, Honda H, Suzuki T, Matsumoto K, Miyazaki K, et al. Effect of self-grown seed layer on thermoelectric properties of ZnO thin films. Thin Solid Films. 2016;**605**:289-294

[5] Hicks LD, Dresselhaus MMS. Effect of quantum-well structures on the thermoelectric figure of merit. Physical Review B. 1993;**47**:12727

[6] Venkatasubramanian R, Siivola E, Colpitts T, O'Quinn B. Thin-film thermoelectric devices with high room-temperature figure of merit. Nature. 2001;**413**:597-602

[7] Ohta H, Kim SW, Mune Y, Mizoguchi T, Nomura K, Ohta S, et al. Giant thermoelectric Seebeck coefficient of a two-dimensional electron gas in $SrTiO_3$. Nature Materials. 2007;**6**:129-134

[8] Tynell T, Giri A, Gaskins J, Hopkins PE, Mele P, Miyazaki K, et al. Efficiently suppressed thermal conductivity in ZnO thin films via periodic introduction of organic layers. Journal of Materials Chemistry A. 2014;**2**:12150

[9] Mele P, Saini S, Tiwari A, Hopkins PE, Miyazaki K, Ichinose A, et al. Thermoelectric and structural characterization of Al-doped ZnO/Y_2O_3 multilayers. Journal of Nanoscience and Nanotechnology. 2017;**17**:1616

[10] Darwish AM, Sagapolutele MT, Sarkisov S, Patel D, Hui D, Koplitz B. Double beam pulsed laser deposition of composite films of poly(methyl methacrylate) and rare earth fluoride upconversion phosphors. Composites B. 2013;**55**:139-146

[11] Darwish AM, Burkett A, Blackwell A, Taylor K, Sarkisov S, Patel D, et al. Polymer-inorganic nano-composite thin film upconversion light emitters prepared by double-beam matrix assisted pulsed laser evaporation (DB-MAPLE) method. Composites B. 2015;**68**:355-364

[12] Darwish AM, Wilson S, Blackwell A, Taylor K, Sarkisov SS, Patel DN, et al. Ammonia sensor based on polymer-inorganic nano-composite thin film upconversion light emitter prepared by double-beam pulsed laser deposition. American Journal of Materials Science. 2015;**5**:8-15

[13] Darwish AM, Moore S, Mohammad A, Alexander D, Bastian T, Sarkisov S, et al. Polymer nano-composite films with inorganic upconversion phosphor and electro-optic additives made by concurrent triple-beam matrix assisted and direct pulsed laser deposition. Composites B. 2017;**109**:82-90

[14] Gross S, Camozzo D, Di Noto V, Armelao L, Tondello E. PMMA: A key macromolecular component for dielectric low-κ hybrid inorganic–organic polymer films. European Polymer Journal. 2007;**43**:673-696

[15] Marquis DM, Guillaume É, Chivas-Joly C. Properties of Nanofillers in polymer. In: Nanocomposites and Polymers with Analytical Methods. Rijeka, Croatia: InTech; 2011

[16] Abdollahi M, Alboofetileh M, Rezaeia M, Behroozb R. Comparing physico-mechanical and thermal properties of alginate nanocomposite films reinforced with organic and/or inorganic nanofillers. Food Hydrocolloids. 2013;**32**:416-424

[17] Ishchenko SS, Budnitskaya VL, Lebedev EV, Kozak NV. Hybrid composites based on organic-inorganic sol-gel systems. Russian Journal of Applied Chemistry. 2011;**84**:1952-1957

[18] Mele P, Kamei H, Yasumune H, Kaname H, Miyazaki K. Development of thermoelectric module based on dense $Ca_3Co_4O_9$ and $Zn_{0\cdot98}Al_{0.02}O$ legs. Metals and Materials International. 2014;**20**(2):389-397

[19] Smith WF, Hashemi J. Foundations of Materials Science and Engineering. New York: McGraw-Hill; 2006. p. 509

[20] Sun Y, Maemoto T, Sasa S. Future of electron devices, Kansai (IMFEDK). In: Proceedings of 2014 IEEE International Meeting for Future of Electron Devices; 19–20 June 2014; Kansai, Kyoto, Japan. IEEE; 2014

[21] Liu Y, Zhao L, Lian J. Al-doped ZnO films by pulsed laser deposition at room temperature. Vacuum. 2006;**81**:18-21

[22] Gondoni P, Mazzolini P, Russo V, Diani M, Amati M, Gregoratti L, et al. Tuning electrical properties of hierarchically assembled Al-doped ZnO nanoforests by room temperature pulsed laser deposition. Thin Solid Films. 2015;**594**:12-17

[23] Schmidt J, Chen X, Chen G. Pulse accumulation, radial heat conduction, and anisotropic thermal conductivity in pump-probe transient thermoreflectance. The Review of Scientific Instruments. 2008;**79**:114902

[24] Yada S, Oyake T, Sakata M, Shiomi J. Filler-depletion layer adjacent to interface impacts performance of thermal interface material. AIP Advances. 2016;**6**:015117

[25] Serway RA, Faugh JS. College Physics. 6th ed. Belmont, CA: Thomson; 2003

[26] Hussien B. The D.C and A.C. electrical properties of PMMA-Al_2O_3 composites. The European Journal of Scientific Research. 2011;**52**:236-242

[27] Ponnambalam V, Varadaraju UV. Observation of variable-range hopping up to 900 K in the $YLa_xBa_2 - xCu_3O_7 - \delta$ system. Physical Review B. 1995;**52**:16213

[28] Cutler M, Mott F. Observation of Anderson localization in an electron gas. Physics Review. 1969;**181**:1336-1340

[29] Kavokin A, Kutrovskaya S, Kucherik A, Osipov A, Vartanyn T, Arakelyn S. The crossover between tunnel and hopping conductivity in granulated films of noble metals. Superlattices and Microstructures. In Press, Corrected Proof. Available online 19 June 2017;1-5

[30] Assaael MJ, Antoniadis KD, Wu J. New measurements of the thermal conductivity of PMMA, BK7, and Pyrex 7740 up to 450K. International Journal of Thermophysics. 2008; **29**:1257-1266

[31] Xu Y, Goto M, Rato R, Tanaka Y, Kagawa Y. Thermal conductivity of ZnO thin film produced by reactive sputtering. Journal of Applied Physics. 2012;**111**:084320

[32] Alvarez-Quintana J, Martinez E, Perez-Tijerina E, Perez-Garcia SA, Rodriguez-Vejo J. Temperature dependent thermal conductivity of polycrystalline ZnO films. Journal of Applied Physics. 2010;**107**:063713

[33] Wikipedia. Wiedemann–Franz law [Internet]. [Updated: 2-July-2017]. Available from: https://en.wikipedia.org/wiki/WiedemannFranz_law [Accessed: 6-July-2017]

[34] Blancet GB, Fincher CR Jr. Laser induced unzipping: A thermal route to polymer ablation. Applied Physics Letters. 1994;**65**(10):1311-1313

[35] Morita N, Shimotsuma Y, Nishi M, Sakakura M, Miura K, Hirao K. Direct micro carbonization inside polymer using focused femtosecond laser pulses. Applied Physics Letters. 2014;**105**:201104

[36] Molenda M, Swietoslawski M, Drozdek M, Dudek B, Dziembaj R. Morphology and electric conductivity of carbon nanocoatings prepared from pyrolized polymers. Journal of Nanomaterials. 2014;**2014**:7/103418

Chapter 11

Fabrication of a Cell Electrostimulator Using Pulse Laser Deposition and Laser Selective Thin Film Removal

Angel Luis Aragón Beloso,
María del Carmen Bao Varela,
Alejandro Fernández Rodríguez, Gerard O'connor,
Eliseo Pérez Trigo, Antonio Pazos Álvarez and
Daniel Nieto García

Additional information is available at the end of the chapter

http://dx.doi.org/10.5772/intechopen.70677

Abstract

In this work, we present a laser-based process for fabricating a cell electrostimulator. The fabrication methodology comprises two laser processes: a pulse laser deposition (PLD) of an aluminum thin film on soda-lime glass and a laser-based selectively removal of the thin film. The laser set-up for PLD consist of Nd:YVO_4 Rofin Power line 20E (1064 nm wavelength, 20 ns pulse width) focused by a lens of 160 mm focal length inside a vacuum chamber to strike a target of the deposited material. The same laser is used for selectively removing the thin film but focused by a lens of 100 mm focal length. The geometry design is made in CAD-like software. Before microfabrication, a thin aluminum layer (1 μm thickness) is deposited on soda-lime glass using the PLD method. In order to assemble the device, the electrical stimulator is placed between two polycarbonate sheets of 1.5 mm thickness. To prevent any contact with the electric circuit, a thin silicate glass (100 μm) is placed over the electrostimulator. Simulations were performed using ANSYS Maxwell software, verifying that the induced electrical field achieves the minimum for cell stimulation.

Keywords: electrical stimulator, PLD, thin film removal, laser ablation, thin film

1. Introduction

Electrostimulation is an electric current application method used for excitation and activation of certain organs and systems of the human body [1]. Although many organs and systems can be stimulated by electric currents by adequate methods and techniques applied, the most widely practiced application is heart electrostimulation, being a specific section of medicine, and electrostimulation of motor nerves and muscles [2].

For electrostimulation, direct impulse currents of different impulse shapes are applied at different duration and frequency. This method is widely used for therapeutic purposes, where the electrical stimulation is used to restore the function of damaged motor nerve, to treat paralysis or muscles restoration while causing motor excitation and contraction of the muscles [3].

Aside from these therapies, where the electric pulses are applied directly to the patient, electrostimulators are used in medical investigation as a tool to induce electric signals in cell cultures simulating different body conditions [4]. In this method, electrostimulators must be carefully designed to work in very specific circumstances. Usually, the device consists of two electrodes in direct contact with the cell culture, where the electric stimulus is induced by applying a voltage between them [5]. In this situation, it is necessary to choose the right material for the electrodes. They must be biocompatible to avoid toxic reactions in the culture and they also must efficiently transfer the charge to the medium in order to minimize the electrode degradation [6].

All these circumstances significantly limit the choice of the materials used in the device [7]. With the purpose of avoiding this problem, we present a design of an electrostimulator in which the electrodes are not in contact with the cell culture, solving the biocompatibility issue, and the stimulation is applied by inducing electric fields, avoiding degradation by electrochemical phenomena. Applications study the growth and information processing of neurons [8], capillary electrophoresis chips for the separation of biochemicals such as amino acids and nucleotides [9], and microstructures for the analysis of DNA [10]. Depending on which types of cells are needed to stimulate, the electrical forces induced can be classified into electrophoresis (EP) or dielectrophoresis (DEP) depending on whether they act on a particle's fixed or induced charge, respectively. In both cases, the electrical field needed to induce notable forces in the cell culture depends on the type of cell, but is usually about 10^4 V/m [11].

Different techniques have been used in the fabrication of microelectrodes. Most of them are fabricated using standard photolithography [12] because this technique is more versatile and allows the preparation of electrodes with a broad range of shapes and sizes, both single electrodes and the electrode arrays. However, photolithography technique needs usually about 3 h to fabricate planar electrodes, and it demands the use of several components in various steps and the final use of chemical components to remove the additional film.

But in the past decades, pulse laser deposition (PLD) has emerged as one of the most popular, flexible, and simple technique for depositing a wide range of materials [13]. Because of its inherent versatility, flexibility, and speed, this method can be applied to almost any material, from simple metals to multicomponent high-quality single crystals [14].

The PLD technique is a physical vapor deposition process where a pulsed laser is focused on the surface of the material to be deposited as a thin film over a substrate. During the interaction between a laser beam with enough energy density and a target, each laser pulse ablates a small amount of the target, creating a plasma plume where particles of different sizes (form single atoms to clusters) are extracted at high energies from the material. These ejected particles usually have an initial speed that could reach values of tens of kilometers per second, decreasing gradually while interacting with the ambient atmosphere, providing a material flux for film growth. In order to optimize this flux, the interaction must happen inside a vacuum chamber to avoid dispersion between the particles and the ambient air [15]. For some applications, it might be interested to collect these ablated particles on a particular surface. By doing this, the coating surface can have different properties from the original material: the object surface could become harder, noncorrosive, or conductive, while the rest of the object properties will remain practically the same [16–18].

Pulse laser deposition presents some serious advantages comparing to other techniques. First, it is very easy to implement, and the flexibility in terms of wavelength and power density allows the ablation of almost any material. Second, almost any geometry of the set-up can be chosen with a high degree of freedom because the laser beam is not part of the vacuum system [19]. Third, the use of a pulse laser enables a very precise control over the growth rate. Besides, due to the fact that this technique does not include impurities from the holder or the environment, it is a very clean technique [20]. And finally, the high kinetic energy of the ejected particles in comparison with other methods leads to an increase in the density of the layer and its adherence to the substrate [21]. However, the high laser energy involved in the process (which makes that microscopic or macroscopic particles can be ejected from the target) and an inhomogeneous distribution in the laser beam profile (and, in consequence, in the angular energy distribution) can cause an inhomogeneous layer with slightly different local densities in detriment to the desired properties of the film [22].

The attractive characteristics of PLD in the synthesis of multicomponent thin-film materials make it the perfect technique for a lot of applications [23, 24]. In some cases, it could be the synthesis of a thin-film material or structure [25]. In other cases, the research focused in the development of specific devices, such as in the growth of nitride films in the development of LED devices or in the fabrication of thin waveguiding films [26, 27]. PLD has also proven to be very effective in the growth of crystalline oxides, [28] or PLD-grown of high-temperature superconducting films (HTS), whose applications include high-frequency electronics for radio frequency, microwave communications, and superconducting quantum interference devices (SQUIDs) for the detection of magnetic fields [29]. In the area of wireless communication, radio frequency filters based on HTS thin films presents excellent results by reducing interference from out-of-band signals due to the low loss in the microwave frequency range for epitaxial oxide superconducting films in which PLD is extremely effective at production [30, 31]. Pulsed laser deposition has also played a key role in exploring more radical oxide device concepts, including various electric field-effect devices based on semiconducting oxide materials [32].

Selective laser elimination of thin materials from glass substrates presents some advantages in terms of time and material by being a relatively accessible and nonexpensive technique. While photolithography needs hours to fully fabricate the electrostimulator circuit, laser ablation can fully mark the same surface in a few minutes. Moreover, once the thin film of material is available, only the laser interaction is needed in order to fabricate the electrode, instead of all the chemical components. This technique uses the process of laser ablation, where the interaction of the laser energy with the sample leads to material removal. Usually, this phenomenon depends on the absorption of laser photons by the sample material, which means that the wavelength of the laser has to be chosen carefully for the maximum absorption. However, the use of ultrafast lasers avoids this approach since ablation takes place as a result of multi-photon absorption at high peak intensities, which means that even materials normally transparent to the laser wavelength, which can be processed. In summary, laser ablation is a simple technique, which can be achieved with inexpensive optics and only requires any kind of beam or sample motion which can interface with CAD programs in order to mark complex patterns. The main problem presented is that the marked is limited to a small area, which is not even an issue if the goal is to work in a surface of a few square millimeters.

In this paper, we present a laser-based fabrication process for fabricating an electrostimulator on a 1 micron aluminum film deposited by PLD over a soda-lime glass. By using laser techniques, the aluminum is selectively removed to obtain a pre-designed electric circuit. In order to avoid direct contact between the circuit and the cell culture, a 100 μm glass is placed between them. Section 2 introduces materials and methods. In Section 3, we describe the electrical stimulator fabrication procedure and results, and Section 4 is devoted to discussion and conclusions.

2. Materials and methods

2.1. Laser set-up

The laser used in this experiment was a Rofin Power Line 20E operating at the fundamental wavelength of 1064 nm and a pulse with of 20 ns. The experimental set-up consists of a galvanometer system coupled to the laser source, with a flat field lens at the output with a spot size at focus of 15 μm. During the PLD process, the lens used has an effective focal length of 160 mm, providing a uniform distribution of irradiance in an area of $120 \times 120\ mm^2$. For the ablation process, the selected lens has 100 mm of effective focal length, with an area of uniform distribution of irradiance of $80 \times 80\ mm^2$.

2.2. Materials

The glass used as a substrate for fabricating the electrostimulator was a commercial soda-lime glass, provided by a local supplier. The composition of this glass (O 50.25%, Na 9.08%, Mg 2.19%, Al 0.54%, Si 33.08% and, Ca 4.87%) was determined by EDX analysis using a scanning electron microscope (SEM) Zeiss FESEM-ULTRA Plus. The material used for pulsed laser deposition of the thin film was an aluminum target of dimensions (10 cm × 10 cm × 1 mm) with a purity of 99.98%, provided by Goodfellow.

2.3. Thin layer deposition

Before PLD deposition, a six-step cleaning process was used for cleaning glass substrates. The samples were first brush-scrubbed in an aqueous and soap bath. They were then ultrasonically pulsed in a second deionized water and soap bath heated to 35°C for 30 minutes. Then, the water was changed and the process repeated. The fourth and fifth baths contained isopropyl alcohol, heated to 35°C for 30 minutes. Finally, the samples were dried using air pressure. Once the cleaning process of the glass substrate was finalized, aluminum layers in the range of 600 nm to 1 m were deposited by PLD on the soda-lime substrate.

2.4. Characterization methods

The chemical composition of the soda-lime glass was determined by using a scanning electron microscope (SEM) Zeiss FESEM-ULTRA Plus issued with EDX analysis. The optical transmission of glass substrate was measured using a Perkin Elmer Spectrometer (type Lambda 950 UV/Vis). Fabricated samples were examined using an optical microscope Nikon MM-400. The characterization of the transmission spectrum across the sample was made using a BLUE-Wave Miniature Spectrometer, from Stellarnet Inc., which can measure in the range from 250 to 1150 nm. The characterization of the topographic profile was made using a Dektak3 profilometer from Veeco.

The electrical characterization of the electrostimulator is made using a Tektronix MSO 1 GHz 5GS/s oscilloscope. The oscilloscope was connected to a coil placed on a 100 μm glass over the surface of the circuit, in order to check if the electrical field is strong enough to pass through the glass to induce an electrical current in the coil.

2.5. Software control and simulation

In order to estimate the intensity and the homogeneity of the electrical field above the electrostimulator, we used the software ANSYS Maxwell. This program was used to simulate the parameters of the circuit and measure the electrical field in different planes over the surface of the circuit. The hardware selected for applying electrical signals to the electrostimulator is the NI USB-6501 portable digital I/O device, from National Instruments. It provides 5 V by default and up to 8.5 mA. The hardware was programmed using LabVIEW, from National Instruments, to apply a square signal whose parameters can be chosen by the final user to stimulate cells.

3. Results

The soda-lime glass samples were covered with an aluminum layer using a PLD process, and then the electrical tracks were fabricated by selective elimination of the thin metal film. **Figure 1** shows the diagram of the PLD process used in this work. The vacuum chamber was custom fabricated by Trinos Vacuum-Projects. It has a primary pump, which can provide a simple vacuum of 10^{-4} mbar and a secondary turbomolecular pump, which can achieve a high vacuum of 10^{-6} mbar.

Figure 1 illustrates the vacuum chamber experimental set-up. The laser source is coupled to a source galvanometer head, with a flat field lens at the output of effective focal length 160 mm and a spot size at focus of 15 μm. The laser beam is focused on the surface of an aluminum layer placed at 45° (see **Figure 1a**). The glass substrate is placed in front of the target, parallel to it, and separated by 1 cm. The optimal pressure inside the vacuum chamber is 4×10^{-5} mbar.

One of the most significant characteristics of the PLD technique is the high kinetic energy of the ejected particles. By raising this energy, the density of the layer increases and the adherence between the particles and the substrate improves. However, one of its problems is the high directionality of the ejected particles, which leads to an inhomogeneous density of the layer. This deposition, in the case of the PLD technique, follows a distribution with the form of $\cos^n\theta$, where θ is the angle between the direction of the ejected particles and the normal to the surface of the target, and n is a number depending on the conditions of the experiment, which can achieve values higher than 30 [20]. To solve this, inconvenient small area of the target (28 × 2 mm) were scanned with the laser beam instead of focussing it in a fixed point.

As consequence of these issues, rising the distance between the target and the substrate the homogeneity of the layer increases, so it is necessary to find the optimal separation in order to have a dense but homogeneous layer. After performing experiments over 60 samples, we have identified the optimal deposition parameters for laser operating at a fluence of 121 J/cm^2, while marking rectangles of 28 × 2 mm, with a separation between the target and the substrate of 1 cm, and with a combination of frequency and scan speed according to **Table 1**.

Figure 2 shows the microscope images of the samples selected in **Table 1** illustrating the edge between the base glass material and the aluminum tracks deposited using PLD.

As it can be seen in **Figure 2**, the homogeneity of the layers, in terms of aluminum particles covering, vary according to the frequency and scan speed. We can observe that the samples 2 and 4 are the most homogeneous and present a uniform distribution of aluminum particles

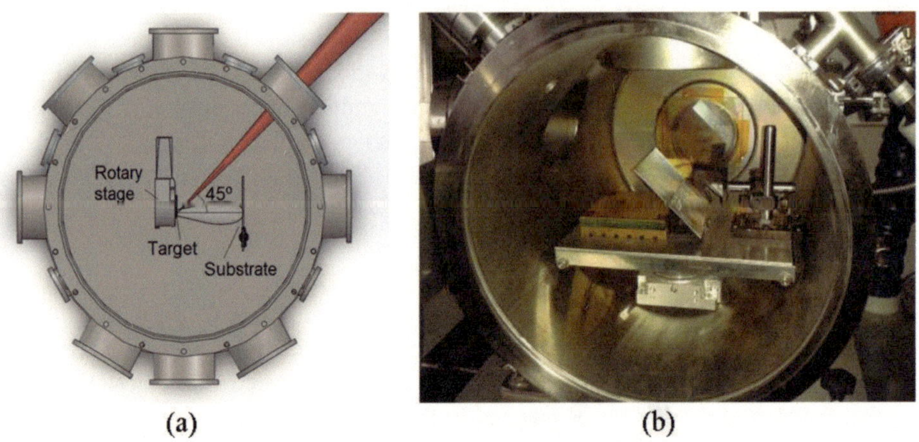

Figure 1. (a) Diagram of the PLD set-up for the fabrication process and (b) image of the PLD system.

Sample	1	2	3	4
Repetition rate (KHz)	1	5	7	3
Scan speed (mm/s)	50	250	350	150

Table 1. Ratio of frequency and scan speed.

Figure 2. Microscope images of the deposited layers obtained with an objective 20×. (Each image corresponds with the laser parameters showed in **Table 1**).

along the surface. Conversely, **Figure 2a, c** shows spaces not covered by aluminum particles. The characterization of the transmission spectrum and the topographic profile of samples 2 and 4 are shown in **Figure 3**.

The percentage of light transmitted across the layer is shown in **Figure 3a, c**. In both cases, the percentage increases from the center to the borders of the aluminum layer due to the inhomogeneity of the deposition. In the central region, the layer is thicker than in the borders and reflects all the light. In **Figure 3b, d**, the topographic profile of the aluminum layer along a distance of 600 μm can be observed. The layer in the sample 2 shows an average height value of 1274 nm and a roughness average of 514 nm. The layer in the sample 4 shows an average height value of 653 nm and a roughness average of 389 nm. From now, the parameters of the sample 4 will be used, due to its less roughness: laser fluence of 121 J/cm^2, frequency of 3 KHz, scan speed of 150 mm/s, while marking rectangles of 28 × 2 mm, and with a separation between the target and the substrate of 1 cm.

3.1. Laser selective thin film removal

The laser direct-write technique for fabricating the electrostimulator system is based on the ablation of an aluminum layer deposited over the soda-lime glass substrate using a physical vapor deposition method. The laser set-up for selective aluminum layer removal is described in Section 2.1. The beam spot size was estimated to be 15 μm. For fabricating the

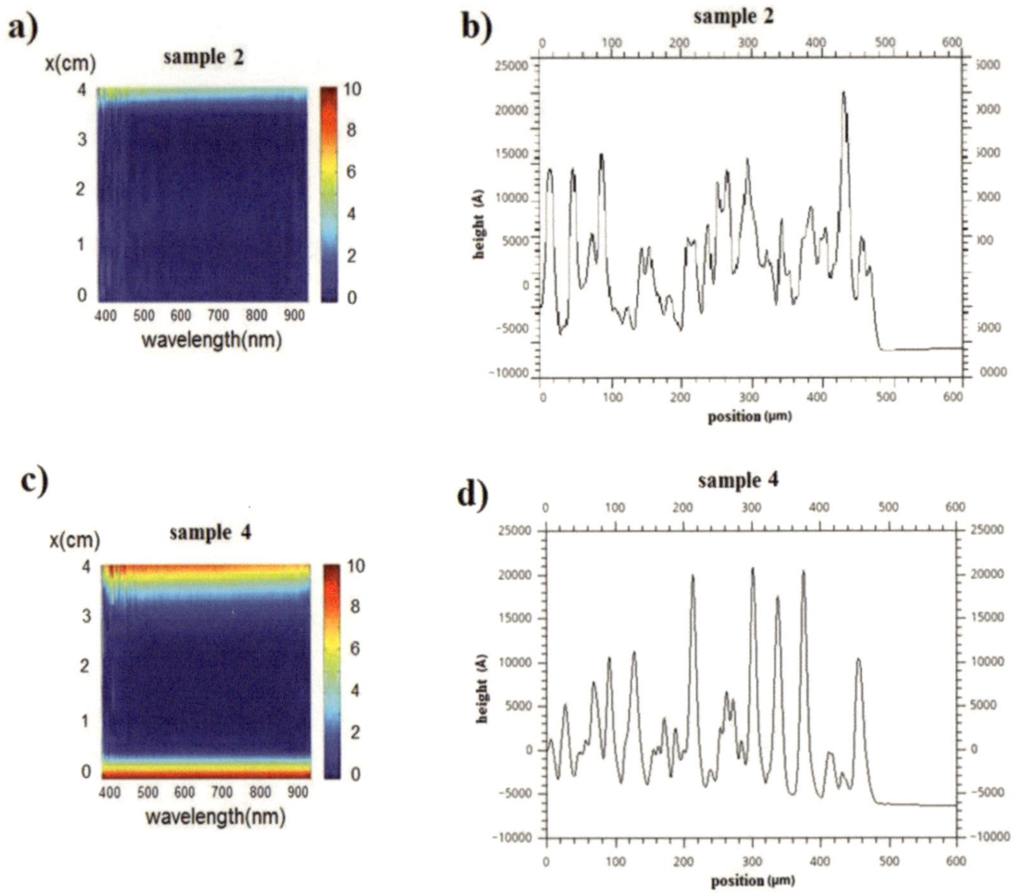

Figure 3. (a) Spectrum of the transmitted light across the layer in sample 2, (b) topographic profile of the surface of the layer in sample 2, (c) spectrum of the transmitted light across the layer in sample 4, and (b) topographic profile of the surface of the layer in sample 4.

electrical stimulator, the laser beam is focused on the top of the 1 micron aluminum layer deposited at the top of the glass substrate. **Figure 4a** shows the laser set-up for fabricating the electrostimulator. **Figure 4b** shows the CAD design used to fabricate the electrical stimulator. The laser parameters used were: average power 1.3 W, repetition rate 16 kHz, and scan speed 60 mm/s.

At laser fluence value below the damaged threshold of glass (920 J/cm^2) and above the ablation threshold of the target (2.4 J/cm^2), we were able to eliminate the aluminum layer selectively, according with the electrostimulator design (see **Figure 4b**). Strong interactions between the laser beam, the plasma, and the aluminum layer take place below the ablation threshold of soda-lime glass, resulting in high-quality elimination of the aluminum layer.

The calculations for determining applied threshold fluence (ϕ_{th}) and (ϕ_0) are obtained according to the method of Liu et al. [27]. The spatial fluence, ($\phi(r)$) for a Gaussian beam is given by:

$$\phi(r) = \phi_0 e^{-2r^2/\omega_0^2} \tag{1}$$

where ϕ_0 is the peak fluence in the beam, r is the distance from the centre of the beam, and ω_0 is the Gaussian spot radius ($1/e^2$). The maximum fluence and the pulse energy, E_p, are related by:

$$\phi_0 = \frac{2E_p}{\pi\omega_0^2} \tag{2}$$

The peak fluence is related to the diameter of the ablated spot

$$D^2 = 2\omega_0^2 \ln\left(\frac{\phi_0}{\phi_{th}}\right) \tag{3}$$

where D^2 is the maximum diameter of the damaged region zone. It is possible to determine the beam radius using the value for ω_0 from the plot of D^2 versus the logarithm of the pulse energy. Once ω_0 is calculated, fluence values can then be found using Eq. (3). By plotting D^2 versus the natural log of the applied laser fluence and extrapolating the D^2 line to zero, ϕ_{th} can be calculated.

Due to the thinness of the aluminum layer, it is necessary to determinate the appropriate laser parameters, which allow to ablate the aluminum without causing damage in the glass substrate. These parameters are the fluence of the beam, the frequency of the pulses, and the speed of the spot along the sample. In order to determinate the ratio between the frequency and the speed, we calculated the degree of pulse overlap between the consecutive spots. This factor is set in the next equation

$$O_d = 1 - \frac{v}{2\omega f} \tag{4}$$

where v and f are the speed and frequency, and ω is the width of the spot, in this case 15 μm. Pulse overlapping is a crucial parameter for fabricating a homogeneous electrical track. Too

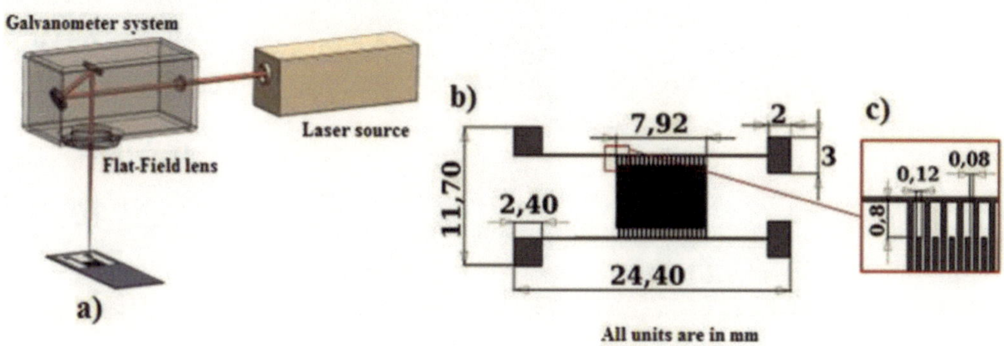

Figure 4. (a) Laser set-up for aluminum layer ablation, (b) electrostimulator design, and (c) extended view of the tracks dimensions.

overlapping will deliver too much energy over the glass, resulting on damage at surface, while low overlapping will result on inefficient material removal. In **Figure 5** are presented some samples of tracks of a 200 nm aluminum layer ablated with different frequencies (10, 12, and 14 KHz) and scan speeds (60 and 100 mm/s), and therefore with different pulse overlapping.

In **Figure 5**, we can observe different tracks created by the laser at different pulse overlapping. All of them were made with a power of 700 mW. Results show how the ratio of 12 KHz and 60 mm/s is the one with the most regular track in terms of width, with an overlap degree of 0.66. In other cases, the pulses are either too separated or too overlapped. Based on the above mentioned results, electrical tracks were fabricated using an overlapping factor of 0.66.

In order to adjust the optimal ratio between power and frequency, different tracks were made with different values of these parameters. We measured the diameter of the tracks related to the energy of each pulse (**Figure 6**).

Figure 6 shows the diameter of the mark after ablation of a single line related to the energy per pulse, obtained by using different combinations of frequency (6–20 KHz) and power (700–2000 mW), where it can be observed the linear relation between the width of the line and the energy per pulse. This width will be taking into account during the aluminum layer removal process. Selected parameters were: Pulse energy: 90 μJ (which correspond to a frequency of 12 KHz, a power of 1.05 W) and a scan speed of 60 mm/s.

In **Figure 7**, you can see microscope images with the result of fabricating the electrostimulator with the previous parameters. The desired aluminum was successfully removed, without interferences between the tracks. Besides, the glass substrate has not been damaged. Tracks have an average length of 60.4 μm and are separated by a distance of 135 μm.

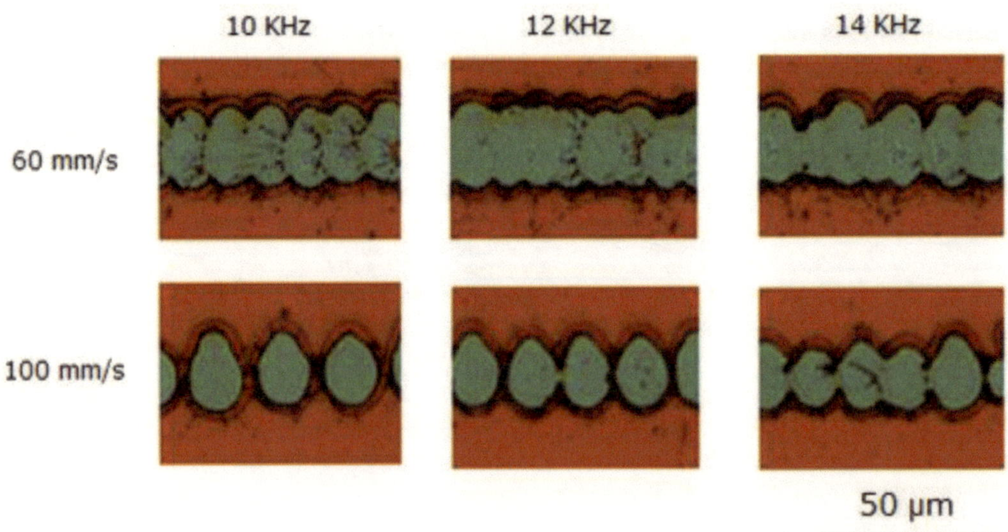

Figure 5. Aluminum layer ablated with different frequencies and scan speeds (thickness: 600 nm).

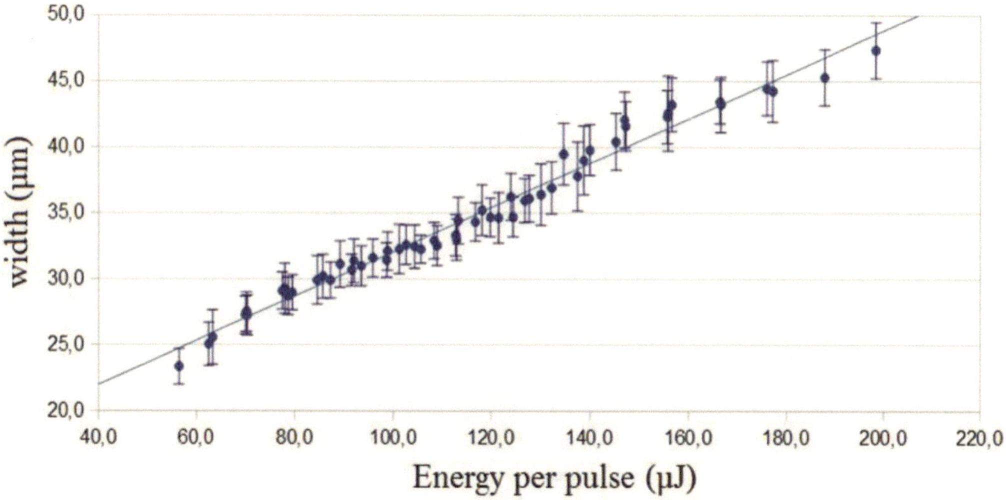

Figure 6. Diameter of the laser mark related to the energy per pulse.

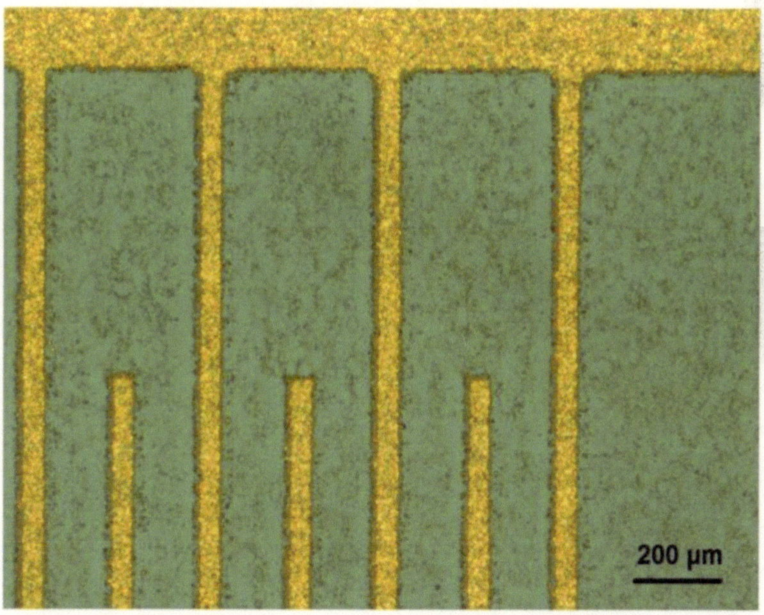

Figure 7. Microscope image of the electrical tracks after selective laser removal.

Motivated to maintain a lab-on-a-chip configuration for medical applications, we have created a culture chamber where performing the electrical characterization of fabricated electrical tracks. A polydimethylsiloxane layer (PDMS, Sigma-Aldrich, Saint Louis, MO) was cut from a PDMS-casted layer and placed on the top of the electrical stimulator (**Figure 8a**). Layer dimensions are 40 × 30 × 10 mm, with a hole of 10 × 8 mm. The glass cover slip (thickness: 100 µm) was placed on the top of the electrical tracks to avoid any contamination. Finally, these elements are set between two polycarbonate layers, which will keep the device compact (**Figure 8b**).

Over the electrostimulator and the 100 μm glass is placed on a coil connected to an oscilloscope. In the terminals of the electrostimulator is induced a square signal with 5 V and 1 Hz, which should induce another signal in the coil (**Figure 9**).

Figure 9a shows a peak voltage induced during the rise time from the low to the high level of the square signal. **Figure 9b** shows an upper view of the set-up. The LED on the right is connected to the margins of the electrostimulator to check that there are no cuts in the circuit. The linear dependence of the voltage peak induced in the coil on the voltage applied to the electrostimulator is presented in **Figure 9c**.

The electrical field induced by the electrostimulator was simulated using the ANSYS Maxwell software. In this simulation, a model of the aluminum circuit in which 5 V was applied between the two terminals was introduced. In order to compare the induced electrical field with and without the glass cover, a second simulation was made keeping the same design of

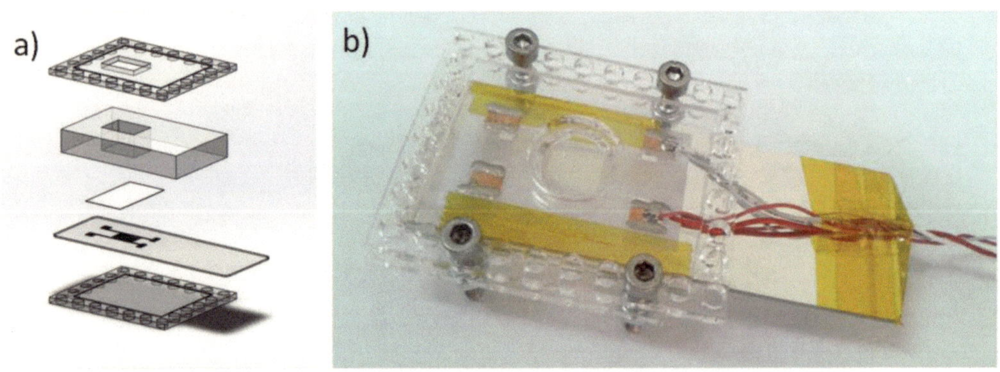

Figure 8. (a) Diagram of the elements of the electrostimulator and (b) picture of the final device.

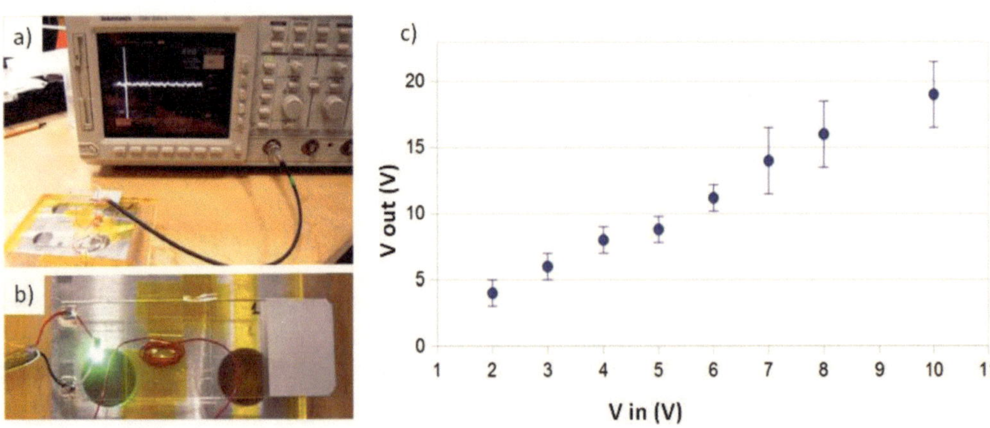

Figure 9. Set-up for checking the presence of an electric field generated above the glass.

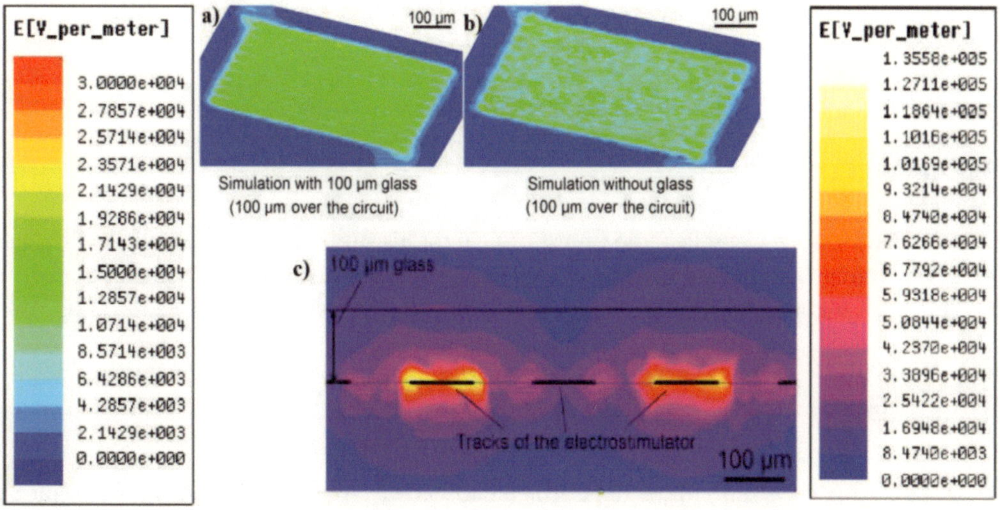

Figure 10. Simulation of the electric field generated above the electrostimulator with 5 V: (a) simulation with a 100 μm glass placed on the circuit, (b) simulation without 100 μm glass, and (c) cross sectional simulation of electrical tracks and glass.

the stimulator and parameters than the first one, but with a 100 μm glass over the electrostimulator. Results are shown in **Figure 10**.

In **Figure 10a, b**, the results show the electric field in a plane parallel to the electrostimulator, 105 μm above. In **Figure 10a**, a 100 μm glass is placed on the circuit, while in **Figure 10b** there is no glass at all. By comparing both results it can be observed that the mean value of the field when the glass is placed is about 1.4 x 10^4 V/m, while in the case without the glass this value varies from 8 x 10^3 V/m to 2 x 10^4 V/m. The presence of the glass makes the electric field slightly weaker but much more homogeneous. As it was previously said, about 10^4 V/m are needed to induce forces in the cell culture, so these results show that the electrical field generated is strong enough to interact with the cells.

4. Conclusions

By combining laser ablation techniques and pulse laser deposition, it fabricated an electrostimulator for medical applications. The layers resulting in the PLD process have been characterized by measuring transmission spectrum, by a profilometer, and by optical microscopy. Results show rough and homogeneous aluminum layers in the central region of the glass substrate. The circuit marked in the laser ablation process was characterized by an optical microscope, verifying that the aluminum was selectively removed without interference between the tracks and without damage in the substrate.

The electrical field generated by the device was simulated using ANSYS Maxwell, verifying that the field is able to pass through the glass with enough intensity to interact with the cells. The presence of the electrical field across the glass was also measured in an experimental way. As distinction regarding to other similar electrostimulator devices, the electrodes have been isolated from the cell culture by using a thin soda-lime glass, solving biocompatibility issues between the material of the tracks and the cell culture.

Acknowledgements

D. Nieto thanks to the Consellería de Cultura, Spain for his support under the Galician Program for Research Innovation and Growth (2011–2015) (I2C Plan). This work has been supported by the Xunta de Galicia under contract Agrupación estratéxica 2015–AEFIS AGRUP2015/11 (PC034).

Author details

Angel Luis Aragón Beloso[1], María del Carmen Bao Varela[1], Alejandro Fernández Rodríguez[2], Gerard O'connor[3], Eliseo Pérez Trigo[4], Antonio Pazos Álvarez[4] and Daniel Nieto García[1,3]*

*Address all correspondence to: daniel.nieto@usc.es

1 Photonics4life Group, University of Santiago de Compostela, Santiago de Compostela, Spain

2 Superconducting Materials and Large Scale Nanostructures, Institut de Ciència de Materials de Barcelona (ICMAB-CSIC), Barcelona, Spain

3 NCLA/Inspire Labs, School of Physics, National University of Ireland, Galway, Ireland

4 Experimental High Energy Physics Group, University of Santiago de Compostela, Santiago de Compostela, Spain

References

[1] Hankey GJ, Pomeroy VM, King LM, Pollock A, Baily-Hallam A, Langhorne P. Electrostimulation for promoting recovery of movement or functional ability after stroke. Stroke. 2006;**37**:2441-2442

[2] Dehail P, Duclos C, Barat M. Electrical stimulation and muscle strengthening. Annales de Readaptation et de Medicine Physique. 2008;**51**:441-451

[3] Boussetta N, Abeldelmalek S, Aloui K, Souissi N. The effect of neuromuscular electrical stimulation on muscle strength, functional capacity and body composition in haemodialysis patients. Biological Rhythm Research. 2017;**48**:157-174

[4] Archer S, Li T, Tudor A, Britland S, Morgan H. Cell reactions to dielectrophoretic manipulation. Biochemical and Biophysical Research Communications. 1999;**257**:687-698

[5] Tandon N, Cannizzaro C, Figallo E, Voldman J, Vunjak-Novakovic G. Characterization of electrical stimulation electrodes for cardiac tissue engineering. Conference Proceedings:

Annual International Conference of the IEEE Engineering in Medicine and Biology Society. 2006;**1**:845-848

[6] Tandon N, Cannizaro C, Chao P, Maidhoe R, Marsano A, Au H, Radisic M, Vunjak-Novakovic MG. Electrical stimulation systems for cardiac tissue engineering. Nature Protocols. 2009;**4**:155-173

[7] Tandon N, Marsano A, Cannizzaro C, Voldman J, Vunjak-Novakovic G. Design of electrical stimulation bioreactors for cardiac tissue engineering, In: Conference Proceedings: Annual International Conference of the IEEE Engineering in Medicine and Biology Society. 2008: pp. 3594–3597

[8] Geremia NM, Gordon T, Brushart TM, Al-Majed A, Vergea VMK. Electrical stimulation promotes sensory neuron regeneration and growth-associated gene expression. Experimental Neurology. 2007;**205**:347-359

[9] Abad-Villar EM, Kubáň P, Hauser PC. Determination of biochemical species on electrophoresis chips with an external contactless conductivity detector. Electrophoresis. 2005;**26**:3609-3614

[10] Woolley AT, Sensabaugh GF, Mathies RA. High-speed DNA genotyping using microfabricated capillary array electrophoresis chips. Analytical Chemistry. 1999;**69**:2181-2186

[11] Voldman J. Electrical forces for microscale cell manipulation. Annual Review of Biomedical Engineering. 2006;**8**:425-454

[12] Yafouz B, Kadri NA, Ibrahim F. Microarray Dot electrodes utilizing dielectrophoresis for cell characterization. Sensors. 2013;**13**:9029-9046

[13] Eason R. Pulsed Laser Deposition of Thin Films: Applications-Led Growth of Functional Materials. Hoboken, New Jersey: John Wiley & Sons, Inc.; 2006

[14] Krebs HU et al. Pulsed laser deposition (PLD)–A versatile thin film technique. Advances in Solid State Physics (Springer). 2003;**43**:505-517

[15] Ojeda A, Schneider CW, Lippert T, Wokaun A. Pressure and temperature dependence of the laser-induced plasma plume dynamics. Journal of Applied Physics. 2016;**120**:225301

[16] Jenniches H, Shen J, Mohan CV, Sundar Manoharan S, Barthel J, Ohresser P, Klaua M, Kirschner J. Structure and magnetism of pulsed-laser-deposited ultrathin films of Fe on Cu.(100). Physical Review B. 1999;**59**:1196-1207

[17] Shen J, Gai Z, Kirschner J. Growth and magnetism of metallic thin films and multilayers. Surface Science Reports. 2004;**52**:163-218

[18] Qiu Z, Gong H, Zheng G, Yuan S, Zhang H, Zhu X, Zhou H, Cao B. Enhanced physical properties of pulsed laser deposited NiO films via annealing and lithium doping for improving perovskite solar cell efficiency. Journal of Materials Chemistry C. 2017;**5**: 7084-7094 *Advance article*

[19] Koinuma H, Nagata H, Tsukahara T, Gonda S, Yoshimoto M. Ceramic layer epitaxy by pulsed laser deposition in an ultrahigh vacuum system. Applied Physics Letters. 1991;**58**:2027-2029

[20] Stafe M, Marcu A, Puscas N. Pulsed Laser Ablation of Solids, Springer Series in Surface Sciences. Vol. 53. Berllin Heidelberg: Springer-Verlag; 2014

[21] Warrender JM, Aziz MJ. Kinetic energy effects on morphology evolution during pulsed laser deposition of metal-on-insulator films. Physical Review B. 2007;**75**:085433

[22] Madsen NR, Gamaly EG, Rode AV, Luther-Davies B. Cluster formation through the action of a single picosecond laser pulse. Journal of Physics: Conference Series. 2007;**59**:762-768

[23] Rosales A, Castañeda-Guzmán R, de Ita A, Sánchez-Aké C, Pérez-Ruiz SJ. Detection of zinc blende phase by the pulsed laser photoacoustic technique in ZnO thin films deposited via pulsed laser deposition. Materials Science in Semiconductor Processing. 2015;**34**:93-98

[24] Kudyakova VS, Shishkin RA, Elagin AA, Baranov MV, Beketov AR. Aluminium nitride cubic modifications synthesis methods and its features. Review. Journal of the European Ceramic Society. 2017;**37**:1143-1156

[25] Lowndes D, Geohegan DB, Puretzky AA, Norton DP, Rouleau CM. Synthesis of novel thin-film materials by pulsed laser deposition. Science. 1996;**273**:898-903

[26] Li G, Wang W, Yang W, Wang H. Epitaxial growth of group III-nitride films by pulsed laser deposition and their use in the development of LED devices. Surface Science Reports. 2015;**70**:380-423

[27] Jelinek M. Functional planar thin film optical waveguide lasers. Laser Physics Letters. 2012;**9**(2):91-99

[28] Wei X, Zhao R, Shao M, Xu X, Huang J. Fabrication and properties of ZnO/GaN heterostructure nanocolumnar thin film on Si (111) substrate. Nanoscale Research Letters. 2013;**8**:112-120

[29] Matthews J, Lee SY, Wellstood FC, Gilbertson AF, Moore GE, Chatraphorn S. Multi channel high-T/sub c/scanning SQUID microscope. IEEE Transactions on Applied Superconductivity. 2003;**13**:219-222

[30] Hontsu S, Sakatani T, Fujimaki A, Nishikawa H, Nakamori M, Kawai T. Mechanically tunable high-temperature superconducting microwave filter with large shift of resonant frequency. Japanese Journal of Applied Physics. 2001;**40**:L1148

[31] Hedge MS. Epitaxial oxide thin films by pulsed laser deposition: Retrospect and prospect. Journal of Chemical Sciences. 2001;**113**:445-458

[32] Ji Y, Qin C, Niu H, Sun L, Jin Z, Bai X. Electrochemical and electrochromic behaviors of polyanilinegraphene oxide composites on the glass substrate/Ag nano-film electrodes prepared by vertical target pulsed laser deposition. Dyes and Pigments. 2015;**117**:72-82

Chapter 12

In Vitro Evaluation of Laser-Induced Periodic Surface Structures on New Zirconia/Tantalum Biocermet for Hard-Tissue Replacement

Alberto Jorge-Mora, Naroa Imaz, Nekane Frutos,
Ana Alonso, Carlota García Santiago,
Rodolfo Gómez-Vaamonde, Jesús Pino-Minguez,
Jose Bartolomé, Gerard O'connor and Daniel Nieto

Additional information is available at the end of the chapter

http://dx.doi.org/10.5772/intechopen.70820

Abstract

This study investigates the biological response of zirconia/tantalum biocermet materials with laser-induced periodic surface structures (LIPSS) generated using a femtosecond laser working at 1030 nm wavelength. LIPSS were formed by laser radiation slightly above the applied threshold fluence. LIPSS features were characterized using techniques such as atomic force microscopy (AFM) and X-ray photoelectron spectroscopy (XPS). LIPSS were generated in this study by applying femtosecond pulses with 500 fs pulse duration at a high-repetition rate to smooth-polished zirconia/tantalum biocermet surfaces, with an original roughness value of 3.8 ± 0.2 and 3.1 ± 0.2 nm, respectively. We have demonstrated in vitro that LIPSS are an efficient option to increase osteoblastic differentiation of human bone marrow mesenchymal stem cells (hBMSCs) in ZrO_2:Ta biocermets. LIPSS created increase cell metabolism statistically (best values in 3-(4,5-dimethylthiazol-2-yl)-2,5-diphenyltetrazolium bromide (MTT) assay) and decrease inflammatory response to the material (IL-6 and TNF-alpha values). Extracellular matrix production (ECM) is produced in more quantity and cells differentiate to osteoblast easily. These differences are seen from the beginning until the endpoint (day 20).

Keywords: laser-induced periodic surface structures, LIPSS, hard-tissue replacement, zirconia/tantalum biocermet

1. Introduction

Orthopaedic surgery has experienced a big development, thanks to the evolution of the materials used, which made it possible to perform procedures, such as total hip replacement with feasibility. These procedures are common in most orthopaedic departments and the survival of these implants is critical to prevent loosening and the need for revision arthroplasty with a cost of more than 15,000 euros in total hip or knee replacement. The ideal surface to interact with bone has not been created. We can find promising results in some materials, for example, hydroxyapatite plasma coating creates early and strong integration, but it seems that this treatment experiments long-term reviews due to the loosening of the coat [1]. Multiple techniques have been proposed to increase osteointegration in metals. Probably the most accepted and used are the creation of a plasma coating with hydroxyapatite in some situations and the texturization of the metal in the microscale [2–6].

Laser processing has gained significant importance in the medical industry and is an integral part of manufacturing, for example, welding endoscopes, drilling holes in hypodermic needles and verification laser marking [2, 3]. Lasers offer high precision and repeatability, high speed and quality, accuracy, cost efficiency, minimal thermal input and noncontact [7–10]. In particular, femtosecond lasers are ideal for surface structuring because they have a minimal heat affected zone, there is rapid heating and cooling and no laser plasma interaction [11–16]. Femtosecond laser-induced periodic surface structures (LIPSS), known as "ripples," have been fabricated on material surfaces such as metals [17], polymers [18], glass, dielectric and semiconductors [19, 20]. There are a number of applications of LIPSS such as increasing the surface area and surface energy, altering the hydrophilic or hydrophobic performance of a materials' surface, improving coating adhesion, optics and tribology [21, 22]. These characteristics are of great interest for biomedical applications, in particular for hard-tissue replacement [23–25].

This study investigates the bioresponse of zirconia/tantalum biocermets (ZrO_2:Ta) with laser-induced periodic surface structures (LIPSS) produced by femtosecond laser pulses. Experiments were carried out using a Yb:KYW chirped-pulse-regenerative amplification laser system (Amplitude Systemes S-pulse HP) that delivered laser pulses with a duration of approximately 500 fs at wavelength of 1030 nm. We investigate the physical and biological response of LIPSS created using a femtosecond laser on an alloy surface. This includes determining the role of incubation and nanostructures, the physical characterization of LIPSS and the biological response of LIPSS. ZrO_2:Ta surfaces were exposed to multiple incident laser shots in air at a repetition rate of 100 kHz at various pulse energies ranging from 2 to 6 mJ. Periodic surface nanostructures were formed by exposure to laser radiation slightly above the applied threshold fluence and observed using techniques such as atomic force microscopy (AFM) and scanning electron microscopy (SEM). Section 2 introduces materials and methods. In Section 3, we describe the LIPPS fabrication procedure and results. Section 4 is devoted to biological results and Section 5 to conclusions.

2. Materials and methods

2.1. Material

We have previously described the manufacturing process of the composite, were ceramic and metal powders are mixed to create the composite [21]. The samples used in this work are discs of 20 mm diameter and 2 mm thickness were cut and machined from cylindrical bars (5 mm length and 2 mm diameter).

2.2. Microstructure and surface characterization

The microstructure of samples was measured on surfaces polished down to 1 mm using a scanning electron microscopy (SEM, Phenom G2). Roughness measurements were performed by using SEM and 3D roughness reconstruction application (Phenom TM Pro Suite). The average surface roughness (Ra) was determined by measuring the surface roughness at 10 different locations over 3 samples of each type.

Morphological observation of the surfaces was undertaken by means of a Carl Zeiss Ultra Plus field emission scanning electron microscope (FE-SEM). The composition was analysed by an Oxford instruments INCA X-sight X-ray electron probe microanalyzer (EDX).The wettability of the studied materials was assessed by measuring the contact angle of deionized water drops deposited onto their surface using a KSV Instruments Cam 200 device.

Laser-induced periodic surface structures are fabricated using a fluence above the damage threshold fluence for the zirconia/tantalum biocermet. Chemical modification was analyzed using a scanning electron microscope (Hitachi S2600N). The topography was measured using an atomic force microscope (Agilent 5500).

During the fabrication process, we used an optical Nikon MM-400 microscope to visually inspect the samples; surface roughness and topography was measured using a SENSOFAR 2300 Plμ confocal microscope. Biological observations were performed using a Zeiss Microscope (Zeiss AxioVert A.1) and Zen software.

2.3. Laser processing

ZrO_2-Ta surfaces were exposed to multiple incident laser shots in air at a repetition rate of 100 kHz at various pulse energies ranging from 2 to 6 mJ. LIPSS were fabricated using a Yb:KYW laser system (Amplitude Systemes S-pulse HP) with a pulse duration of approximately 500 fs and a wavelength of 1030 nm. The beam delivery process for fabricating the LIPSS can be described as follows: the beam is focused on the top of the zirconia/tantalum biocermet surfaces using a flat-field lens (focal length = 100 mm). This kind of lens ensures a homogeneous light distribution in an area of 8×8 cm^2. The beam spot size (21 μm) was determined at $1/e^2$ of the Gaussian profile. The spatial profile of the laser was Gaussian in nature with a nominal M2 value of <1.2. The laser set-up for texturing ZrO_2-Ta surfaces with LIPSS is shown in **Figure 1**.

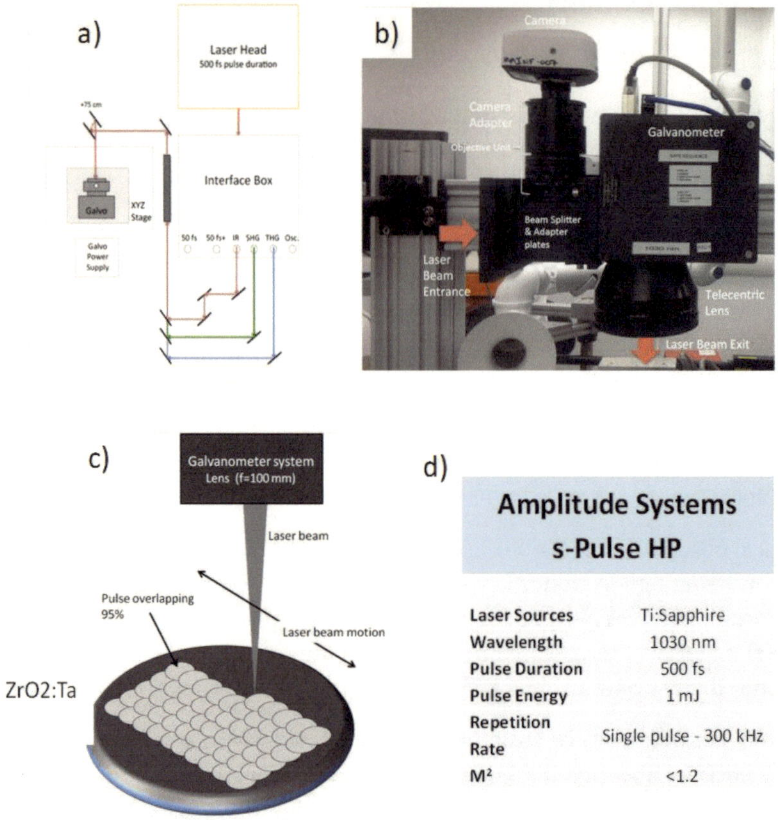

Figure 1. (a) Laser set-up and (b) beam delivery system for generating the LIPSS; (c) schematic of LIPSS fabrication process and (d) femtosecond laser characteristics.

2.4. Biological assay

We created 50 discs of the cermet. Each disc had a diameter of 6.75 mm (± 0.02 mm) and a thickness of 1.25 mm (± 0.02 mm). Once all the discs have been obtained, we proceed to polish those discs to prepare and to apply further treatments. We maintained 25 unaltered discs as a control group. In the remaining 25, we apply a laser treatment, which is explained later in detail, to create regular structures or roughness in the metallic structures of the cermet at disc surfaces. The samples used were treated with an ultrasonic bath for effective cleaning and then autoclaved (Selecta Autotester) at 121°C for 30 min. We used one disc from each group to culture mesenchymal stem cells (MSC) for 48 h to compare under direct vision cell behaviour.

After Institutional Review Board approval, and after giving consent, we obtained mesenchymal stem cells (MSC) from iliac crest of patients who were operated elective, with no previous disease and with an age between 18 and 50 years old. Under spinal anaesthesia, during the main orthopaedic procedure, we performed a puncture with a trocar in the iliac crest to aspirate bone marrow mesenchymal stem cells. We used a low-pressure technique with heparin and saline to prevent haemolysis and the formation of clots.

http://dx.doi.org/10.5772/intechopen.70820

Afterwards, we used a Ficoll-Histopaque (Sigma-Aldrich) gradient to isolate MSC. Then, we cultured the MSC to expand them under sterile conditions until we had obtained a concentration of 25,000 cells/cm^2 as previously reported. We confirmed the desired cell concentration with a Neubauer chamber.

The survival and proliferation of hBMSCs on discs were examined with a MTT assay after 5, 10, 15 and 20 days of culture (groups of 6 discs). The MTT assays were carried out as per manufacturer's protocol (Sigma-Aldrich). Briefly, hBMSCs were seeded on the sample discs (both surfaces were tested at the same time to diminish variability) at a density of 25×10^3 cells per well in a 24-well plate. The medium was removed periodically at each time point, MTT solution was added and cells were incubated overnight. Next step was to remove the MTT solution and dissolve the purple formazan crystals in 100 mL of dimethyl sulphoxide (DMSO) by shaking the plate for 15 min (50 mL solution of each well was added into a new 24-well plate). An automated plate reader (PerkinElmer) was used to quantify the OD value by measuring the absorption at 570 nm.

Afterwards, we proceed to clean the discs to use them for remaining biological assays. We proceed to clean all surfaces chemically (to destroy all biological tissues without modifying the surface) and treated them with an ultrasonic bath for effective cleaning and then autoclaved (Selecta Autotester) at 121°C for 30 min.

To assess the amount of inflammatory cytokine concentration, both IL-6 and TNF-alpha ELISA kits (eBioscience) were performed according to the manufacturer's indications. Kits were read on a spectrophotometer at the wavelength indicated by the manufacturer.

Formation of an ECM implies the deposition of collagen (the main protein of the ECM) on the biomaterial surface. Collagen concentration was assessed by means of a total protein assay (Pierce Labs USA).

Alkaline phosphatase (ALP) protein was assessed every 5 days. Samples were lysed using Cell Lysis Buffer (Cell Signalling), collected, and spun down at 14,000× *g* for 10 min at 4°C. The supernatant was then collected to measure ALP protein levels via LabAssay ALP (Wako); protein levels were normalized by measuring total protein content via a BCA Protein Assay kit (Thermo Fisher Scientific).

Osteocalcin and osteopontin levels were assessed to gain an insight of the mesenchymal cell fate determination. Increased values of osteocalcin and osteopontin are associated with osteoblastic differentiation. Supernatants were collected at different time points and measured by ELISA kits (Thermo Fisher).

We performed a Kolmogorov-Smirnov test to confirm that values have a normal distribution. After we used a T-student test to compare rough and smooth surfaces, we accepted a p value lower than 0.05 as statistically significant.

3. Experimental results

The laser set-up for fabricating the laser-induced periodic surface structures (LIPSS) on ZrO_2: Ta surfaces consists of a Yb:KYW chirped-pulse-regenerative amplification laser system

(Amplitude Systemes S-pulse HP). The laser set-up used for fabricating the LIPSS is described in Section 2.3. **Figure 1** shows the laser set-up for fabricating the LIPSS on ZrO_2:Ta samples.

The laser system for fully texturing materials with LIPSS is shown in **Figure 1c**. For fabricating the LIPSS, the incident direction of the laser was orthogonal to the sample plane. Each sample was fully textured by combining an aligned array of spots at a laser wavelength of 1030 nm. The irradiation conditions for both materials were applied fluence of 0.30 J/cm^2, laser spot of 21 μm, wavelength of 1030 nm, repetition rate of 100 kHz, pulse width of 500 fs and sample scan speed of 200 mm/s; the sample was fully textured using highly over-lapped (95%) pulses. This method is very effective in accelerating the processing times while the obtained results are similar to conventional pulsed-laser treatments for creating LIPSS. Fluence was calculated using Eqs. (2) and (3). The applied threshold fluence for one pulse was previously calculated and it was found that with increasing number of laser pulses, the threshold fluence decreases [26].

The calculations for determining applied threshold fluence (ϕ_{th}) and (ϕ_0) are obtained according to the method of Liu et al. [27]. The spatial fluence, (ϕ_r), for a Gaussian beam is given by:

$$\phi(r) = \phi_0 e^{-2r^2/\omega_0^2} \tag{1}$$

where ϕ_0 is the peak fluence in the beam, r is the distance from the centre of the beam and ω_0 is the Gaussian spot radius (1/e^2). The maximum fluence and the pulse energy, E_p, are related by:

$$\phi_0 = \frac{2E_p}{\pi\omega_0^2} \tag{2}$$

The peak fluence is related to the diameter of the ablated spot

$$D^2 = 2\omega_0^2 \ln\left(\frac{\phi_0}{\phi_{th}}\right) \tag{3}$$

where D^2 is the maximum diameter of the damaged region zone. It is possible to determine the beam radius using the value for ω_0 from the plot of D^2 versus the logarithm of the pulse energy. Once ω_0 is calculated, fluence values can then be found using Eq. (3). By plotting D^2 versus the natural log of the applied laser fluence and extrapolating the D^2 line to zero, ϕ_{th} can be calculated.

Figure 2 shows a polished ZrO_2:Ta sample used for generating ripples that exhibits a roughness average of 3.7 ± 0.2 nm.

LIPPS were fabricated on ZrO_2:Ta samples illustrated on **Figure 2** by irradiation with a p-polarized femtosecond laser. The laser parameters used for generating the LIPSS were repetition rate of 100 kHz, pulse width of 500 fs, sample scan speed of 200 mm/s and 50 passes of highly over-lapped (95%). In **Figure 3**, we can see the typical fabricated LIPSS or ripples observed in the scanning electron microscopy images. Periodical micropatterns were generated in ZrO_2:Ta using the abovementioned laser parameters at 200 mw laser power.

As shown in **Figure 3**, periodic structures were generated on tantalum while zirconia shows irregular structures at surface. The period of the surface ripples is significantly smaller than the

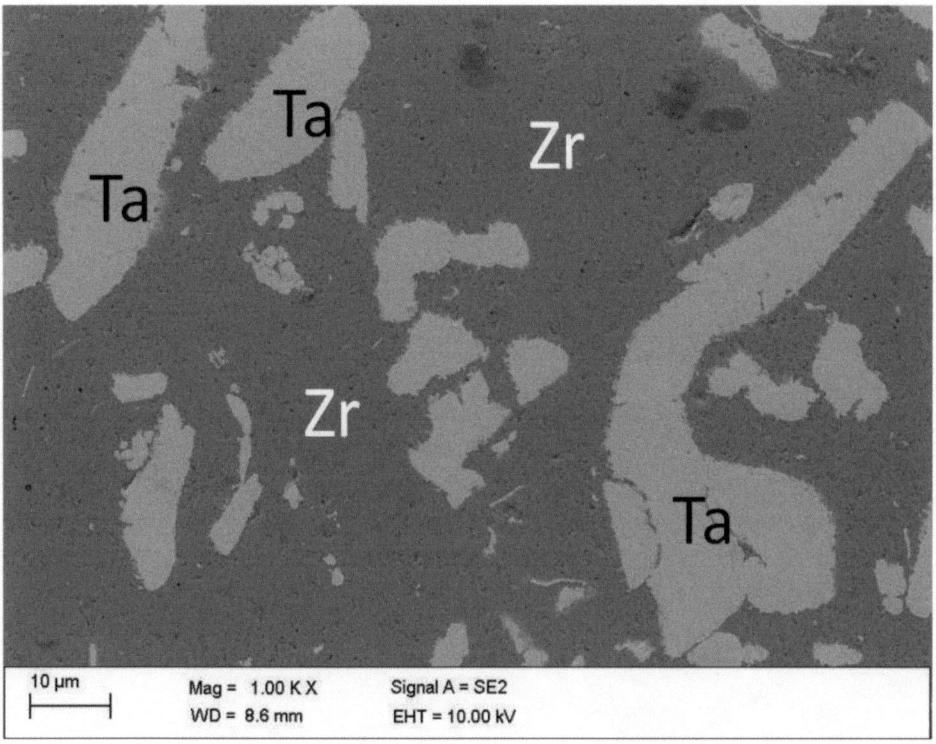

Figure 2. FE-SEM image of the ZrO_2:Ta cermet used for performing experiments.

wavelength of the incident laser beam. The period of ripples is estimated to be 760 ± 48 nm and the depth is 251 ± 7 nm when exposed to a 1030 nm, and the roughness average is 54 ± 2 nm. The depth of LIPSS depends on various parameters such as absorption coefficient, thermal diffusion and radiative cooling of the material. The effective penetration depth (α−1) is proportional to the laser wavelength. After femtosecond laser exposure, the alloy changes its optical penetration depth, as the material become amorphous. After multiple-pulse exposure which incorporates rapid heating and cooling cycles, the zirconia/tantalum material could evolve toward a metallic-like glass which would favor deeper surface structures [28].

Wettability results assessed in aqueous medium are presented in **Figure 4**. Since liquid ZrO_2:Ta interfacial tension is not easy to measure directly, the contact angle of a deionized water droplet on the studied cermets was used to indicate surface wettability. The equilibrium contact angle (θeq) of a liquid drop on an ideal solid surface is defined by Young's equation [29]:

$$\sigma\mathrm{LV}\cos\theta\mathrm{eq} = \sigma SV - \sigma SL \tag{4}$$

where σSV, σSL and σLV are the interfacial tensions between solid-gas, solid-liquid and liquid-gas phases, respectively. Our results indicate that the as-cast cermet shows high wettability with a contact angle value of 81.2°. On the other hand, laser-treated ZrO_2:Ta alloy presents significantly higher contact angle (113.5°), that is, it shows a more hydrophobic character.

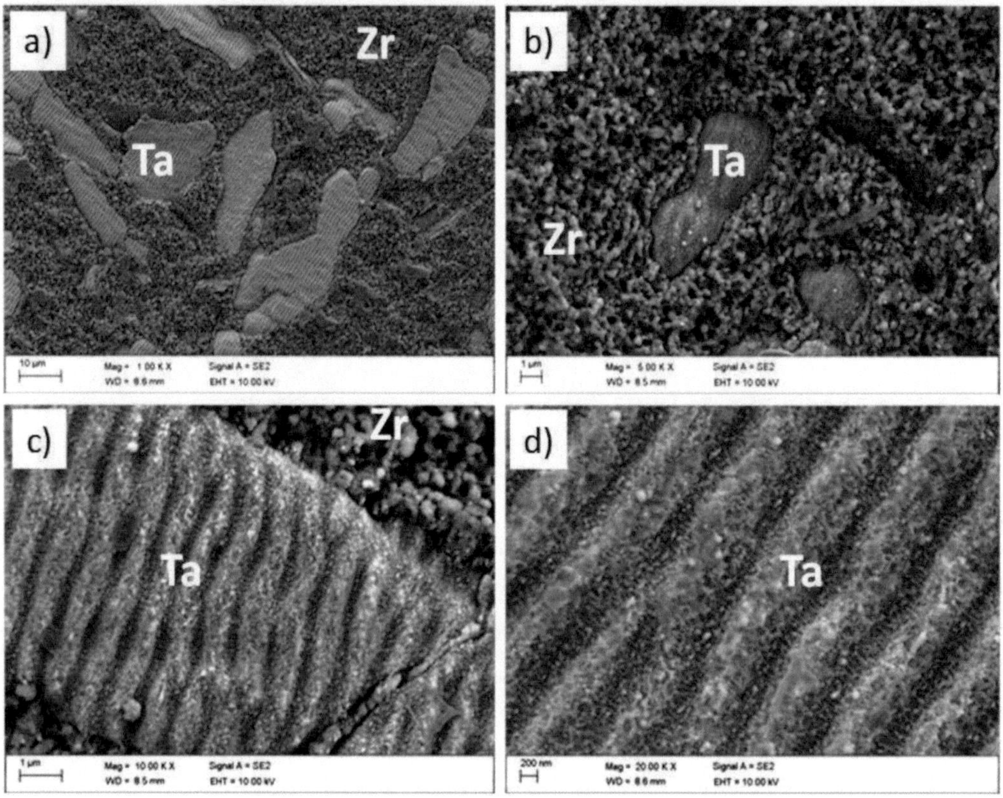

Figure 3. FE-SEM images of the fabricated micropatterned structures on ZrO_2:Ta (Ta: Tantalum; Zr: Zirconia) at different magnifications.

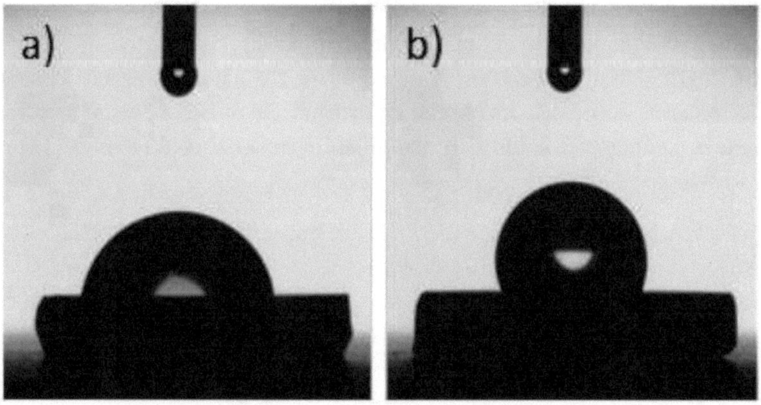

Figure 4. Photographs of as-deposited water droplets onto the surface of (a) as-cast and (b) ZrO_2-Ta with ripples.

Considering γlv constant for a certain liquid in a particular environment, the observed variability in wettability values could be ascribed to variations in γsv and γsl, caused by the hierarchical structures fabricated on ZrO_2:Ta surface by the applied laser process. LIPSS nanopattern may be responsible for the observed hydrophobicity increase, which could also be influenced by the intrinsic change in the surface chemistry with the exposure to air after the laser fabrication process [30].

Cell adhesion and several cellular functions are strongly influenced by the surface properties of a material. It has been proved that surface chemistry, energy, topography and wettability are key factors that affect protein adsorption and blood coagulation, which may have an impact on the adhesion, migration and differentiation of osteogenic cells, playing an important role in implants' osteointegration [31, 32]. In the studied cases, an increase in the protein adsorption and osteoblast differentiation parameters for laser-treated surfaces was observed, in spite of their lower wetting properties. The higher specific surface area offered by the fabricated nanostructures can play a decisive role in providing more adsorption sites for bioactive molecules and proteins, mitigating the effect of surface hydrophobicity [33].

4. Biological results

On the biological side, one of the main challenges in this study was to investigate the cells' behaviour when cultured over nanometric structures (LIPPS) to increase osteoblastic differentiation. After fabricating laser-induced periodic surface structures, we studied the behaviour of

		SMOOTH 5	LIPSS 5	SMOOTH 10	LIPSS 10	SMOOTH 15	LIPSS 15	SMOOTH 20	LIPSS 20
MTT	Mean	0.711	0.722	0.784	0.767	0.868	0.917	0.943	0.980
	Std. deviation	0.046	0.053	0.067	0.051	0.061	0.119	0.079	0.085
IL-6	Mean	0.018	0.019	0.026	0.023	0.031	0.031	0.032	0.030
	Std. deviation	0.002	0.002	0.004	0.003	0.003	0.003	0.002	0.003
TNF-alpha	Mean	0.018	0.018	0.023	0.023	0.024	0.024	0.045	0.044
	Std. deviation	0.001	0.002	0.003	0.002	0.002	0.002	0.003	0.007
ECM	Mean	0.033	0.029	0.044	0.055	0.325	0.452	0.333	0.425
	Std. deviation	0.026	0.015	0.028	0.031	0.044	0.034	0.040	0.052
AF	Mean	0.544	1.037	0.580	1.123	0.626	1.224	0.679	1.317
	Std. deviation	0.038	0.198	0.026	0.145	0.047	0.073	0.033	0.132
OSC	Mean	0.539	0.639	0.575	0.624	0.622	0.778	0.601	0.787
	Std. deviation	0.052	0.043	0.049	0.069	0.063	0.069	0.049	0.050
OSP	Mean	0.716	0.838	0.765	0.825	0.837	0.831	0.922	0.884
	Std. deviation	0.057	0.075	0.077	0.050	0.130	0.123	0.059	0.076

Table 1. Mean value for each assay and time point with the standard deviation.

bone marrow mesenchymal stem cells compared to polished surfaces. We cultured hBMSCs in a concentration of 25,000 cells/cm^2 in the surfaces created for 20 days. We determined MTT, TNF-alpha, IL-6, collagen production, alkaline phosphatase, osteopontin and osteocalcin for every 5 days until day 20. **Table 1** summarizes the average of all measurements for each assay.

MTT values in both surface treatments were similar (**Figure 5**), without reaching the statistical significance at any time point of control. Inflammatory response of both materials described by IL-6 and TNF-alpha values was also similar, with no difference between surfaces. We found very low values for both surfaces. **Figure 5** shows the data obtained for MTT assay with the standard deviation.

We found a significant difference in ECM production (**Figure 6**) and osteoblast differentiation parameters (FA, OSC and OSP). ECM production was statistically increased in LIPSS surfaces at days 15 and 20. Alkaline phosphatase activity was increased at all time point controls. Osteocalcin levels were also elevated at every control, but at day 10 there was no statistical difference. Osteopontin values were also superior to control in every control, but only reached statistical difference at day 5. **Figure 6** shows the data obtained for ECM assay with the standard deviation.

Confocal microscopic image (**Figure 7**) showed a polarization of cells in the metal area (where LIPSS are generated). In the surface where the ceramic is present, we can see a disordered pattern seen in most of the smooth surfaces.

We can see a transition between LIPSS and the ceramic surface. In the upper left corner, we can see a polarization of the long axis of the cell, which is not seen in the ceramic surface

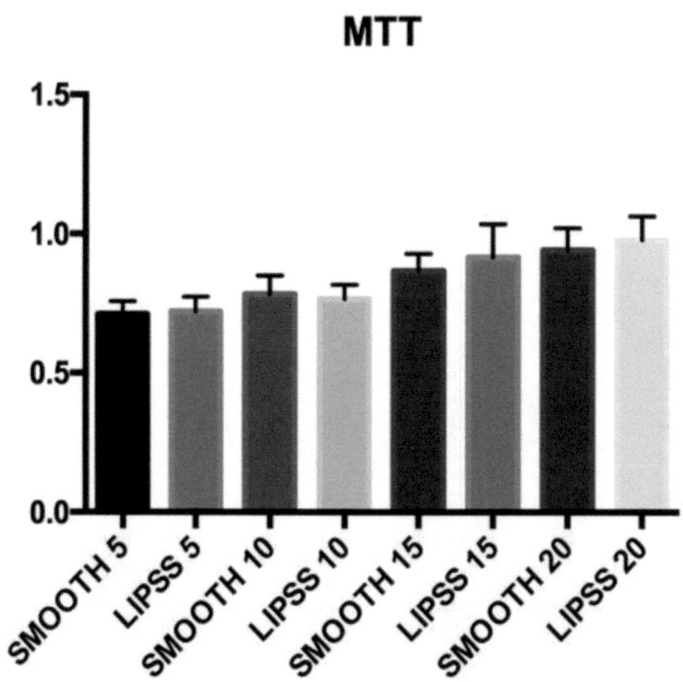

Figure 5. Data obtained for MTT assay with the standard deviation.

(lower right corner). **Figure 8** shows the images obtained on smooth surfaces, which showed a nonorganized mess of cells without polarization.

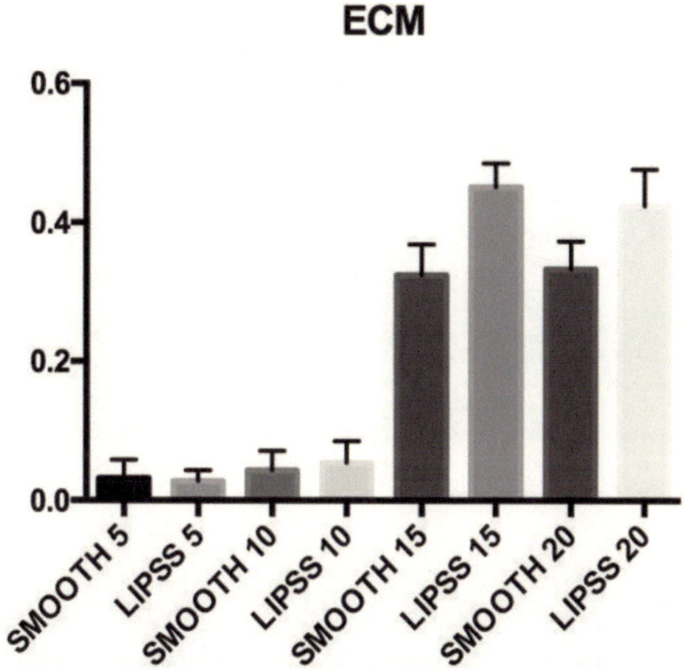

Figure 6. Data obtained for ECM assay with the standard deviation.

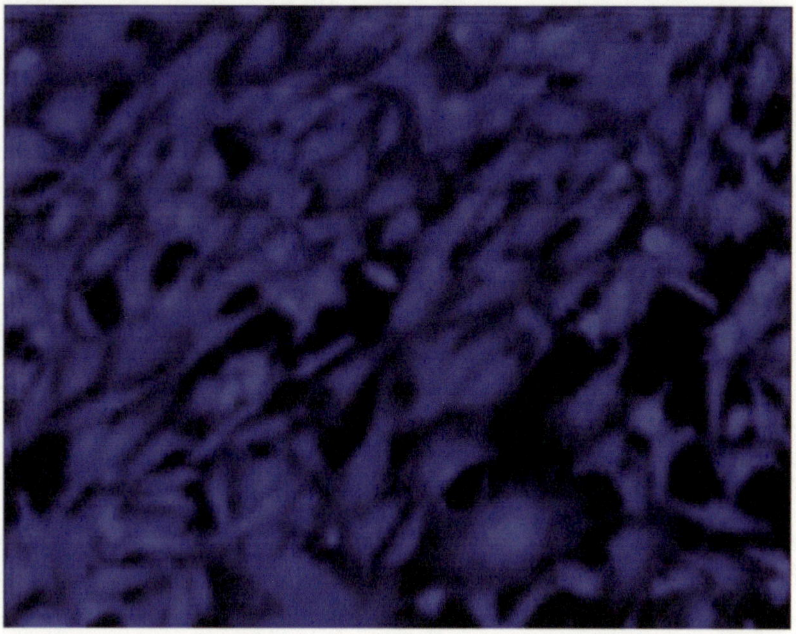

Figure 7. Confocal image of a treated surface.

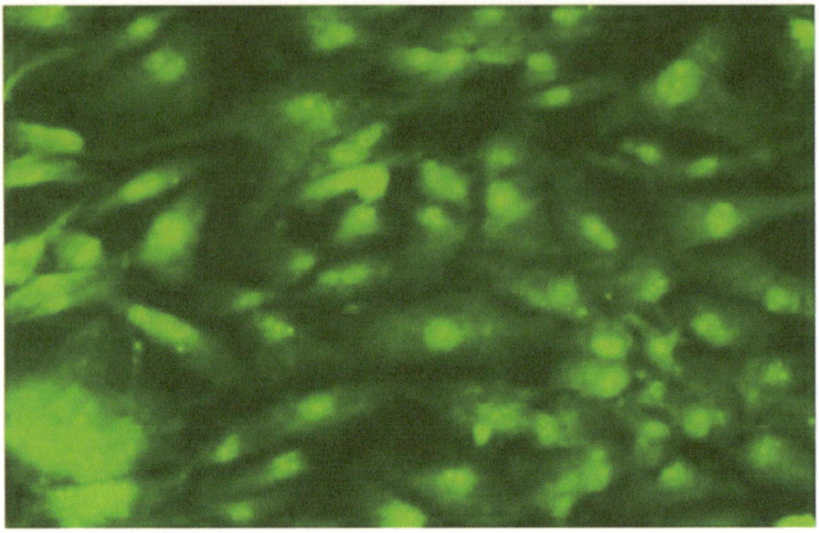

Figure 8. Confocal image of the smooth surface.

5. Conclusions

We have demonstrated in vitro that LIPSS are an efficient option to increase osteoblastic differentiation of hBMSCs in ZrO_2:Ta biocermets. When we create a laser-induced periodic surface with a femtosecond laser, we found that proliferation and inflammatory response are not increased when compared to a smooth surface. We know that LIPSS in metals decreases hydrophobicity compared to smooth surfaces, but in cermet this situation is not seen. We believe that we found a dual phenomenon. In the metal area of the cermet, the contact angle is decreased, but in the ceramic area, the transformation phase seen with this laser treatment increases the contact angle and the hydrophobicity. The optimization of the laser treatment probably will improve the wettability minimizing the transformation phase.

We can see an increase in the ECM in treated surfaces at days 15 and 20. These values add more significance to the values obtained for osteoblastic differentiation.

With values of proliferation similar in both materials, we found an increase in osteoblastic differentiation in LIPSS created over ZrO_2:Ta compared to smooth surfaces. This increase is demonstrated by higher values in the studied assays of alkaline phosphatase, osteocalcin and osteopontin production at every control point. This behaviour is seen in metals such as titanium and tantalum. This situation is similar to previous studies that demonstrate a more biological activity in LIPSS surfaces compared to smooth ones.

We can conclude that creating laser-induced periodic surfaces is a promising and cheap modification to increase osteoblastic response to a material, without adding chemicals or new materials. Optimization of the laser fluence and parameters will increase this effect, minimizing the transformation phenomenon in the ceramic area, but creating LIPSS in the metal islands.

Acknowledgements

D. Nieto is grateful to the Consellería de Cultura, Spain for the support under the Galician Program for Research Innovation and Growth (2011–2015) (I2C Plan).

Author details

Alberto Jorge-Mora[1,2], Naroa Imaz[3], Nekane Frutos[3], Ana Alonso[2], Carlota García Santiago[4], Rodolfo Gómez-Vaamonde[1,2], Jesús Pino-Minguez[1,5], Jose Bartolomé[6], Gerard O'connor[6] and Daniel Nieto[7,8]*

*Address all correspondence to: daniel.nieto@usc.es

1 SERGAS (Galician Health Service), Universidad de Santiago de Compostela, Santiago de Compostela, Spain

2 IDIS (Musculoskeletal Pathology Group), Santiago de Compostela, Spain

3 Surface Engineering Division, IK4-CIDETEC, San Sebastián, Spain

4 Department of Pharmacology, Universidade de Santiago de Compostela, Spain

5 IDIS (NEIRID Group), Orthopaedic Department, Santiago de Compostela, Spain

6 Instituto de Ciencia de Materiales de Madrid (ICMM), Consejo Superior de Investigaciones Científicas (CSIC), Madrid, Spain

7 NCLA / Inspire Labs, School of Physics, National University of Ireland, Galway, Ireland

8 Photonics4Life Research Group, Departamento de Física Aplicada, Facultad de Física, Universidade de Santiago de Compostela, Santiago de Compostela, Spain

References

[1] Arifin A, Sulong AB, Muhamad N, Syarif J, Ramli MI. Material processing of hydroxyapatite and titanium alloy (HA/Ti) composite as implant materials using powder metallurgy: A review. Materials and Design. 2014;**55**:165-175

[2] Coathup MJ, Blunn GW, Mirhosseini N, Erskine K, Liu Z, Garrod DR, et al. Controlled laser texturing of titanium results in reliable osteointegration. Journal of Orthopaedic Research. 2016

[3] Kurella A, Dahotre NB. Review paper: Surface modification for bioimplants: The role of laser surface engineering. Journal of Biomaterials Applications. 2005;**20**(1):5-50

[4] Rokkum M, Reigstad A, Johansson CB. HA particles can be released from well-fixed HA-coated stems: histopathology of biopsies from 20 hips 2–8 years after implantation. Acta Orthopaedica Scandinavica. 2002;**73**(3):298-306

[5] Han CD, Shin KY, Lee HH, Park KK, Yang IH, Lee WS. The results of long-term follow-up of total hip arthroplasty using hydroxyapatite-coated cups. Hip & Pelvis. 2015;**27**(4):209-215

[6] Martinez-Calderon M, Manso-Silvan M, Rodriguez A, Gomez-Aranzadi M, Garcia-Ruiz JP, Olaizola SM, et al. Surface micro- and nano-texturing of stainless steel by femtosecond laser for the control of cell migration. Scientific Reports. 2016;**6**:36296

[7] Nieto D, Delgado T, Flores-Arias MT. Fabrication of microchannels on soda-lime glass substrates with a Nd:YVO_4 laser. Optics and Lasers in Engineering. 2014;**63**:11-18

[8] Nieto D, Couceiro R, Aymerich M, Lopez-Lopez R, Abal M, Flores-Arias MT. A laser-based technology for fabricating a soda-lime glass based microfluidic device for circulating tumour cell capture. Colloids and Surfaces B: Biointerfaces. 2015;**134**:363-369

[9] Chen M-F, Chen Y-P, Hsiao W-T, Gu Z-P. Laser direct write patterning technique of indium tin oxide film. Thin Solid Films. 2007;**515**:8515-8518

[10] Nieto D, McGlynn P, de la Fuente M, Lopez-Lopez R, O'connor GM Laser microfabrication of a microheater chip for cell culture outside a cell incubator Colloids Surf B Biointerfaces. 2017;**154**:263-269

[11] Mannion PT, Magee J, Coyne E, O'Connor GM, Glynn TJ. The effect of damage accumulation behaviour on ablation thresholds and damage morphology in ultrafast laser micro-machining of common metals in air. Applied Surface Science. 2004;**233**:275-287

[12] Hwang TY, Vorobyev AY, Guo CL. Ultrafast dynamics of femtosecond laser-induced nanostructure formation on metals. Applied Physics Letters. 2009;**95**

[13] Hwang TY, Guo CL. Femtosecond laser-induced blazed periodic grooves on metals. Optics Letters. 2011;**36**:2575-2577

[14] Bonse J, Kruger J, Hohm S, Rosenfeld A. Femtosecond laser-induced periodic surface structures. Journal of Laser Applications. 2012;**24**

[15] Gurevich EL, Gurevich SV. Laser induced periodic surface structures induced by surface plasmons coupled via roughness. Applied Surface Science. 2014;**302**:118-123

[16] Sedao X, Maurice C, Garrelie F, Colombier JP, Reynaud S, Quey R, Blanc G, Pigeon F. Electron backscatter diffraction characterization of laser-induced periodic surface structures on nickel surface. Applied Surface Science. 2014;**302**:114-117

[17] Huynh TTD, Petit A, Semmar N. Picosecond laser induced periodic surface structure on copper thin films. Applied Surface Science. 2014;**302**:109-113

[18] Perez S, Rebollar E, Oujja M, Martin M, Castillejo M. Laser-induced periodic surface structuring of biopolymers. Applied Physics A: Materials Science & Processing. 2013; **110**:683-690

[19] Kumar B, Soni RK. Pulsed-laser-induced photochemical growth of the periodic surface structure on InP. Semiconductor Science and Technology. 2009;**24**

[20] Hohm S, Herzlieb M, Rosenfeld A, Kruger J, Bonse J. Formation of laserinduced periodic surface structures on fused silica upon two-color double-pulseirradiation. Applied Physics Letters. 2013;**103**

[21] Bartolome JF, Moya JS, Couceiro R, Gutierrez-Gonzalez CF, Guiti F, Martinez-Insua A. In vitro and in vivo evaluation of a new zirconia/niobium biocermet for hard tissue replacement. Biomaterials. 2016;**76**:313-320

[22] Daminelli G, Krüger J, Kautek W. Femtosecond laser interaction with silicon under water confinement. Thin Solid Films. 2004;**467**:334-341

[23] Suchanek W, Yoshimura M. Processing and properties of hydroxyapatite-based biomaterials for use as hard tissue replacement implants. Journal of Materials Research. 1998;**13**:94-117

[24] Karageorgiou V, Kaplan D. Porosity of 3D biomaterial scaffolds and osteogenesis. Biomaterials. 2005;**26**:5474-5491

[25] Vasconcellos LMRd, Oliveira MVd, Graça MLdA, Vasconcellos LGOd, Carvalho YR, Cairo CAA. Porous titanium scaffolds produced by powder metallurgy for biomedical applications. Materials Research. 2008;**11**:275-280

[26] Nieto D, Arines J, O'connor GM, flores-Arias MT. Single-pulse laser ablation threshold of borosilicate, fused silica, sapphire, and soda-lime glass for pulse widths of 500 fs, 10 ps, 20 ns. Applied Optics. 2015;**54**(29):8596-8601

[27] Liu X-L, Cheng W, Petrarca M, Polynkin P. Measurements of fluence profiles in femtosecond laser filaments in air. Optics Letters. 2016;**41**(20):4751-4754

[28] McDaniel C, Gladkovskaya O, Flanagan A, Rochev Y, O'Connor GM. In vitro study onthe response of RAW264.7 and MS-5 fibroblast cells on laser-induced periodic surfacestructures for stainless steel alloys. RSC Advances. 2015;**5**:42548-42558

[29] Young T. Philosophical Transactions of the Royal Society of London. 1805;**95**:65

[30] Martínez-Calderón M, Rodríguez A, Dias A, Morant-Miñana MC, Gómez-Aranzadi M, Olaizola SM. Femtosecond laser fabrication of highly hydrophobic stainless steel surface with hierarchical structures fabricated by combining ordered microstructures and LIPSS. Applied Surface Science. 2015;**374**:81-89

[31] Wennerberg A, Jimbo R, Stübinger S, Obrecht M, Dard M, Berner S. Nanostructures and hydrophilicity influence osseointegration – A biomechanical study in the rabbit tibia. Clinical Oral Implants Research. 2014;**25**:1041

[32] Kenar H, Akman E, Kacar E, Demir A, Park H, Abdul-Khaliq H, AktaCs KE. Femtosecond laser treatment of 316L improves its surface nanoroughness and carbon content and promotes osseointegration: An in vitro evaluation. Colloids and Surfaces B. 2013;**108**:305-312

[33] Kopf BS, Ruch S, Berner S, Spencer ND, Maniura-Weber K. The role of nanostructures and hydrophilicity in osseointegration: In-vitro protein-adsorption and blood-interaction studies. Journal of Biomedical Materials Research Part A. 2015;**2015**(103A):2661-2672

Chapter 13

Laser Ablation of Energetic Materials

Ruiqi Shen, Lizhi Wu, Wei Zhang and
Haonan Zhang

Additional information is available at the end of the chapter

http://dx.doi.org/10.5772/intechopen.71892

Abstract

The initiation of explosives by laser is a new initiation method. Compared with traditional initiation methods, laser initiation has the characteristics of high reliability and high safety. It can be used as one of the alternative technologies for future initiation device. A microscopic understanding of the complex physical and chemical processes involved in the reaction process is essential for laser initiation. Shock initiation technology of laser-driven flyer was studied. Several typical laser-driven flyer targets researches were introduced. Some significant characteristics including velocity and impact stress of flyers were tested via photonic Doppler velocimetry and polyvinylidene fluoride pressure sensor, respectively. Some types of flyers including Al and Cu single-layer flyers and CuO/Cu, CuO/Al, and CuO/Al/Cu multilayer flyers with relatively high velocities were used to initiate PETN explosive. In order to give a better understanding of the mechanism of laser interaction with typical energetic materials (RDX, HMX, TNT, and HNS), a time of flight mass spectrometer (TOFMS) was used to detect the positive ions and the negative ions were produced in the laser-induced dissociation processes. The influences of laser wavelength, the laser fluence, and the delay time of the decomposition process have been studied as well. The results may throw some light on the laser interaction mechanism of energetic materials.

Keywords: laser ablation, PDV, laser-driven flyer, TOFMS, nitramines, aromatic nitro compounds

1. Introduction

Laser ablation of energetic materials, such as pyrotechnics, explosives, and propellants, is different from other materials because of the interaction with chemical reaction and laser beam. The interaction processes between energetic material and laser beam are accompanied with thermal physics process irradiated by laser (photo-thermal, melting, and gasification),

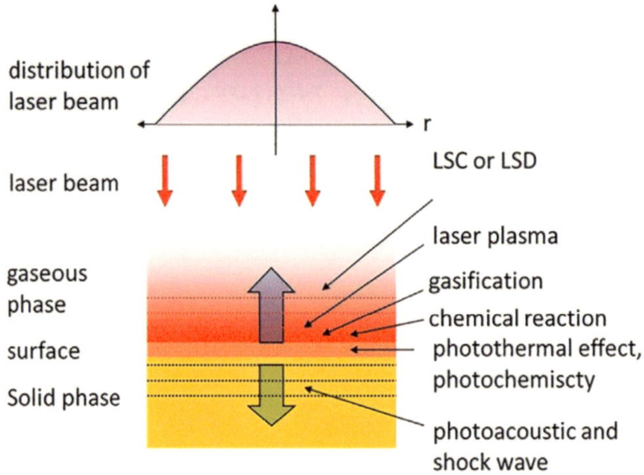

Figure 1. Interaction mechanism between laser beam and energetic materials.

chemical reaction in condensed phase and gaseous phase, generation of plasma, and laser-supported combustion (LSC) and laser-supported detonation (LSD) depending on laser power and material characteristics, seen in **Figure 1**.

Laser ablation of energetic materials is different from other materials' in chemical reaction, in which the chemical reactions in condensed and gaseous phase can induce ignition, combustion, and/or detonation. The research results show the reaction irradiated by laser is attributed to the category of thermochemistry at low power ($<10^3$W/mm^2), but it is a complex reaction including laser ablation and laser plasma at high power ($>10^3$W/mm^2).

2. Shock initiation technology of laser-driven flyer

As a type of driving energy, laser is able to ablate the target to generate substances with high temperature and pressure to propel objects up to high velocity. There exist mainly two modes of laser-driven flyers including inward driving and linear driving as shown in **Figure 2**.

Figure 2(a) shows an inward driving structure with free outward spherical surface as target. After being ablated by intense laser (with intensity of light over 10^{12} W·cm^{-2}), the target will be disaggregated as plasmas with high temperature, high pressure, and high density to promote the rest part of target shell to be a flyer with high velocity. Immediately, the high-velocity flyer will cause implosion of deuterium and tritium medium inside the shell to achieve compressed fusion ignition (CFI). In linear driving structure shown in **Figure 2(b)**, the incident surface is covered by the transparent substrate or the optical fiber. If the intensity of incident laser is not that high (10^9–10^{11} W·cm^{-2}), the transparent substrate can block the expansion of plasmas temporarily to enhance the impact stress and velocity of flyer (the remainder of target plate after ablation). This chapter mainly focuses on laser-driven flyer technology based on transparent substrate, which is gradually becoming a significant dynamic high pressure loading technology.

The process of laser-driven flyer is very complex and can be roughly divided into several steps as follows. First, the laser beam penetrates into the transparent substrate and irradiates on the

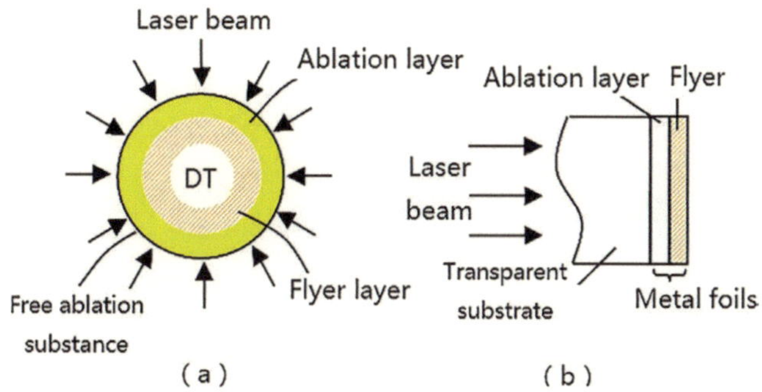

Figure 2. Two modes of laser-driven flyers: (a) inward driving (b) linear driving.

surface of the metal film, so that the film substance is heated and evaporated. Then, the plasmas formed at the interface between substrate and metal absorb the laser energy continuously to heat up and expand. Next, the rest part of metal film is released out to form high-velocity flyer.

Compared with conventional high-velocity driving methods including gas cannon, detonation, and electromagnetic driving, laser-driven flyer shows advantages of greatly enhancing the velocity of flyer (up to tens of km/s) and impact stress (up to TPa order), owning simple structure and low cost. Therefore, laser-driven technology is widely used in high-pressure physics, space science, material science, and research of rapid detonation for explosives [1].

Laser-driven flyer impact initiation technology is an application of laser-driven flyer, which is the process of initiating high energy explosives with high-velocity flyer driven by laser. As shown in **Figure 3**, the process of laser-driven flyer initiating explosive can be divided into three main steps:

a. The laser beam penetrates into the transparent substrate and interacts with the film on the back of substrate so that a part of film material is ablated instantaneously to form plasmas, which can absorb the laser energy unceasingly,

b. The plasmas with high temperature and high pressure can only expand in the direction of the film material due to the confinement of substrate so that the remaining nonablated films are sheared down to form high-velocity flyer, which is accelerated in accelerating chamber immediately,

c. The flying flyer at high velocity impacts on explosive surface and causes detonation to initiate explosive (e.g. HNS and PETN).

It is the precise time control, rapid response, strong electromagnetic interference resistance, and low sensitivity that makes laser-driven flyer impact initiation technology as a good potential for the application to modern fuze system.

2.1. Laser-driven flyer target

At present, the flying target with transparent substrate is mainly used in laser-driven flyer impact initiation technology due to its great performance of enhancing the absorption of laser energy by plasma and increasing the impact momentum of flyer. Usually, BK7, K9, or

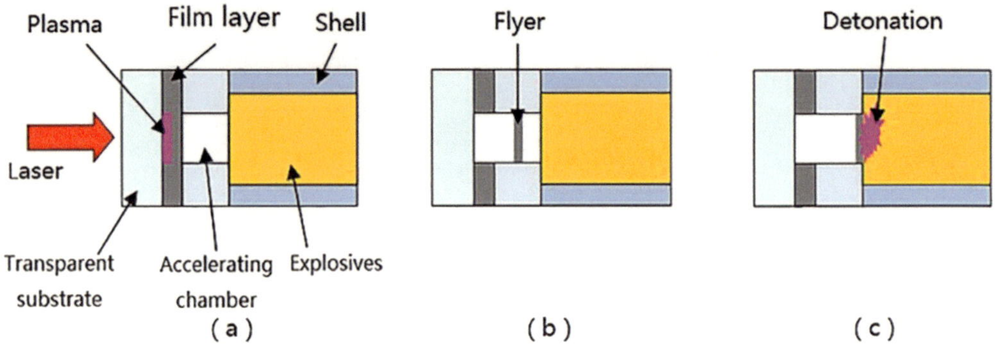

Figure 3. Process of laser-driven flyer impact initiation.

sapphire glass with high transmittance of the corresponding wavelength are adopted as the transparent substrate.

In general, thickness of the film deposited on the transparent substrate is typically a few microns to tens of microns and the diameter is about 1 mm. Paisley et al. [2] found that the impact stress of flyer must be higher than the rarefaction wave's reflected from the free surface of flyer during plasma-driving laser ablation. Only under this condition the flyer continue to be accelerated to reach the final velocity. In addition, the laser pulse width must be 2.5 times larger than round-trip time of sound in the flyer in order to ensure that the flyer will not be fully impacted, ablated, and chipped during the acceleration.

$$R = T/\left[\sim 2.5\left(2t/C_1\right)\right], \text{that is } R = \frac{0.2\left(TC_1\right)}{t} \tag{1}$$

In Eq. (1), the dimensionless quantity R must be greater than 1, T (s) is the laser pulse width, t (mm) is the thickness of flyer, C_1 (mm/s) is the radial sound velocity of flyer.

2.1.1. *Structure of flyer target*

The structure of flyer target can be divided into two types: single-layer flyer and multilayer flyer. **Figure 4(a)** shows the structure of single-layer flyer. As we can see, single-layer flyer is prepared by depositing metal film on the transparent substrate. After the interaction between laser and metal film, the plasma with high temperature and high pressure produced through ablation will drive the nonablated film and shear it out to form a flyer.

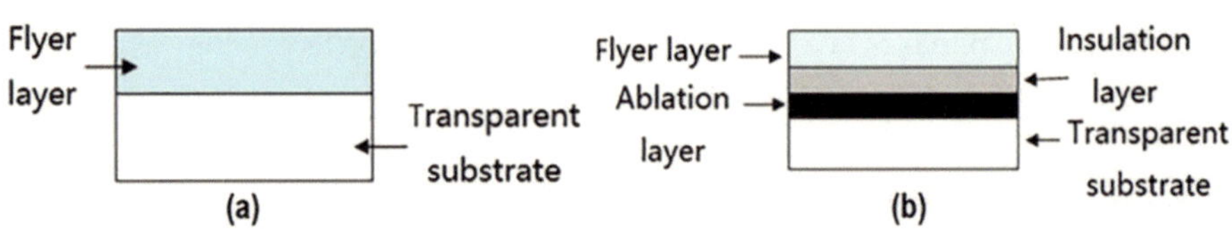

Figure 4. Structure of (a) single-layer flyer and (b) multilayer flyer.

Multilayer flyer is developed according to the formation process of laser-driven flyer. Generally, multilayer flyer owns three-layer structure, as shown in **Figure 4(b)**. The main function of ablation layer is to absorb laser energy and be ablated to form plasma with high temperature and high pressure to drive the flyer. Thermal insulation layer is to restrain the formation and diffusion of plasma in order to reduce the ablation of flyer and ensure its integrity. Sometimes multilayer flyer can be divided into four layers. An absorption layer is added between ablation layer and transparent substrate to improve the absorption of laser energy.

2.1.2. Typical flyer targets

Several typical flyer targets are listed in **Table 1** in detail. In laser-driven flyer impact initiation technology, the research on flyer has been developed from single-metal flyer to multilayer flyer.

According to **Table 1**, the main materials of single-layer flyer are Al and Cu. As for multilayer flyer, the ablation layer materials are mostly C, Al, Mg, Ge, Ti, Zn, etc. Al_2O_3, MgF_2, ZnS, and TiO_2 are commonly good choices for thermal insulation materials. Al, Ti, and Cu are usually used to prepare flyer layers.

2.2. Characteristics of laser-driven flyer

In the study of laser-driven flyer impact initiation technology, the detonation of explosive depends mainly on the impact stress of flyer and pulse width of impact stress. The impact stress of flyer depends on the velocity, structure, and material of flyer while the pulse width

Structure	Author	Flyer layer	Insulation layer	Ablation layer	Absorption layer	Transparent substrate
Single-layer flyer	S. A. Sheffield [3]	Al	—	—	—	—
		Al	—	—	—	BK7
	D. L. Paisley [4]	Al, Cu	—	—	—	glass
Multilayer flyer	D. L. Paisley [5, 6]	Al, Cu	Al_2O_3	—	—	
	D. B. Stahl [7]	Al	—	C	—	
	W. M. Trott [8, 9]	Al	Al_2O_3	Al	—	or
	D. J. Hatt and J. A. Waschl [10]	Al, Cu	Al_2O_3, MgF_2, ZnS	Al, Mg	—	
	J. L. Labaste [11]	Al	Al_2O_3	Ge, C, Ti, Al	—	K9 glass
	D. L. Paisley [12, 13]	Al	Al_2O_3	Al	C	
	M. D. Bowden and S. L. Knowles [14]	Al, Ti	Al_2O_3, TiO_2	C, Mg, Ge, Al, Ti	C, Ti	or
	L. Wu [15]	Al	Al_2O_3	Al	CuO	
	H. R. Brierley [16]	Al	Al_2O_3	Ge, Ti, Zn		Sapphire glass

Table 1. Typical laser-driven flyer targets researched.

of impact stress depends on the planarity and integrity of flyer. Therefore, it is necessary to select flyer with high velocity, strong impact stress, and good planarity and integrity to develop laser-driven flyer impact initiation technology.

2.2.1. Velocity of flyer

2.2.1.1. Measurement of velocity

The velocity of flyers can be measured via photonic Doppler velocimetry (PDV). **Figure 5** shows a schematic view of PDV. It is mainly based upon two physical processes: Doppler effect and optical mixing, that is, heterodyning.

Figure 6 shows schematic view of the measuring setup of flyer velocity. Velocity information of flyer is collected and recorded by PDV measurement system and time history of flyer is obtained by computer program analysis.

2.2.1.2. Regularity of velocity

According to the research of Wu [15] in Nanjing University of Science and Technology, four types of flyers including $Al/Al_2O_3/Al$ multilayer, $C/Al/Al_2O_3/Al$ multilayer, $Mg/Al/Al_2O_3/Al$ multilayer and Al single-layer were tested with the laser energy of 36.4–195.1 mJ by PDV measurement system. The result is indicated in **Figure 7**.

According to **Figure 7**, the analysis to result are as follows. At the same laser energy, $Mg/Al/Al_2O_3/Al$ multilayer flyer owns the maximum velocity, which is little bit better than Al single-layer flyer. The velocity of $C/Al/Al_2O_3/Al$ multilayer flyer is higher than that of $Al/Al_2O_3/Al$ multilayer flyer when the laser energy is less than 150 mJ. Within the range of 150–195.1 mJ of laser energy, the velocity of $C/Al/Al_2O_3/Al$ multilayer flyer is comparable to that of $Al/Al_2O_3/Al$ multilayer flyer. As the ablation layer, Mg layer can greatly improve the velocity of flyer. The addition of Al_2O_3 insulation layer increases the mass of flyer, which makes the velocity

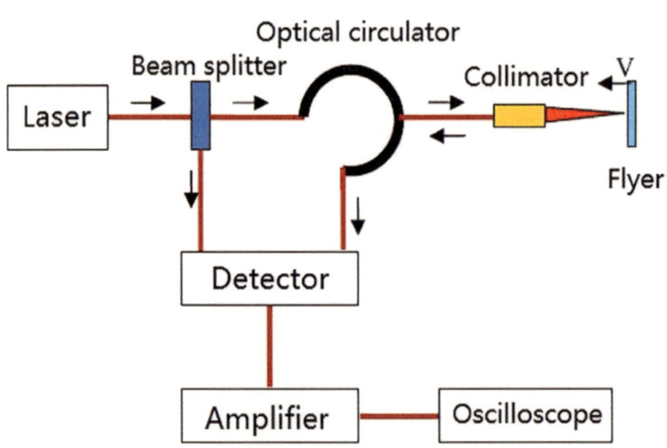

Figure 5. Schematic of PDV.

http://dx.doi.org/10.5772/intechopen.71892

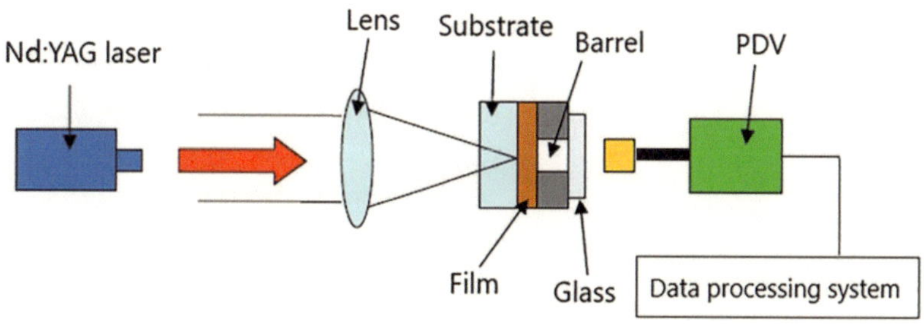

Figure 6. Schematic of measuring setup of flyer velocity.

of $Al/Al_2O_3/Al$ multilayer flyer lower than that of Al single-layer flyer. The addition of C absorption layer increases the velocity of $C/Al/Al_2O_3/Al$ multilayer flyer at low laser energy. However, C layer is easy to be broken down at high laser energy, which reduces the restraining effect of substrate and accelerating chamber to plasma. As a result, the velocity of $C/Al/Al_2O_3/Al$ multilayer flyer is lower than that of $Al/Al_2O_3/Al$ multilayer flyer.

2.2.2. Impact stress of flyer

2.2.2.1. Measurement of impact stress

The impact stress can be measured by a polyvinylidene fluoride (PVDF) pressure sensor (shown in **Figure 8**), and the current mode of PVDF pressure sensor is used to measure the impact stress of flyer.

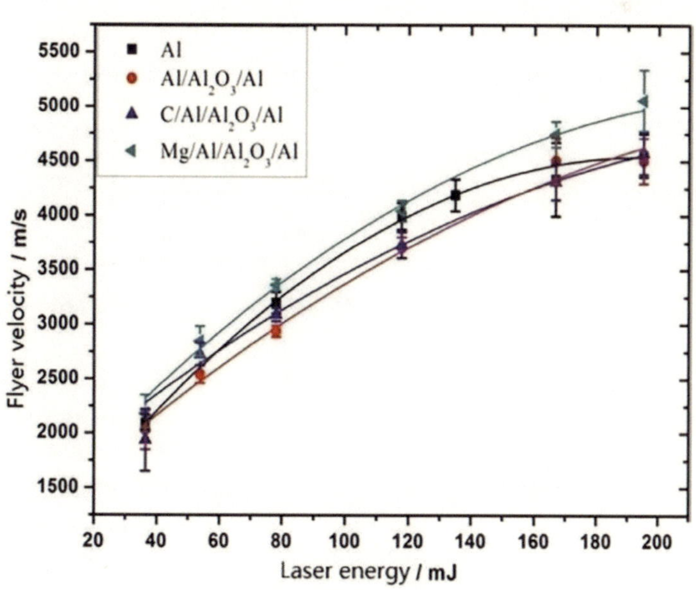

Figure 7. Final velocities of four flyers in different laser energy.

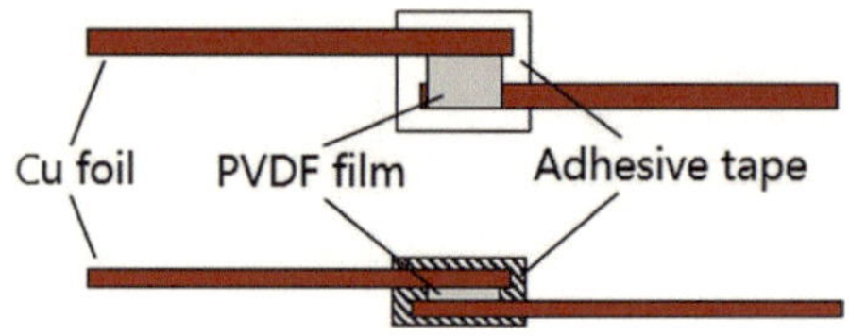

Figure 8. Structure of PVDF sensor.

2.2.2.2. Regularity of impact stress

Figure 9(a) shows the impact stress of six types of flyers at 175, 212, 248, 282, 315, and 362 mJ laser energy, respectively. With the increase of laser energy, the impact stresses of flyer tend to rise up gradually. At the same laser energy, the impact stress of Al single-layer flyer is lower than $Al/Al_2O_3/Al$ multilayer flyer. When the laser energy is not more than 248 mJ, the impact stress of $CuO/Al/Al_2O_3/Al$ multilayer flyer is comparable to that of $CuO/Al_2O_3/Al$ multilayer flyer. As the energy continues to increase, the impact stress of former will exceed that of the latter. In most cases, the impact stress of $C/Mg/Al/Al_2O_3/Al$ multilayer flyer is greater than $C/Al/Al_2O_3/Al$ multilayer flyer.

In 1969, Walker and Wasley [17] proposed the criterion for impact initiation of heterogeneous explosive

$$E_C = P^2 \tau \tag{2}$$

In Eq. (2), P is the shock wave pressure entering into explosive, τ is the time that shock wave travels back and forth in flyer, and E_c is a constant determined by the specific explosive. Only when the value of $P^2\tau$ is higher than that of E_c, the explosive can be initiated. Ignoring the area effect of impact initiation, the criterion for it can be written as

$$P^2 \tau \geq E_C \tag{3}$$

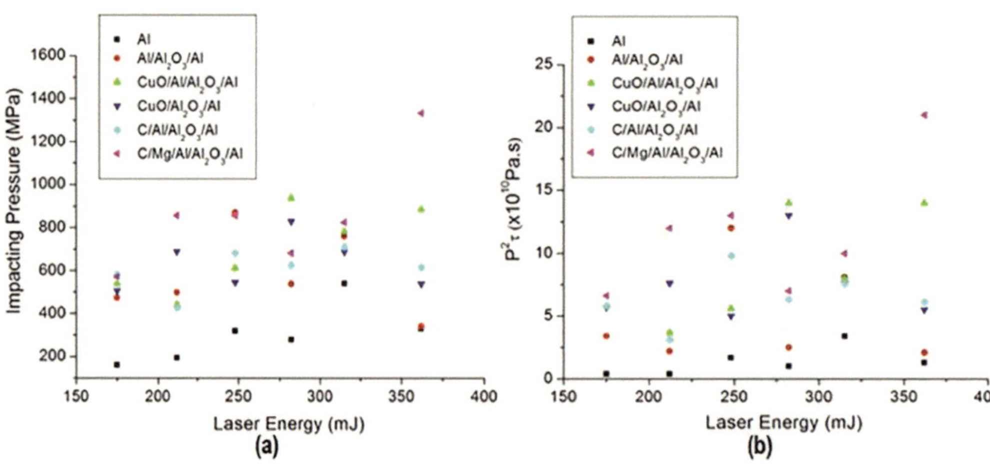

Figure 9. Impact stresses (a) and value of $P^2\tau$ for flyers at different laser energy (b).

http://dx.doi.org/10.5772/intechopen.71892

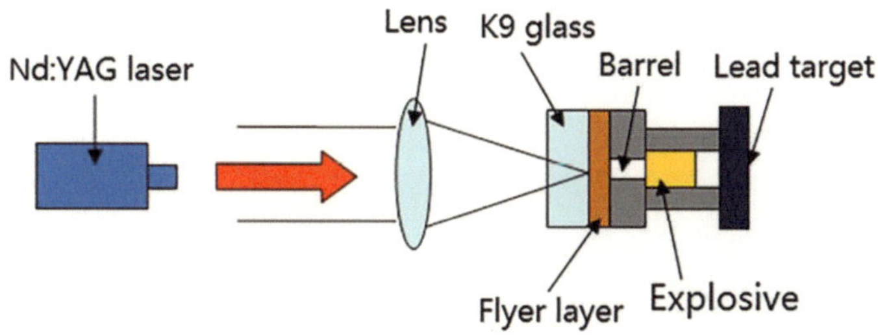

Figure 10. Schematic view of measuring setup for impact initiation.

Value of $P^2\tau$ for six types of flyers are shown in **Figure 9(b)**. The value of $P^2\tau$ indicates the flyer's ability to initiate explosive. It is obvious from **Figure 10** that $P^2\tau$ of multilayer flyers is higher than that of single-layer flyers. Especially, C/Mg/Al/Al_2O_3/Al multilayer flyer obtains the maximum value of $P^2\tau$, which means the best ability to initiate explosive.

2.3. Characteristics of laser-driven flyer impact initiation

2.3.1. Theory of initiating explosives

Impact initiation of agglomerate explosive is a significant initiation method besides thermal initiation. Due to the press fitting of solid agglomerate explosive, voids are retained around the crystal grains with the degree of porosity of 1–4%. After the shock wave entering explosive, the air voids and bubbles are compressed adiabatically under impact. The specific heat of gas is lower than that of the explosive crystal. Therefore, the temperature of bubble is higher than that of the crystal, which will cause the appearance of hot spots.

2.3.2. Measuring setup for impact initiation

The schematic view of measuring setup for impact initiation is shown in **Figure 10**. The laser focused by lens irradiates on flyer target and ablates a part of the film to form plasma. Immediately, the remaining part of the film is sheared along the edge of accelerating chamber and is driven out at high velocity to form a flyer. Then, the flyer is accelerated in accelerating chamber and hits explosive cylinder causing the explosive to be struck and detonated.

2.3.3. Initiation of PETN

Five types of flyers including Al and Cu single-layer flyers and CuO/Cu, CuO/Al and CuO/Al/Cu multilayer flyers with relatively high velocities were used to initiate PETN explosive [15]. The results are shown in **Table 2**.

Compared to the results in **Table 2**, it is easy to find that Al single-layer flyer shows the best performance of initiation in the same laser energies, which succeeded in initiating as long as the laser energy was higher than 161 mJ. CuO/Cu multilayer flyer owned four successful detonations as well. Besides, the success rate of initiation for other flyers is arranged from high to low, followed by Cu single-layer flyer, CuO/Al and CuO/Al/Cu multilayer flyers. The results indicate

Structure of flyer	Laser energy/mJ	67	109	161	203	264	336.7	375.7
Cu	Flyer velocity/m·s^{-1}	1965	2402	2597	3389	3389	4020	3065
	$P^2\tau$/×10^{10} Pa·s	0.2	0.2	1.3	6.3	10.4	6.3	3.0
	Detonation	×	×	×	√	√	√	×
Al	Flyer velocity/m·s^{-1}	3162	3433	4145	4324	4761	5194	5000
	$P^2\tau$/×10^{10} Pa·s	0.2	0.9	5.4	5.6	9.4	7.6	6.6
	Detonation	×	×	√	√	√	√	√
CuO/Cu	Flyer velocity/m·s^{-1}	2105	2500	2649	2909	3149	3252	3305
	$P^2\tau$/×10^{10} Pa·s	2.1	1.5	2.6	5.8	6.8	5.4	7.1
	Detonation	×	×	×	√	√	√	√
CuO/Al	Flyer velocity/m·s^{-1}	3478	3773	4597	4878	5031	4848	5673
	$P^2\tau$/×10^{10} Pa·s	0.6	2.4	3.2	3.7	11.0	8.7	4.9
	Detonation	×	×	×	×	√	√	×
CuO/Al/Cu	Flyer velocity/m·s^{-1}	2352	2507	2702	3100	3265	3791	3625
	$P^2\tau$/×10^{10} Pa·s	0.9	1.1	1.9	3.3	5.1	2.0	2.0
	Detonation	×	×	×	×	√	×	×

Table 2. Initiation results of PETN for different types of flyers.

that a certain range of laser energy is necessary in the process of laser-driven flyer impact initiation, which can ensure enough velocity and impact stress of flyer.

2.4. Analysis of laser-driven flyer initiating explosives

There are two main factors affecting the impact initiation of flyer including velocity and shape of flyer, which are indispensable. Al flyer shows the best performance in impact initiation while CuO/Al and CuO/Al/CuO multilayer flyers show the worst performance. It is analyzed that the oxidation reaction between Al and CuO will release a lot of energy to melt or vaporize some part of flyer and change the shape of flyer, which can reduce the impact effect of it and result in the decrease of flyer velocity. The velocity of Cu flyer is higher than Al's, but its impact initiation ability is not as good as Al flyer. This is because Cu flyer is more brittle than Al flyer and is more likely to be broken during flight, which will affect the initiation effect. A large number of experimental results show that a certain range of laser energy is needed to drive the flyer. Thus, the velocity and impact stress can be high enough to initiate the explosives successfully.

3. Laser-induced time of flight mass spectroscopy of typical energetic materials

Laser-material interaction is essentially based on thermal and photonic processes, where as a consequence of laser irradiation, different phenomena occur on a microscopic or macroscopic scale [18]. It will populate an excited state that can either couple to a reaction pathway, resulting in chemical reactions, or undergo internal conversion back to the ground state of the starting molecule. A microscopic understanding of the complex physical and chemical processes involved in the combustion and decomposition of energetic material is essential for the development of predictive detonation and reliable model for performance, stability, and hazard analysis of explosives [19]. Many investigations have been conducted concerning the mechanism of decomposition of explosives under laser irradiation [20–22].

In order to ensure the reliability and safety of laser initiation system, it is necessary to reveal the microscopic mechanism of the reaction of energetic materials under laser irradiation. Spectroscopic method is an effective means to reveal the mechanism of laser-induced dissociation of energetic materials [23]. In the present work, time of flight mass spectrometry (TOFMS) was employed to study the ion fragments produced by explosives after laser excitation with the purpose of understanding the effect of different experimental parameters on the resultant mass spectra and laser-induced reaction.

3.1. Detection of laser-induced decomposition products

The explosive samples were pressed into an aluminum holder (6 mm diameter and 11 mm depth) using a pressing tool to ensure a flat surface and a consistent filling density with a pressure of 80 MPa. The apparatus used in this study was a homemade reflection time of flight mass spectrometer. The time of flight mass spectrometer consists of four parts: ion source, ion acceleration zone, flight area, reflector and detector. The detector is made from a pair of microchannel plates (MCPs) [24]. It was combined with a vacuum chamber, pumping down to 4×10^{-4} Pa with two turbomolecular pumps. The output from a Q-switched Nd:YAG laser (LOTIS TII, LS-2147, 450 mJ@532 nm/15 ns, 800 mJ@ 1064 nm/15 ns) was focused to the sample surface by a quartz focus lens with focus length of 500 mm) and spot size of about 0.8 mm^2. The laser fluence was measured by an energy meter (Ophir, Model 30A). After ionization, the laser generated ion fragments were first driven by the initial kinetic energy into the acceleration zone, and then accelerated by a pulsed electric field, and finally detected by the MCPs [25]. The ions were mass analyzed by its flight time. In order to reduce the noise, the weak signal measurement technique with multiple signal superposition was adopted. The TOF signals were accumulated 128 times by a digital oscilloscope (Tektronix, DPO7140) to improve the signal-to-noise ratio. The trigger of each instrument was controlled by a pulse delay generator (University of Science and Technology of China, GH024). **Figure 11** shows the schematic diagram of the experimental setup [18].

3.2. Laser-induced TOF mass spectroscopy of nitramines

3.2.1. RDX

Typical negative and positive ion mass spectra of RDX obtained by 532 nm laser dissociation are shown in **Figure 12**. The laser energy was 4.8 J/cm^2 for negative ions with delay time of 50 μs and 5.2 J/cm^2 for positive ions with delay time of 50 μs. For negative mode, the peak with the highest intensity in the negative-ion spectrum are at m/z = 46 and corresponds to the NO_2 fragment. Other relatively strong peaks appear at m/e = 26, 106, and 134 and can be assigned to CN, $(CNC)C(NN)_2$, and $H_2C(N\text{-}NO_2)_2$, respectively. Finally, some low intensity

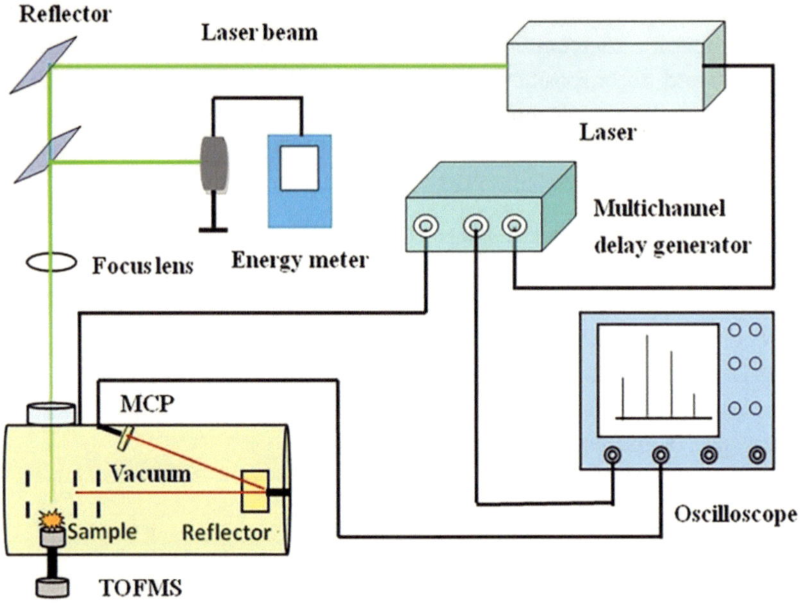

Figure 11. Schematic diagram of the TOFMS system.

peaks are observed at m/z = 42,60,84,88,93,130, and 176, possible assignments for major peaks are N(CH_2)N, N-NO_2, $(H_2CN)_3$,$(CH_2)_2$N–NO_2, (CH)(NCNC)(NN), $(CH_2)_3$N(N-NO_2)N and $(CH_2)_3$(N-NO_2)$_2$N, respectively. For positive mode, positive ions have three series, m/z = 17 can be attributed to OH. m/z = 26, 27, 28, 29, 30, and 32 ions can be classified as CN, HCN, CO/N_2/H_2CN, HCO, NO, and O_2, respectively. The third series with m/z of 42, 44, and 47, and their corresponding attributions are C_2H4N/ CH_2N_2/N (H_2C) N, N_2O/CH_2NO, and HONO, respectively [18, 26].

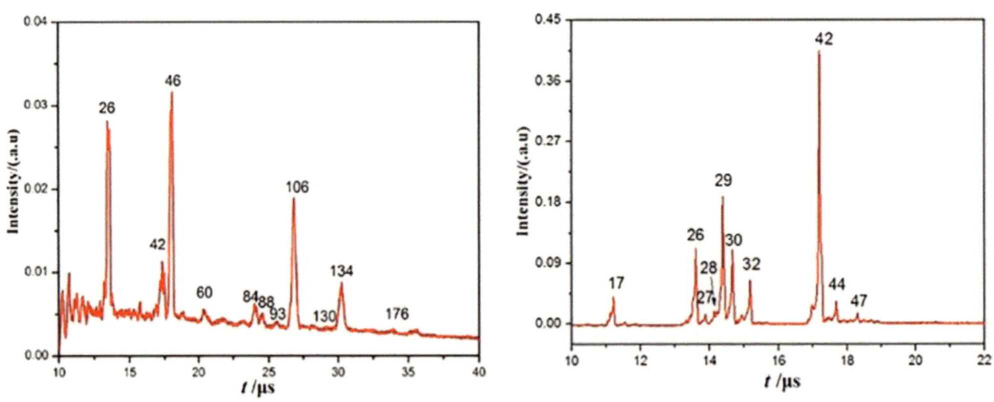

Figure 12. Typical TOF mass spectra of negative ions (left) and positive ions (right) for RDX produced by 532 nm laser ablation.

Typical negative and positive ion mass spectra of RDX obtained by 1064 nm laser dissociation are shown in **Figure 13**. The laser energy was 34.9 J/cm^2 for negative ions with delay time of 80 μs and 49.4 J/cm^2 for positive ions with delay time of 70 μs. The position of the peaks of negative ion mass spectrum under 1064 nm is similar to that under 532 nm. The difference lies in the intensity of the major peak. From the negative ions mass spectrum, it can be found that the highest peak with the appears at *m/z* = 106 and corresponds to $CH_2(NO_2)_2$ fragment. Different with the results at 532 nm, two small new peaks produce at *m/z* = 98 and 100, which can be assigned to $(CH_2)_3N_2(NN)$ and $(CH_2)_3N_2(NO)$. The relative strong peaks at *m/z* = 26, 46, 88 and 134 may be CN or C_2H_2, NO_2, $(CH_2)NNO_2$ and $H_2C(N\text{-}NO_2)_2$. Moreover, the peaks with low intensity can be observed at *m/z* = 42, 64, and 84 are most likely the CNO or $(CH_2)_2N$ or CH_2N_2 and $(CH_2N)_3$ fragments, respectively. In the positive ions mass spectrum, the peak with the highest intensity can be found at *m/z* = 30, corresponding to the NO or CH_2O species. The second strong peak appears at *m/z* = 18, and can be ascribed to H_2O or NH_4, respectively. A series of peaks are observed at *m/z* = 40, 42, 43, 44, 46, and 47, and can be ascribed to CN_2 or C_2H_2N, $N(CH_2)N$ or $(CH_2)_2N$, NCOH, N_2O or CO_2 or CH_2NO, NO_2 and HONO, respectively. The thermal decomposition results of the isotopically labeled RDX confirmed that the formula for the ion signal at *m/z* = 45 was H_2NCHO and which might be formamide. Likewise, the isotopically labeled results also showed that *m/z* = 59 was corresponding to CH_3NHCHO and was most likely N-methylformamide. Some small peaks can be observed at *m/z* = 57, 58, 59, and 60, these ions can be assigned to $(CO)NH(CH_2)$ or N(COH)N, CNO_2 or $(COH)NH(CH_2)$, $CHNO_2$ and CH_2NO_2, or NNO_2, respectively. Some other peaks can be found at *m/z* = 16, 19, 24, 27, 28, 74, 79, and 86, which may be from CH_4 or O, H_3O, C_2, C_2H_3, or CHN, N_2 or CO or C_2H_4 or CH_2N, CH_2NNO_2, $(HC)N(CN)_2$, and $(CH)_2NNO_2$, respectively [18, 25].

3.2.2. HMX

Typical negative and positive ion mass spectra of HMX obtained by 532 nm laser dissociation are shown in **Figure 14**. The laser energy was 8.6 J/cm^2 for negative ions with delay time of 75 μs and 7.8 J/cm^2 for positive ions with delay time of 70 μs. It can be seen from the negative

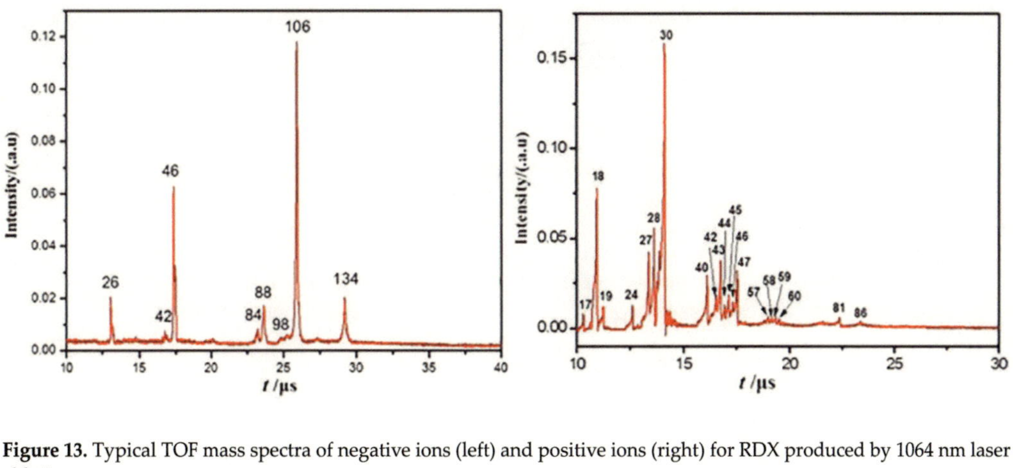

Figure 13. Typical TOF mass spectra of negative ions (left) and positive ions (right) for RDX produced by 1064 nm laser ablation.

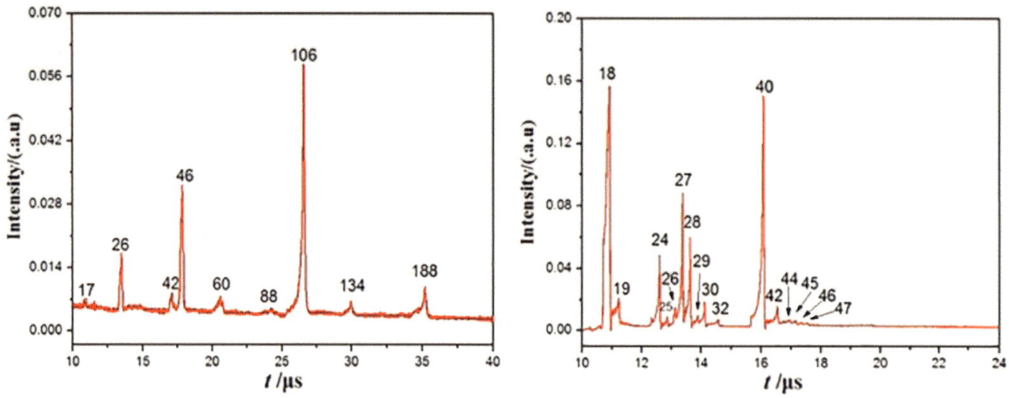

Figure 14. Typical TOF mass spectra of negative ions (left) and positive ions (right) for HMX produced by 532 nm laser ablation.

ion spectrum that the ion distribution of HMX has some similarity with that of RDX. HMX has three strong ion peaks with *m/z* of 26, 46, and 106. According to the molecular structure and decomposition properties of HMX, it can be deduced that these three peaks are CN_2 and $NO_2(NO_2)_2/N(NO_2)_2$, respectively. There are some relatively weak peaks at *m/z* of 17, 42, 60, 88, 134, and 188, which correspond to OH, $CH_2N_2/C_2H_2O/CNO$, CH_2NO_2/N_2O_2, $(CH_2)_2NNO_2$, $CH_2(NNO_2)_2$, and $CNN(CH_2NNO_2)_2$, respectively. The distribution of positive ions is obviously different from that of negative ions, which is mainly distributed in three regions: *m/z* of 18 and 19 were in the first region, which may be due to H_2O and H_3O, respectively. The *m/z* of the second distribution region is 24–32, the possible attributions are C_2(*m/z* = 24), C_2H(*m/z* = 25), CN(*m/z* = 26), HCN(*m/z* = 27), $CO/N_2/H_2CN$ (*m/z* = 28), HCO (*m/z* = 29), NO/CH_2O (*m/z* = 30), and O_2 (*m/z* = 32). The *m/z* of the third distribution region is 40–47, which can be assigned to CN_2/C_2H_2N (*m/z* = 40), $CH_2N_2/C_2H_2O/CNO$ (*m/z* = 42), $N_2O/CO_2/CH_2NO$(*m/z* = 44), HN_2O (*m/z* = 45), NO_2 (*m/z* = 46), and HONO (*m/z* = 47), respectively [18].

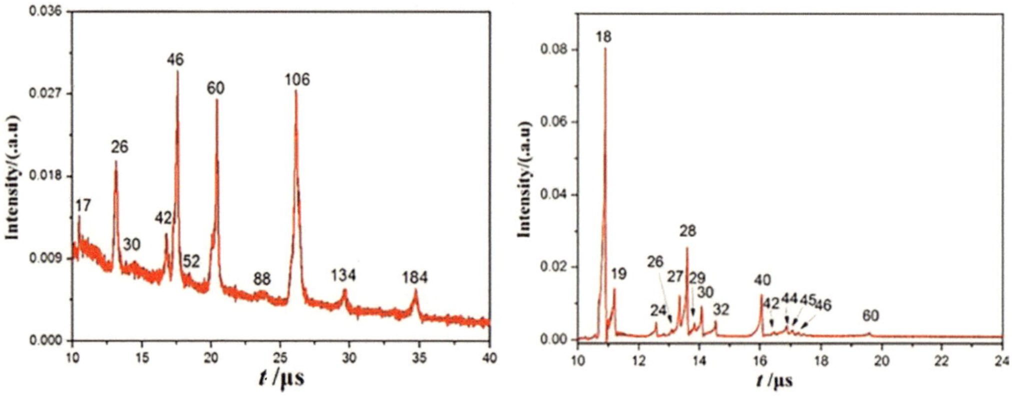

Figure 15. Typical TOF mass spectra of negative ions (left) and positive ions (right) for HMX produced by 1064 nm laser ablation.

Typical negative and positive ion mass spectra of HMX obtained by 1064 nm laser dissociation are shown in **Figure 15**. The laser energy was 45.0 J/cm² for negative ions with delay time of 80 μs and 41.4 J/cm² for positive ions with delay time of 70 μs. The distribution of negative ion peaks of HMX at 1064 nm is similar to that of 532 nm laser, and all of them have more obvious ion peaks at *m/z* of 17(OH), 26(CN), 42 ($CH_2N_2/C_2H_2O/CNO$), 46(NO_2), 60(CH_2NO_2/N_2O_2), 88($(CH_2)_2NNO_2$), 106 ($N(NO_2)_2/CH_2(NO_2)_2$), and 134 ($CH_2(NNO_2)_2$), except that the relative intensity of *m/z* of 60 and 46 increased significantly and a new ion peak $CNN(CNNO_2)_2$ is generated at *m/z* of 184. The distribution of positive ions is almost the same as that of 532 nm. However, there are differences in the intensity for the ions, among which the relative intensity of 24 (C_2), 27(HCN), and 40(CN_2/C_2H_2N) reduced significantly. The similarity of the distribution of ions indicates that the decomposition mechanism of HMX under the irradiation of two kinds of laser wavelengths is similar [18, 24].

3.3. Laser-induced TOF mass spectroscopy of aromatic nitro compounds

3.3.1. TNT

Typical negative and positive ion mass spectra of TNT obtained by 532 nm laser dissociation are shown in **Figure 16**. The laser energy was 24.6 J/cm² for negative ions with delay time of 70 μs and 25.8 J/cm² for positive ions with delay time of 70 μs. For the negative ions, the strongest peak appeared at *m/z* = 26, which might correspond to CN. Two relatively strong peaks appeared at *m/z* = 42 and 46, which might be due to CH_2N_2/C_2H_2O/CNO and NO_2. A series of small peaks were observed at *m/z* = 50, 66, 77, 137, 152, 167, 197, 212, 225, and 238, these ions could be ascribed to CHCCHC, CCHCOCH, CHCHCHCHCHC, CH_3CHCOCHCOCCO, $(CHCO)_2CCNO_2$, $CH_3(CHCO)_2CCNO_2$, $CH_3CHCOCHCNO_2CCNO_2$, $C_6H_2(NO_2)_3$, $C_6H_2CH(NO_2)_3$, and $C_6H_2CHCH(NO_2)_3$. The mass charge ratio of $C_6H_2CHCH(NO_2)_3$ (*m/z* = 238) is larger than TNT (*m/z* = 227), it can be conjectured that TNT may get some atoms to form larger fragments. For the positive ions, the dominant peaks were appeared at *m/z* = 18, 19, 24, 27, 28, 30, and 40, these ions could be correspond to H_2O, H_3O, C_2, HCN, $CO/N_2/CH_2N$, NO,

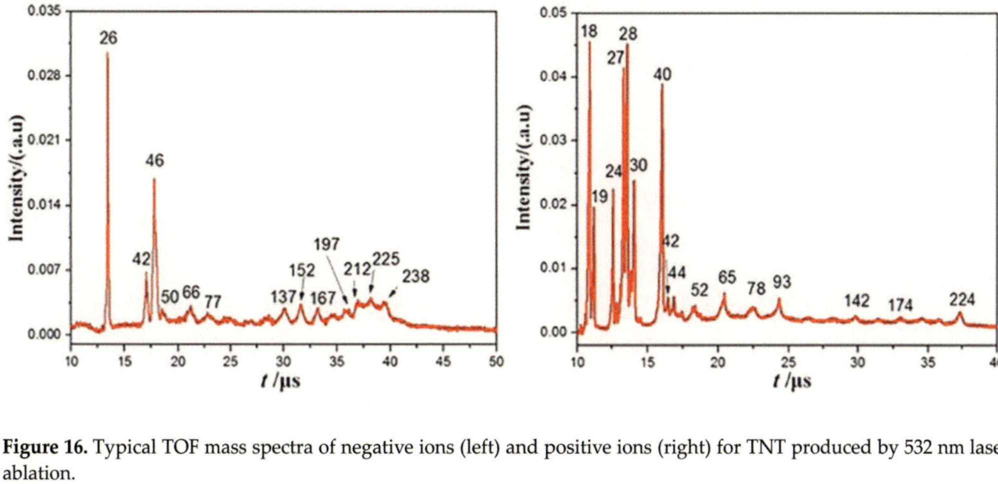

Figure 16. Typical TOF mass spectra of negative ions (left) and positive ions (right) for TNT produced by 532 nm laser ablation.

and C_2H_2N. Some low intensity peaks were observed at 42, 44, 52, 65, 78, 93, 142, 174, and 224, possible assignments for these peaks were CH_2N_2/C_2H_2O/CNO, N_2O/CO_2/CH_2NO, C_2N_2, CNCHCN, C_6H_6, C_6H_5O, $(CCN)_2C$ (CCNO), $C(CCNO)_3$, and $C(CHNO_2)_2(CCNO_2)$ [18, 27].

Typical negative and positive ion mass spectra of TNT obtained by 1064 nm laser dissociation are shown in **Figure 17**. The laser energy was 15.9 J/cm^2 for negative ions with delay time of 70 μs and 13.3 J/cm^2 for positive ions with delay time of 70 μs. Compared with previous studies at 532 nm laser irradiation, we find that the distribution of both the negative and positive ion fragments has a certain similarity. The difference is that three strong peaks appear at $m/z = 26$, 212, and 238 for the negative ion mode after 1064 nm irradiation, which can be attributed to CN, $C_6H_2(NO_2)_3$, and $C_6H_2(NO_2)_3$(CHCH), as shown in **Figure 3**. At the same time, some weak negative ion fragment peaks are produced at $m/z = 92$, 103, 116, 146, 158, 176, and 197, these peaks are probably attributable to $(CCH)_2$(CHCOH), $(CCH)_2$CO(CCH), $(CCO)_2C_3$, $(CCO)_2$C(CCON), $(CCON)_2$CC(CN), (CCON)($CCNO_2$) C(CCO), and CH_3(CHCO) ($CHCNO_2$) $CCNO_2$. For the positive ion mode, different with 532 nm, these peaks at $m/z = 24$, 27, 30, 40, 42, 65, 78, 93, and 174 are observed, these ion fragments may be due to C_2, HCN, NO, C_2H_2N, CH_2N_2/C_2H_2O/CNO, CNCHCN, C_6H_6, C_6H_5O, and $C(CCNO)_3$ [18, 28].

3.3.2. HNS

Typical negative and positive ion mass spectra of HNS obtained by 532 nm laser dissociation are shown in **Figure 18**. The laser energy was 8.6 J/cm^2 for negative ions with delay time of 60 μs and 8.6 J/cm^2 for positive ions with delay time of 55 μs. For the negative mode, the main peaks appear at $m/z = 26$, 42, 46, and 62, which corresponds to the CN, CH_2N_2/C_2H_2O/CNO, NO_2, and NO_3 fragments. A series of relatively weak peaks can be found at $m/z = 50$, 66, 77, 93, 107, 142, 172, 196, 212, 225, and 238. According to the structure of HNS, these peaks can be assigned to $(CHC)_2$, CCHCOCH, C_6H_5, C_6H_5O, C_6H_5NO, $(CCN)_2$C(CCNO), $(CCN)_2$CNO (CCNO), $C_6H_2NO(NO_2)_2$, $C_6H_2(NO_2)_3$, $C_6H_2CH(NO_2)_3$, and C_6H_2CHCH $(NO_2)_3$, respectively. For the positive mode, the peaks with high intensity are mainly distributed in $m/z = 28$, 30,

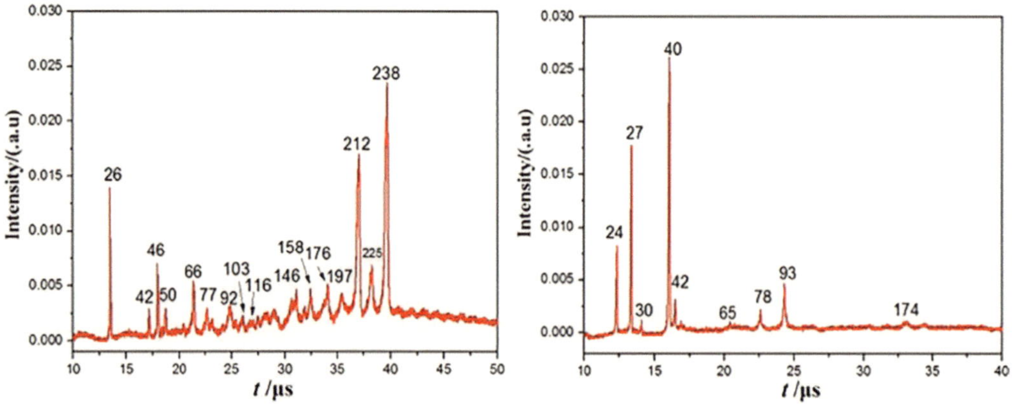

Figure 17. Typical TOF mass spectra of negative ions (left) and positive ions (right) for TNT produced by 1064 nm laser ablation.

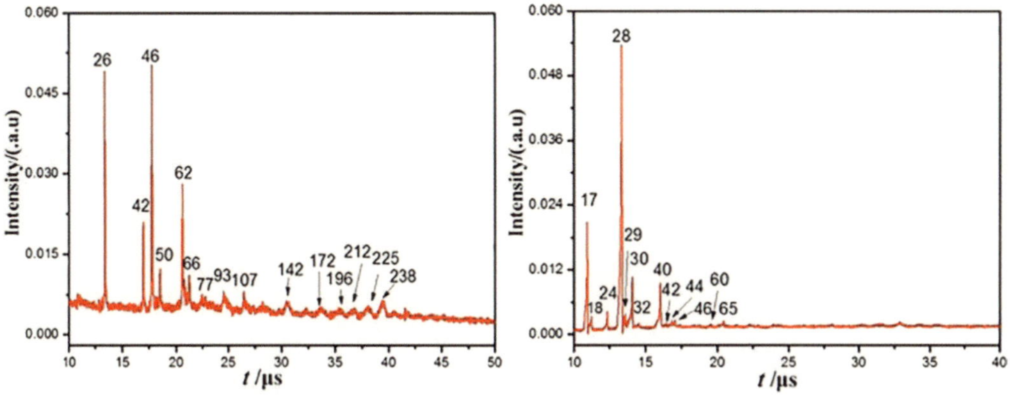

Figure 18. Typical TOF mass spectra of negative ions (left) and positive ions (right) for HNS produced by 532 nm laser ablation.

17, and 40, respectively, corresponding to OH, CO/N_2/CH_2N, NO, and CH_2CN. In addition, at m/z = 18, 24, 29, 32, 42, 44, and 65, the strength of the ions is weak, which can be attributed to H_2O, C_2, CHO, O_2, CH_2N_2/C_2H_2O/CNO, N_2O/CO_2/CH_2NO, and CNCHCN, respectively [18].

Typical negative and positive ion mass spectra of HNS obtained by 1064 nm laser dissociation are shown in **Figure 19**. The laser energy was 8.4 J/cm^2 for negative ions with delay time of 80 μs and 13.1 J/cm^2 for positive ions with delay time of 70 μs. For the negative ions, a strong peak is observed at m/z = 26, which can be attributed to CN fragment. A series of regular peaks with a mass to charge ratio step of 4 appear at m/z = 42, 46, and 50, according to the molecular structure of HNS, possible assignments for these peaks are CH_2N_2/C_2H_2O/CNO, NO_2, and CHCCHC. Fragments like CH_2N_2, C_2H_2O, and CNO that have similar mass to charge ratio are difficult to distinguish from the instrument due to the limited mass resolution, which mainly caused by the large broadening of the kinetic energy distributions of the species produced in the excitation process. Another series

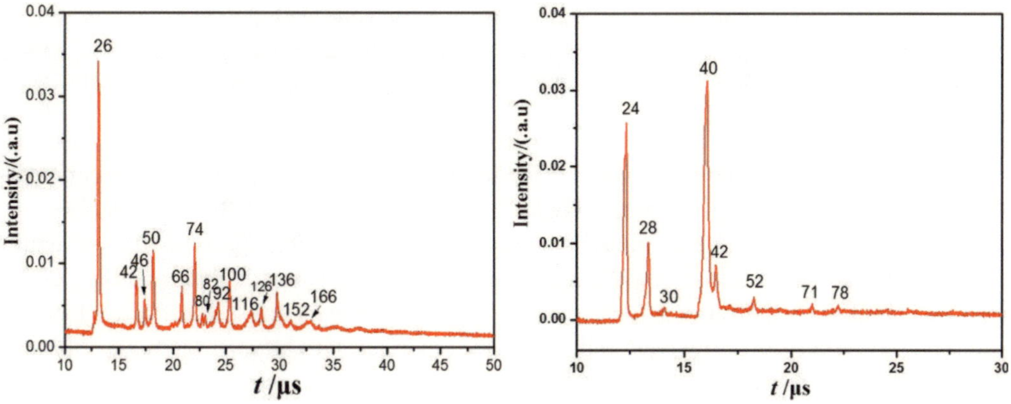

Figure 19. Typical TOF mass spectra of negative ions (left) and positive ions (right) for HNS produced by 1064 nm laser ablation.

of peaks can be found at *m/z* = 50, 66, 74, 80, 82, 92, 100, 116, 126, 136, 152, and 166, these peaks can be assigned to CHCCHC, CCHCOCH, C_6H_2, CHCCHCNO, CHNCHCNO, CHCCHCCNO, C_6H_2CN, C_6H_2CNO, $CHCNOCHCNO_2$, $C_6H_2ONO_2$, $C_6H_2O_2NO_2$, and $C_6H_2ONONO_2$ respectively. Unlike the spectra in the positive mode, there are less ion lines in the negative mode. The dominant peaks appear at *m/z* = 24, 28, 40, and 42, which corresponded to C_2, $CO/N_2/CH_2N$, CH_2CN, and $CH_2N_2/C_2H_2O/CNO$, respectively. Some relatively weak peaks can be found at *m/z* = 30, 52, 71, and 78, which may be due to $(CN)_2/CHCNCH$, $CHCNO_2$, and C_6H_2N, respectively [18, 29].

4. Conclusions

The structure of flyer target can be divided into two types: single-layer flyer and multilayer flyer. The main materials of single-layer flyer are Al and Cu. As for multilayer flyer, the ablation layer materials are mostly C, Al, Mg, Ge, Ti, Zn, etc. Al_2O_3, MgF_2, ZnS, and TiO_2 are commonly good choices for thermal insulation materials. Al, Ti, and Cu are usually used to prepare flyer layers. Four types of flyers including $Al/Al_2O_3/Al$ multilayer, $C/Al/Al_2O_3/Al$ multilayer, $Mg/Al/Al_2O_3/Al$ multilayer, and Al single-layer were tested with the laser energy of 36.4–195.1 mJ by PDV measurement system. At the same laser energy, $Mg/Al/Al_2O_3/Al$ multilayer flyer owns the maximum velocity. Mg layer can greatly improve the velocity of flyer as the ablation layer. The addition of Al_2O_3 insulation layer increases the mass of flyer, which makes the velocity of $Al/Al_2O_3/Al$ multilayer flyer is lower than that of Al single-layer flyer. The addition of C absorption layer increases the velocity of $C/Al/Al_2O_3/Al$ multilayer flyer at low laser energy. The current mode of PVDF pressure sensor was used to measure the impact stress of flyer. With the increase of laser energy, the impact stresses of flyer tend to rise up gradually. $P^2\tau$ of multilayer flyers are higher than that of single-layer flyers. Especially, $C/Mg/Al/Al_2O_3/Al$ multilayer flyer obtains the maximum value of $P^2\tau$, which means the best ability to initiate explosive. Besides, five types of flyers including Al and Cu single-layer flyers and CuO/Cu, CuO/Al and CuO/Al/Cu multilayer flyers with relatively high velocities were used to initiate PETN explosive. Al single-layer flyer shows the best performance of initiation in the same laser energies, which succeeded in initiating as long as the laser energy was higher than 161 mJ. A certain range of laser energy is necessary in the process of laser-driven flyer impact initiation, which can ensure enough velocity and impact stress of flyer.

Laser ionization time of flight mass spectrometry has been applied to study the decomposition process of typical energetic materials (RDX, HMX, TNT, and HNS). Both the negative and positive ion fragments of these energetic compounds were detected. Based on the structure of explosives, possible attributions of the ion fragments were obtained. The attribution of the ions was similar for the same explosive at two different wavelengths, it indicated that the similar dissociation paths involved in the processes of explosives after irradiated by 532 and 1064 nm laser. For the same category of energetic compounds, the attribution of the ions was similar, that means the dissociation paths were similar. The results might give some help for the further understanding of the process of laser initiation of explosives.

Acknowledgements

This work was supported by the National Natural Science Foundation of China (No. 11604149).

Author details

Ruiqi Shen*, Lizhi Wu, Wei Zhang and Haonan Zhang

*Address all correspondence to: rqshen@njust.edu.cn

Department of Applied Chemistry, School of Chemical Engineering, Nanjing University of Science and Technology, Nanjing, China

References

[1] Goujon J, Musset O, Marchand A, Bigot C. Synchronous initiation of optical detonators by Q-switched solid laser sources. Proceedings of SPIE. 2008;**7115**:71150P-71150P-11

[2] Paisley D, Swift D, Tierney IV T, Munson C, Johnson R. Diagnostics for confined plasma ablation for plate launch and shock generation. Proceedings of SPIE. 2005;**5580**: 868-872

[3] Sheffield S, Rogers J Jr, Castaneda J. Velocity measurements of laser-driven flyers backed by high impedance windows. Shock Waves in Condensed Matter. 1986;541-546. https://link.springer.com/chapter/10.1007/978-1-4613-2207-8_78#citeas

[4] Paisley DL, Montoya NI, Stahl DB, Garcia IA. Interferometry and High Speed Photography of Laser-Driven Flyer Plates. 33rd Annual Techincal Symposium. 1990;245-249

[5] Paisley D. Laser-Driven Miniature Flyer Plates for Shock Initiation of Secondary Explosives, APS Meeting. 1989;14-17

[6] Paisley D. Laser-Driven Flyer Plate. US Patent. US 5046423;1991

[7] Stahl DB, Paisley DL. Carbon-Assisted Flyer Plates, US Patent. US 5301612;1994

[8] Trott WM. Investigation of the Dynamic Behavior of Laser-Driven Flyers. AIP Conference Proceedings. 1994;**309**(1):1655-1658

[9] Trott WM. High-speed optical studies of the driving plasma in laser acceleration of flyer plates. AIP Conference Proceedings. 1996;**370**(1):921-924

[10] Hatt D, Waschl J. A Study of Laser-Driven Flyer Plates. AIP Conference Proceedings. 1996;**370**(1):1221-1224

[11] Labaste JL, Brisset D, Doucet M. Investigation of driving plasma materials for laser acceleration of flyer plates. AIP Conference Proceedings. 2000;**505**(1):1189-1192

[12] Paisley DL. Confined plasma ablation for shock physics, plate launch and material dynamics. High-Power Laser Ablation VI. 2006;**6261**:62611Y

[13] Paisley DL, Luo SN, Greenfield SR, Koskelo AC. Laser-launched flyer plate and confined laser ablation for shock wave loading: Validation and applications. Review of Scientific Instruments. 2008;**79**(2):023902

[14] Bowden M, Knowles S. Optimisation of laser-driven flyer velocity using photonic Doppler velocimetry. Proceedings of SPIE. 2009;**7434**

[15] Wu L. Shock Initiation Technology of Explosives by Laser-Driven Metal Flyer. Nanjing University of Science & Technology; 2010

[16] Brierley H, Williamson DM, Vine T. Improving laser-driven flyer efficiency with high Absorptance layers. AIP Conf. Proc. 2012;**1426**(1):315-318

[17] Walker F, Wasley R. Critical energy for shock initiation of heterogeneous explosives. Explosive Stoffe. 1969;**1**:9-13

[18] Zhang W. Laser-Induced Decomposition Mechanism of Typical Energetic Compounds. Nanjing University of Science & Technology; 2014

[19] Glascoe EA, Zaug JM, Armstrong MR, Crowhurst JC, Grant CD, Fried LE. Nanosecond time-resolved and steady-state infrared studies of photoinduced decomposition of Tatb at ambient and elevated pressure. The Journal of Physical Chemistry A. 2009;**113**(20): 5881-5887

[20] Kosmidis C, Ledingham K, Kilic H, McCanny T, Singhal R, Langley A, Shaikh W. On the fragmentation of nitrobenzene and nitrotoluenes induced by a femtosecond laser at 375 nm. The Journal of Physical Chemistry A. 1997;**101**(12):2264-2270

[21] Civiš M, Civiš S, Sovová Kn, Dryahina K, Španěl P, Kyncl M. Laser ablation of FOX-7: Proposed mechanism of decomposition. Analytical Chemistry. 2011;**83**(3):1069-1077

[22] Delgado T, Alcántara JF, Vadillo JM, Laserna JJ. Condensed-phase laser ionization time-of-flight mass spectrometry of highly energetic nitro-aromatic compounds. Rapid Communications in Mass Spectrometry. 2013;**27**(15):1807-1813

[23] Zhang W, Ma X, Shen R, Wu L, Ye Y, Hu Y, Zhu P. Progress on laser-induced decomposition of explosives investigated by spectroscopic methods. Applied Spectroscopy Reviews. 2014;**49**(7):550-563

[24] Zhang W, Shen R, Ye Y, Wu L, Zhu P, Hu Y. Distribution and formation of particles produced by laser ablation of cyclotetramethylene tetranitramine. Laser and Particle Beams. 2017;**35**(3):1-6

[25] Zhang W, Shen R, Ye Y, Wu L, Hu Y, Zhu P. Dissociation of cyclotrimethylenetrinitramine under 1064-nm laser irradiation investigated by time-of-flight mass spectrometer. Spectroscopy Letters. 2014;**47**(8):611-615

[26] Zhang W, Shen R, Wu L, Qin Z, Ye Y, Hu Y, Zhu P. Dissociation of RDX at 532nm laser. Chinese Journal of Explosives & Propellants. 2012;**35**(1):39-42

http://dx.doi.org/10.5772/intechopen.71892

[27] Zhang W, Shen R, Ye Y, Wu L, Hu Y, Zhu P. Photodissociation of 2, 4, 6-trinitrotoluene with a Nd: YAG laser at 532nm. Proceedings of SPIE. 2015;**9543**:95431A

[28] Zhang W, Shen R, Ye Y, Wu L, Hu Y, Zhu P. Ion fragments of TNT produced by 1064 nm laser dissociation. Journal of Beijing Institute of Technology (English Edition). 2016;**25**(1): 103-106

[29] Zhang W, Wang F, Shen R, Ye Y. Time of flight mass spectroscopy of 2, 2′, 4, 4′, 6, 6′-hexa-nitrostilbene under 1064 nm excitation. Proceedings of SPIE. 2017;**10173**:1017315